Preface

The papers presented here are taken from the annual conference of the Association for Environmental Archaeology held in Edinburgh in 1992. The conference, entitled '*On the edge – human settlement in marginal areas*', set out to examine the concept of marginality and its application to archaeology.

As designed, the conference set out to explore the environmental limits to human activity, but what became clear was that archaeologists of all persuasions are increasingly committed to more complex models of person-land interaction and that a variety of social and economic factors are of central importance to any discussion of marginality. In consequence, the running order of the papers here differs from that seen at the conference with papers of similar approach or related themes being grouped together.

It is perhaps appropriate that the theme of a conference held in Edinburgh should have been marginality. After all where else has such a variegated Fringe Festival?

Post-script: Since going to press, the editors were greatly saddened to hear of the death in May 1998 of Camilla Dickson, who wrote a paper for this collection. The editors wish to acknowledge the immense contribution made by Camilla to the field of environmental archaeology in Scotland, and to support the decision of the AEA to dedicate the Proceedings of the Bradford conference to her.

Editors' acknowledgements

The editors wish to thank the many people who have assisted with the production of this volume and those who contributed to, and helped to run, the conference from which it originates. In particular, we are grateful for the generous support of our colleagues from the University of Edinburgh, especially the Department of Archaeology and the Department of Geography, and from AOC (Scotland) Ltd. We are also most grateful for the support of the AEA committee, for practical help and advice from a number of AEA members and for the patience of the contributors.

We wish to thank especially Finbar McCormick, who was a co-organiser of the conference, and Andrew Dugmore for arrangement of the venue. Thanks also to Rod McCullagh and Jerry O'Sullivan for undertaking slide projection at the conference.

Many of the illustrations are as provided by the contributors, but additional work was undertaken by Andrew Heald and Mark Roughly. We are grateful to Alan Braby for the front cover illustration.

Contents

LIFE ON THE EDGE:
HUMAN SETTLEMENT AND MARGINALITY

Symposia of the Association for Environmental Archaeology No. 13

Edited by C M Mills and G Coles

Oxbow Monograph 100

Published by
Oxbow Books, Park End Place, Oxford OX1 1HN

ISBN 1 900188 57 0

This book is available direct from
Oxbow Books, Park End Place, Oxford OX1 1HN
(Phone: 01865–241249; Fax: 01865–794449)

and

The David Brown Book Company
PO Box 511, Oakville, CT 06779, USA
(Phone: 860–945–9329; Fax: 860–945–9468)

and

via our website
www.oxbowbooks.com

Printed in Great Britain at
The Alden Press
Oxford

Clinging on for grim life: an introduction to marginality as an archaeological issue

Geraint Coles and Coralie M Mills

Abstract

Three overlapping definitions of marginality are currently used in archaeology. They are often vaguely expressed but may be summarised as environmental, economic and social/political marginality. In each case different criteria are used to define marginality and thus different criteria are proposed for the archaeological recognition of the condition. Historic and ethnographic sources indicate the complexity of the issue and suggest that a single explanation of marginality is unlikely to be satisfactory. Archaeological and palaeoecological findings appear confused, with some lines of evidence indicating the potential importance of extreme environmental 'events' as triggers for the collapse of systems already operating at the limits of growth, while other studies suggest that the perception of land as marginal has more to do with social and socio-economic factors than any inherent quality of environmental marginality. In part, this confusion appears to relate to the different definitions of marginality used. Current research is emphasising a greater attention to the inter-relationship of environmental, economic and social systems.

INTRODUCTION:
MARGINAL LAND, MARGINAL PEOPLE?

The simple observation that some people live in areas perceived to be less favourable than others prompts the obvious speculation that such patterns of 'centre and periphery' can be found in antiquity. Groups thought to be living 'on the edge' are called marginal and their plight is termed marginality. The use of these terms is now so common that archaeologists rarely take time to consider whether they have any underlying basis as a concept at all. Indeed, one might argue that marginality has become a 'fuzzy catchall' which, while apparently offering explanation, actually serves only to disguise our ignorance of the complex environmental systems and economic and social choices facing past human groups.

The concept that some human groups live in more, or less, favourable environments than others is not, of course, new. Indeed the concept that external, environmental, factors especially climate, determine the 'success' of a particular human group may be traced back to the ancient Greeks, who naturally defined the ideal, vigorous, intellectually stimulating, climate as being that of Greece. The Romans were converts to the idea and moved the ideal west and north. Rediscovered in the Renaissance and with the 'ideal' again moved north, such ideas came, in time, to be viewed as explanations of the material precociousness, and helped to justify the colonial expansion, of Western Europe (Simmons 1989). Where we stand now, the notion has become so laden with conceptual baggage that it takes only the slightest suggestion that environmental archaeologists are unduly 'deterministic' to reduce them collectively to tears. While such crude determinism is now discredited, the interaction of the environment and human groups through time continues to be widely discussed (eg Macinnes & Wickham-Jones 1992) particularly with regard to human groups perceived by archaeologists as living in 'marginal situations'. The linkage of certain types of environment with constraints upon economic and social development is most clearly seen in the context of British archaeology.

The concept that certain environments are inherently marginal is one that has had an inordinate, almost subliminal, influence on British archaeology since the 19th century. The publication of Cyril Fox's 'Personality of

Britain' in 1924 cemented into place the central tenet that the British Isles could be divided into a fertile lowland zone and an unforgiving and marginal highland zone. This supposition is one which continues to play to packed archaeological houses (cf Evans 1975, Limbrey & Evans 1978, Bintliff *et al* 1988).

In such environmental models of marginality the limits to economic growth of a particular human group are determined by the scarcity or absence of certain critical resources necessary for the group's survival. Scarcity or absence of critical resources is seen as resulting either from an inherent property of the landscape itself (such as the underlying geology or soils) or through environmental changes which result in the diminishing effectiveness of the subsistence base to supply the fundamental needs of the society in question.

The main objectives of the meeting from which the papers presented in this volume are derived were, firstly, to examine a long-standing issue in the 'determinism debate', that is the concept that some regions or areas within the landscape are inherently 'marginal' for human settlement, and, secondly, to consider whether the issue of marginality can be competently investigated through archaeology.

DEFINING AND EXPLAINING MARGINALITY

Marginal: at the outer edge, peripheral, on the fringe, with low returns

The first problem facing examination of the concept of marginality is with the definition of the term itself. The terms marginal and marginality have several concurrent and often overlapping meanings in archaeological discussions. It has been rare for the exact meaning employed to be more than implied by the context of use. Nevertheless, one may distinguish a number of strands of thought with regard to the definition of 'marginal' and thus the presumed causes of marginality. These might be broadly characterised as environmental, economic, and social/political marginality.

It is obvious that these factors do not stand in isolation since each interacts with the others. In consequence, explanations as to why groups become marginal or utilise locations which are less that ideal for a given subsistence strategy usually involve interaction between all these factors. What differs is the view of the authors as to the primary causal factor in marginality and the order in which changes cascade through the system.

(a) Environmental marginality

Here the environment is thought of as a limiting factor. The concept that organisms are adapted to their environments and that their tolerances of different environmental conditions vary is fundamental to ecology (Mayr 1982). Considerable effort has been put into establishing the

optimum conditions and limits to the growth of organisms. In ecological terms we may consider any environment where at least one of the requirements for the growth of an organism is either restricted in supply or is sporadic in occurrence as 'marginal'.

Essentially all such environmental explanations of marginality are variants upon Liebig's 'Law of the Minimum' in that growth and reproduction are constrained by whatever is in shortest supply (Wild 1988). Marginal situations may therefore arise either because a critical environmental resource, such as good quality soil, is absent or in short supply or because a critical environmental variable, such as climate, changes. In the former case the environment is seen as a series of potential constraints upon population growth which may be overcome by the adoption of new technologies or subsistence methods (Dickson, this volume). In the latter case environmental change may be seen as a trigger for the collapse of societies which are already operating at the limits of their particular ecological niche. Often these views are combined and gradual, long term, environmental change is seen as leading to the creation of a situation in which the returns from subsistence activities become increasingly borderline for continued survival leading to a situation in which sudden, relatively short term, environmental change may result in the failure to gain sufficient subsistence returns for effort invested (eg Baillie; Gratton, this volume). Famine and starvation result in population collapse and either the complete reorganisation of the subsistence strategy and/ or the abandonment of settlement in particular ecological niches.

Several factors have been cited as probable causes of environmental marginality including climate, soils, biota and disease. The essential link is that they all relate to factors which are beyond the control of human populations and which therefore may be regarded as inherent to the landscape. What does lie within human control is the economic and political response of human groups to these inherent factors.

(b) Economic marginality

Here marginality is the result of the way in which the subsistence economy is structured. This differs from environmental marginality in that the environment is not seen as *inherently* marginal. Instead, marginality is a property of the economic system and may be created or ameliorated by changes in the subsistence base, the technological base and/or the organisation of the economy itself. In other words, economic marginality is the result of a fundamental mismatch between the means by which resources are procured from the environment and the resources available in the environment.

This associates the term 'marginal' with the viability or otherwise of a landscape for a particular form of subsistence exploitation. Marginal lands may be those which are seen in terms of modern commercial agriculture to have either too low a rate of return for the effort invested or which have inherent high risk defined as a high return rate of potential crop failure. The term marginal is therefore associated with excessive input of agricultural labour, and the development of intensive crop cultivation systems such as so called 'lazy bedding' is commonly cited as an example. It is also associated with the cultivation of crops in a high risk environment, one where the probability of crop failure exceeds a given figure; one failure in five years is commonly cited as commercially unsustainable. How 'acceptable agricultural risk' can be quantified in the context of subsistence economies remains problematic. While it is tempting to point to recent historical events such as the Great Famine of the 1840s in Ireland to underline the consequences of repeated crop failures, it is clear that the response to the famine was highly complex and included increased expenditure of cultivation effort in regions previously considered unsuitable for agriculture rather than general land abandonment (Bell, this volume). Such efforts emphasize the control of social and political factors in determining the nature of the response to environmental or economic pressure.

Since the economy may change it is apparent that changes in subsistence base may result in the transformation of the economic potential of a particular environment; for example, heavy clay soils which are difficult to till with hoe and ard may become prime agricultural land with the introduction of the mouldboard plough and heavy draught animals.

A further obvious corollary of the economic argument is that environments or locations that are marginal to one economy may well be perfectly suited to another. The example of Norse Greenland is possibly pertinent. Here one human population, the Greenland Norse, died out sometime in the 14th century while surrounded by marine resources which were exploited by another group, the Inuit, which shows no evidence for population reduction or ecological stress at this time (Fredskild 1988; Buckland *et al* 1996).

(c) Social and political marginality

This strand of thought associates marginality with the political and cultural isolation or near isolation of communities living on the edge of larger groupings. Here the main reason for marginality is held to be the social and/or political position which a community or group occupies within the wider social and political system (polity).

Isolation may occur because of the geographic remoteness of the community from the centre of power, the presence of religious, ethnic or linguistic differences or the status of the group within the polity. In many cases these are explicitly linked and the marginal status of a group may be reinforced by a combination of these factors leading to economic backwardness or underdevelopment. The Western and Northern Isles of Scotland during the 18th and early 19th century are commonly seen as typical of such 'marginal communities' (cf Armit, this volume). These communities are held to possess entrenched and

largely closed (read static or inflexible) social systems and to have a backward economy based upon 'primitive' agricultural practices. As Bond (this volume) notes, however, the perception of the backwardness has more to do with the position and motives of the observers than with the fitness of the economy or society to provide the material means of survival.

Changes in the organisation of social and political systems, in particular the movement of the centre of organisation coupled to the growth of larger polities, will tend to lead to political and social marginalisation of those on the geographic periphery of the enlarged polity. Ian Armit suggests that this is the case in the Hebrides where apparent economic or environmental marginality did not prevent this area forming the core of the Norse Kingdom of the Isles and which he argues only lapsed into marginality when the Isles were incorporated into the larger polity of the Kingdom of Scotland after the 14th century AD.

It should be noted, however, that social and/or political marginality does not necessarily have to be measured in terms of *geographic* remoteness since there are potential cases of groups living in the heartland of a polity but only possessing either restricted access to resources or access to areas of land considered to be marginal by other communities within the population. One possible example of social and political marginality within a population is given by Watson (this volume) in his discussion of the recent historic Irish practice known as booleying. This illustrates the potential complexity of the situation with different communities within a single population and polity having different perceptions as to the value or marginality of different parts of the landscape.

Conversely, we have examples of communities in peripheral geographical positions with considerable economic, social and political importance as points of pilgrimage, exchange and trade. Such communities may well be located in areas of extreme environmental marginality and may only be maintained by the actions of the polity of which they are part. The ability of large polities to mobilise huge resources to overcome environmental marginality is illustrated by Hamilton-Dyer (this volume) with reference to trade routes across the Eastern Desert of Egypt and the supply of food and materials to the Roman quarry settlement of Mons Claudianus. Ten centuries later the development of pilgrimage routes to Mecca provides similar evidence for the provision of facilities for travellers in locations where no subsistence activity is possible (Al-Resseeni *et al*, this volume). In general, however, we might envisage communities in such situations as the exception rather than the rule and that usually communities remote from political centres will have restricted access to some resources.

Explaining marginality

The concept that certain landscapes are inherently marginal is common to several working definitions of marginality. However, what became increasingly clear during debate was that the range of conditions which bring about the marginalisation of a human group has very little to do with the inherent qualities of the land itself and much more to do with the way in which that landscape is both perceived and exploited. While we may attempt to define marginality using discrete criteria it is clear that few workers now seek mono-causal explanations derived from these definitions. In many cases these uses of the term marginality are combined and, for example, the economic impoverishment of a group is seen as being the direct cause of its relative position in the political or social system. Conversely the social position of a group may be seen to lead to its underdevelopment and consequent economic marginalisation.

Such polythetic models, in which marginality occurs because of a combination of factors, are increasingly common and reflect a wider interest in the behaviour of complex, and possibly chaotic, systems. As we noted above, the actual explanation of why a group is marginal or becomes marginal inevitably involves discussion of the inter-relationship of all elements of the environment, economy and society.

ARCHAEOLOGICAL RECOGNITION OF MARGINALITY

How we might identify marginal situations and marginality in the archaeological and palaeoecological record depends to a great extent upon the definition of marginality in use. Where environmental and economic factors are seen as key players then it should be possible to use proxy records to identify changes in the environment using established palaeoecological techniques. The principal problem remains the precision with which these proxy records of environmental change can be dated and the danger that changes occurring in an imprecisely dated archaeological record will be 'sucked in' and bracketed with an equally imprecisely dated environmental record (Baillie 1991). Even where both proxy climate and dating evidence of a high quality is available from sources such as tree-rings (eg Baillie, this volume) it is still difficult to correlate this to the archaeological record with sufficient precision to be able to establish whether the changes noted in the proxy environmental record precede or follow the changes noted in the archaeological record. In other words, can causality be demonstrated? Baillie attempts to overcome this by treating trees as a 'parallel biological system' and by equating the periods of poor tree growth recorded by tree-rings with periods of adverse climatic conditions and hence periods in which human activity was likely to be under severe environmental stress.

An alternative attack on the question of causality is seen in the work of Tipping (this volume) which sets out to examine the relationship between periods of known climatic deterioration and the abandonment of upland arable agriculture and settlement. The idea that settlement in the

highland zone occurs at higher altitude during periods of climatic amelioration and retreats from the margins with climatic deterioration was put forward by Parry (1975; 1978) in his study of the Lammermuir Hills south of Edinburgh. In this study Parry suggested that Medieval settlement occurred at altitudes now far above the conventional limits for arable cultivation and settlement. Following the climatic downturn of the 'Little Ice Age' this upland settlement was abandoned. Such ideas have become important tenets in prehistoric archaeology where the ebb and flow of upland settlement is now widely held to be climatically controlled and has closely influenced the design of many research projects (eg, de Rouffignac, this volume). Tipping has used palynological data to establish that upland arable agriculture (and by inference settlements) were not abandoned with the onset of the 'Little Ice Age'. Where abandonment and retreat did occur it appears to have been a drawn out, sporadic process, more probably connected with changes in the social and economic system than with environmental factors. This has important implications for the 'Parry hypothesis' and its use in prehistoric archaeology.

Since Tipping fails to find evidence for land abandonment in the face of climate change in the Historic period it is inevitable that we should question the role of climate as a forcing factor in prehistory. Indeed Armit noted that the major downturn in climate recorded in the Western Isles is coincident with the construction of the major ceremonial monuments. Clearly adaptive strategies were at least initially able to cope with deteriorating conditions sufficiently well to produce the surpluses widely recognised as necessary for the mobilisation of sufficient people for large scale undertakings such as the erection of the Calanais stone circles.

Nonetheless, since shifts in settlement patterns and land-use do appear to have taken place over time and appear to be well documented by both the archaeological record (see Cowley, this volume) and the palaeoecological record (see Edwards & Whittington, this volume), the question as to the causes of these phases of expansion and contraction must now be regarded as more open than previously thought. The emerging consensus that the most marked climatic downturn during the later Holocene took place around 4000 BP (Blackford 1993; Chambers 1993), well before the archaeologically attested upland expansion of the Middle Bronze Age, is clearly paradoxical in terms of the 'Parry hypothesis' and suggests that explanations involving agricultural extensification in the face of reduced crop yields throughout the Highland zone may be as valid as the more established view. One is tempted to point to possible parallels with the historic and ethnographic examples of extensification noted by Bell and we would be wise to look for integrated environmental, economic and social explanations of this paradox.

The recognition of social and political marginality from archaeological data is, as Watson's paper well demonstrates, more difficult. Nevertheless, Stallibrass (this volume) is able to point to detailed differences in the livestock seen in Romano-British settlements in Northern England as possibly reflecting the relative social and economic positions of different regions within the Northern Frontier of the Roman Empire and hence the degree to which the local population had become Romanised. Potential evidence for economic adaptation to the changing position of a community within a wider social network is noted by Bond (this volume) who recorded changes in the arable crops grown following the transition from the Pictish to the Norse periods in Sanday, Orkney Islands.

ADAPTATIONS AND RESPONSES TO MARGINALITY

Whatever the cause of marginality the fate of a human population faced with changes in the environment, the economy or the social and political system will largely depend on their ability to adapt. Several workers have pointed to the adoption of particular environmental, economic and social strategies as evidence of a possible response to marginality. A number of possible responses have been recognised including economic and social innovation, especially the adoption or use of novel resources, the adoption of new technologies and subsistence methods and changes in settlement pattern. Conversely, a failure to respond adequately to changes in the situation of the group are perceived to result in social fragmentation, famine, death and large scale outward population migration.

Economic and social innovation

Economic and social responses to marginality are closely interlinked. Two temporal trajectories of marginality are commonly asserted in the archaeological literature. The first involves periodic shortages in critical subsistence resources and invokes social mechanisms for coping with natural environmental variability (year by year changes in weather, crop disease, etc). The second involves long term changes in the environment over periods of hundreds of years which fundamentally affect the available resource base and its reliability and is widely seen to result in more fundamental changes in the subsistence base.

Periodic marginality is thought to occur in settings where 'marginal conditions' (perceived as reduced agricultural yield) occur irregularly either over short time periods (three to ten years) or over a restricted geographic range (within the transport capabilities of the group concerned). In such circumstances it has been argued that such periodic marginality may favour the growth of redistributive economies and hence is closely tied to the centralisation of political control and the emergence of ranked or hierarchical social systems (eg Halstead & O'Shea 1982). Such social innovations leave the subsistence base essentially unchanged but provide mechanisms for coping with short term oscillations around mean conditions.

Conversely, change in the subsistence base has been widely perceived as a possible response to long term changes in the environment. In Assendelft, Netherlands, changes in economic strategy from the Late Bronze Age to the Iron Age have been linked to changes in local environment, in particular to marine transgression and increases in the area of coastal wetlands. This is thought to have driven a general shift away from mixed arable agriculture towards pastoralism (Wijngaarden-Bakker, this volume). Elsewhere the adoption of particular economic strategies, such calf slaughter, has been held to reflect adaptations to specific economic and environmental circumstances (McCormick, this volume).

The adoption of novel resources in terms of food, fuel and building materials (McLaren; Carter; Dickson, this volume) may also reflect specific economic adaptations to shortages. Whether these shortages were inherent to the landscape concerned or stemmed from economic or social factors is unknown; to what extent such adaptations can be attributed directly to marginality is problematic since many economic innovations coincide with changes in material culture over long time periods (Bond, this volume).

Adoption of new technologies and subsistence methods

The adoption of new technologies and methods of gaining a living may also be stimulated by marginality. The adoption of manuring may reflect one such methodological innovation, as might the creation of 'man-made' or plaggen soils (Simpson; Acott, this volume). In both cases additional human energy input is required to create a greater crop return and it might be argued that this is in response to increased pressure on the agricultural system either in response to declining yields (caused by climate deterioration, declining soil fertility, crop disease) or increased demand (rising population, presence of non-productive elites and craft specialists) or some combination of both. At the lowland site of Tentsmuir, Fife, evidence for repeated firing of the vegetation cover of sandy soils may indicate the use of fire to enrich nutrient depleted soils and may also be a response to marginal conditions (Whittington & McManus, this volume).

Changes in settlement pattern

Change in settlement pattern in the archaeological record has been observed on a number of spatial scales, ranging from abandonment of isolated upland areas involving movement of only a few kilometres to complete abandonment of a region covering several hundred square kilometres. These changes are often thought to be in response to increasing marginality of particular settlement locations and while the evidence for such shifts in settlement is now widespread the degree to which changes in environmental conditions can be correlated with settlement changes is poor (Cowley, this volume). More detailed investigation of the chronology of settlement within restricted geographic areas is necessary if progress is to be made in understanding the causal factors of settlement pattern changes and several such studies are in progress (eg, de Rouffignac; O'Connor, this volume).

CONCLUSIONS

The definitions of marginality discussed here reflect the development of ideas about the concept, from one of simple determinism to increasingly complex polythetic explanations of cultural change. They also reflect a widespread realisation that many models of marginality are uncomfortably simplistic and that in many cases ancient human groups are characterised as marginal because of their position in the modern world system rather than by any objective criteria.

From the papers presented here, there appears to be an emerging consensus that 'marginality can only be a meaningful idea in relation to a particular economic and social system' (Brown *et al*, this volume) and that explanations of marginality must embrace the complexity of human societies, the way that they perceive their position and their decision making processes.

ACKNOWLEDGEMENTS

Thanks to Dr Andrew Dugmore for commenting on an earlier draft of this paper.

REFERENCES

Baillie, M G L 1991 'Suck in and smear: two related chronological problems for the 1990s', *Journal of Theoretical Archaeology,* 2, 12–16.

Bintliff, J N, Davidson, D A & Grant, E G (eds) 1988 *Conceptual issues in environmental archaeology.* Edinburgh: Edinburgh University Press.

Blackford, J J 1993 'Peat bogs as sources of proxy climatic data: past approaches and future research', *in* Chambers, F M (ed) *Climate change and human impact on the landscape.* London: Chapman and Hall, 47–56.

Buckland, P C, Amorosi, T, Barlow, L K, Dugmore, A J, Mayewski, P A, McGovern, T H, Ogilvie, A E J, Sadler, J P & Skidmore, P 1996 'Bioarchaeological and climatological evidence for the fate of the Norse farmers in medieval Greenland', *Antiquity* 70, 88–96.

Chambers, F M 1993 'Late Quaternary climatic change and human impact: commentary and conclusions', *in* Chambers, F M (ed) *Climate change and human impact on the landscape.* London: Chapman and Hall, 247–260

Evans, G J (ed) 1975 *The effect of man on the landscape: the highland zone.* CBA Research Report 11. London: Council for British Archaeology.

Fredskild, B 1988 'Agriculture in a marginal area: South Greenland AD 985–1985', *in* Birks, H H, Birks, H J B, Kaland, P E & Moe, D (eds) *The cultural landscape past, present and future.* Cambridge: Cambridge University Press, 381–393.

Fox, C 1924 *The personality of Britain* (1st Edition). Cardiff: National Museum of Wales.

Halstead, P & O'Shea, J 1982 'A friend in need is a friend indeed: social storage and the origins of social ranking', *in* Renfrew, C & Shennan, S (eds) *Ranking, resource and exchange – aspects of the archaeology of early European society.* Cambridge: Cambridge University Press, 92–99.

Limbrey, S & Evans, J G (eds) 1978 *The effect of man on the landscape: the lowland zone.* CBA Research Report 21. London: Council for British Archaeology.

Macinnes, L & Wickham-Jones, C 1992 *All natural things: archaeology and the green debate.* Oxbow Monograph 21. Oxford: Oxbow Books.

Mayr, E 1982 *The growth of biological thought: diversity, evolution and inheritance.* Cambridge (Mass): Belknap Press.

Parry, M L 1975 'Secular climate change and marginal agriculture', *Transactions of the Institute of British Geographers,* 64, 1–13.

Parry, M L 1978 *Climate change, agriculture and settlement.* Folkstone: Dawson.

Simmons, I G 1989 *Changing the face of the earth.* London: Basil Blackwell.

Wild, A (ed) 1988 *Russell's soil conditions and plant growth* (11th Edition). Harlow: Longman Scientific and Technical.

1. Cereal cultivation on the Anglo-Scottish Border during the 'Little Ice Age'

Richard Tipping

Abstract

A high-resolution pollen diagram from 365 m OD in the northern Cheviot Hills has allowed the identification of a vegetational response to the 'Little Ice Age' climatic deterioration. The dating and nature of this change is discussed. The altitude of the site is critical in terms of Parry's (1975; 1978) 'retrodictive' model of enhanced marginality and agricultural retreat during the 'Little Ice Age', developed for the adjacent Lammermuir Hills. In the Cheviot Hills, the evidence suggests that no such 'retreat from the margins' took place, and that changes in land-use in the historic period occurred only through socio-economic incentives. The relevance of this finding, and attendant uncertainties, are discussed.

INTRODUCTION

The effect of deteriorating climate on 'marginal' human settlement and agriculture has been explored in a series of publications since the mid 1970s by Martin Parry (1973; 1978; 1981a & b; Parry & Carter 1985), based on work in the Lammermuir Hills, in south-east Scotland (Figure 1.1a). This upland massif, rising to over 500 m OD and today mostly given over to rough grazing and heather moorland, has extensive traces of 'medieval' rig-and-furrow, and the timing and cause of abandonment of this agricultural system in the uplands was the focus of Parry's investigation.

The methods employed have been elaborated in his publications, and will not be considered here. Broadly, the aim was two-fold: firstly to understand how the changing climate from the late medieval to modern times could be expected to have influenced the growing of the staple arable crop of this region, oats, in order to determine periods when the risk of crop failure at critical altitudes might be expected to have been too great to maintain the agricultural system; and secondly, to establish whether times of inferred climatically-induced stress on the productivity of this crop coincided with periods of settlement and land abandonment, derived from historical and archaeological evidence.

The results of this analysis suggested to Parry that deteriorating climate in the 'Little Ice Age' had a role to play in the abandonment of upland arable on the Lammermuir Hills in the early modern period. How significant this role was in comparison with socio-economic pressures is hard to gauge, and on a number of occasions Parry has sought to lessen the significance of climate change and to emphasize the complex web of interacting forces on agricultural decision-making.

Parry's more cautionary comments may in part have originated from the many assumptions and apparent weaknesses in his methods, of which he was aware. These include the need to assume one particular aspect of the agricultural economy (the expected frequency of crop failure) to have been predominant in decision-making amongst farmers, and the need arising from this to arbitrarily infer 'critical frequencies' of crop failure, beyond which farmers would be 'forced' to give up. The suggested correlation between areas of land expected on palaeoclimatic grounds to have become too risky for the maintenance of crop-growing and those known from historical/archaeological evidence to have been abandoned, which remains Parry's principal argument for suggesting climate stress to be a causal factor in abandonment, is hampered by weak and imprecise dating of farm and land abandonment. At best these two aspects are only

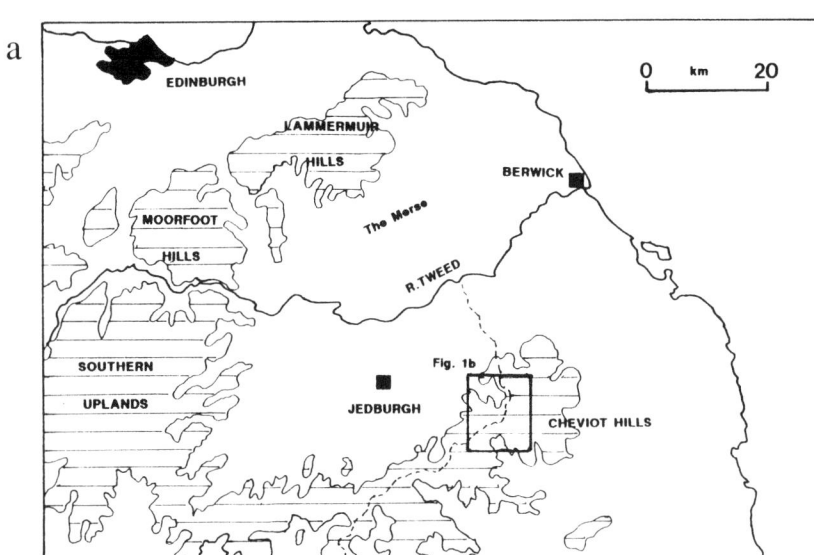

Figure 1.1 (a) South east Scotland, showing land over 300 m OD, the location of the study area (outline of Figure 1.1b) and its proximity to the Lammermuir Hills

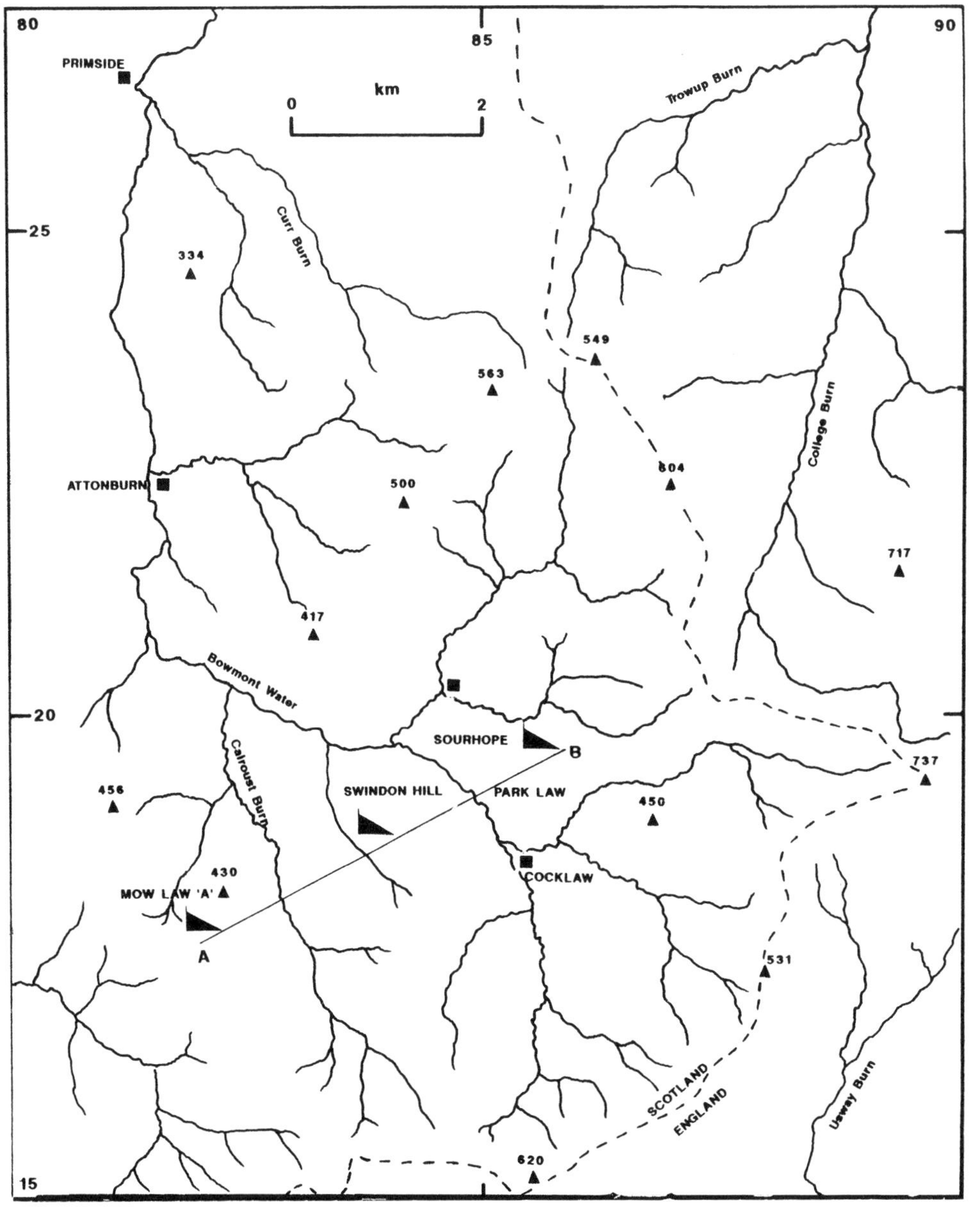

Figure 1.1 (b) The location of the three pollen sites in the upper part of the Bowmont Valley, northern Cheviot Hills, in relation to the principal hills

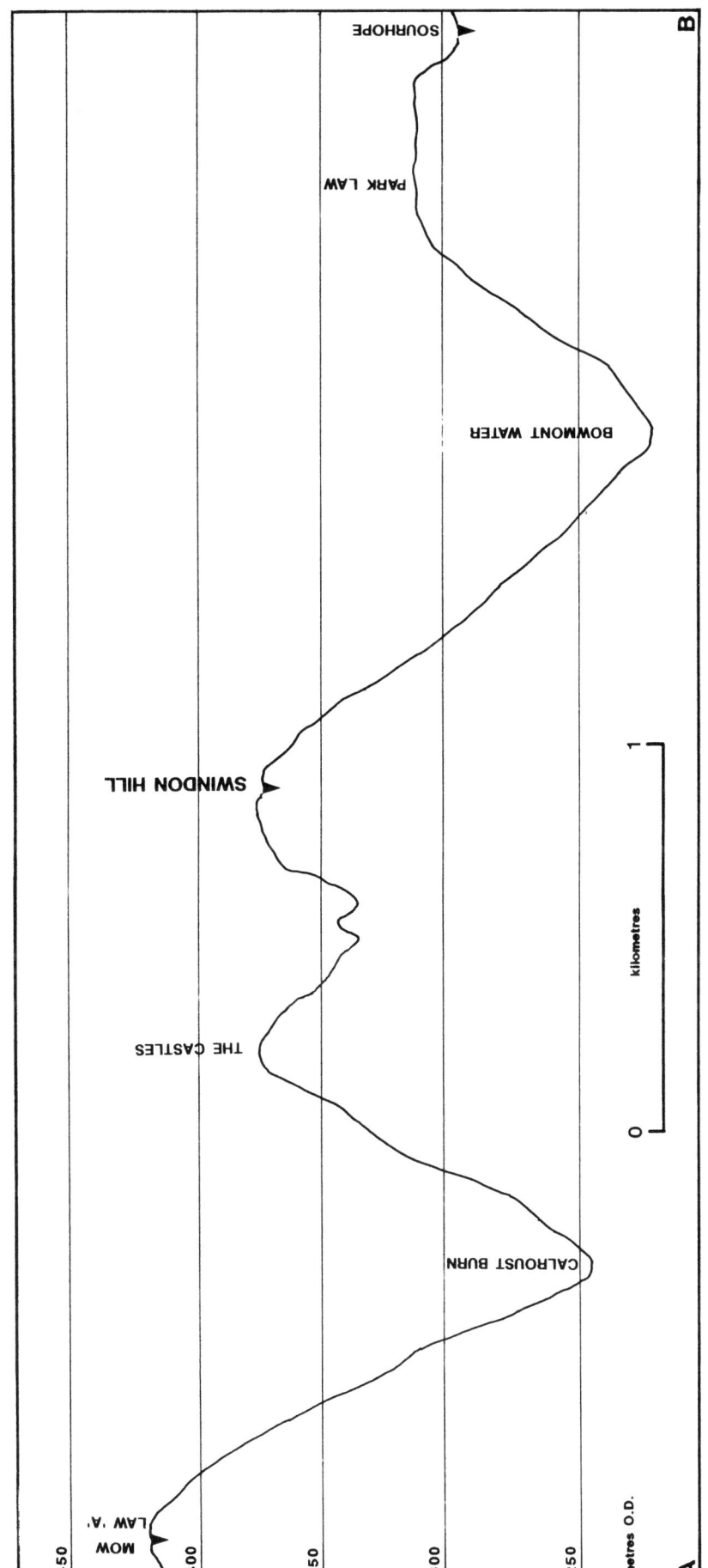

Figure 1.1 (c) Transect A–B (see Figure 1.1b) showing the spatial and altitudinal relations of the three pollen sites

spatially associated and are not clearly causally connected.

Yet there are more direct ways than those adopted by Parry of measuring the presence and persistence of cereal cultivation at high altitudes. This paper will describe the methods, results and assumptions of exploring this problem by applying radiometrically-dated (^{14}C and ^{210}Pb) high-resolution pollen analyses from carefully chosen sites in an area of south east Scotland close to and comparable in nearly all respects with Parry's study area. The intention is to test by independent and, it is argued, better means Parry's original conclusion that climatic deterioration in the 'Little Ice Age' was in part responsible for the cessation of arable production, and thus the abandonment of previously farmed land, in 'marginal' areas of southern Scotland.

A PALYNOLOGICAL TEST OF PARRY'S MODEL

Hypothesis testing

The hypothesis to be explored here is simple: pollen sites dated to the 'Little Ice Age' lying close to and above critical altitudinal cultivation limits (derived from Parry's model) will show a cessation of cereal pollen representation coinciding in time with the onset or intensification of climatic deterioration.

Within this hypothesis are a number of research-design requirements which need to be satisfied in order for the hypothesis to be tested. These together with some assumptions will be considered next, and other assumptions will be evaluated following presentation of the results.

The study area and sites analysed

Geographical concerns

The study area is located in the northern Cheviot Hills, within the catchment of the Bowmont Water (Figure 1.1a). This study forms only part of a much larger palaeo-environmental investigation within the valley (Tipping in prep.). The results do not apply directly to the Lammermuir Hills, but these two areas are closely similar in nearly all respects, and sufficiently close to allow comparison. They lie only 40 km apart, and are comparable in climate and land-use history.

Within the Bowmont Valley, sites were sought that lay in close proximity, so that any changes in land-use would be explicable by altitudinal and not spatial factors. Three peat basins were located, not on the same hillside, but within 4 km of each other. The sites are Sourhope, Swindon Hill and Mow Law 'A' (Figures 1.1b and 1.1c). Mow Law 'A' is adjacent to Mow Law 'B', but this site lacks the palynological and temporal resolution to be considered here.

Altitudinal concerns

Crucial in site selection is the altitudinal distribution of sites. Parry (1975; 1978) reasoned that the risk of crop failure can be expected to increase exponentially rather than linearly with increasing altitude. He defined critical altitudes at which he deemed the inferred frequency of crop failure to be important in farmers' decision-making. Clearly the risk at any one altitude will change as climatic stresses change, but as an average over the time period for which annual climatic records are available (from 1659; Manley 1974), which is generally thought to include the nadir of the 'Little Ice Age' (Grove 1988), an altitude of *circa* 275 m OD can be shown to have a risk of oat crop failure of one year in fifty (1:50). At 310 m OD the risk is increased to 1:10, and at 420 m OD it is 1:2, ie a predicted crop failure once in two years.

The altitudes of the three pollen sites are: Sourhope (297 m OD), Swindon Hill (375 m OD) and Mow Law 'A' (420 m OD). Assuming the same critical altitudes for the northern Cheviots, a reasonable assumption given their similar climates, the expected risk of crop failure at each site can be calculated as: Sourhope (1:25); Swindon Hill (1:5); and Mow Law 'A' (1:2). From this it can be seen that all three sites can be regarded as potentially sensitive to inferred climatically-induced changes in the cultivation limit, and the higher sites can be expected to have been particularly exposed to increasing risk of crop failure.

Pollen-representational concerns

There is a need for the pollen sites chosen to reflect agricultural practices related specifically to the landscape and altitude of the site. Sites which receive a substantial component of their pollen from distances further than a few hundred metres from the basin cannot be used. A recently developed model of pollen recruitment to sites suggests a strong relation between basin diameter and size of pollen recruitment area (Jacobson & Bradshaw 1981), with the smallest sites having a greater local component. Quantifying this relation is difficult, but a site less than 20 m in diameter can be expected to receive in excess of 90% of its pollen from within a few hundred metres (called 'extralocal' pollen).

Using this model, the basin diameters and suggested pollen recruitment areas are: Sourhope (*circa* 350 m – 33% extralocal pollen); Swindon Hill (*circa* 60 m – 85% extralocal pollen); and Mow Law 'A' (*circa* 40 m – 90% extralocal pollen). Each of the sites is a small valley peat, and none has a stream catchment substantially greater than the actual peat basin diameter. From these observations it can be seen that the higher sites satisfy the requirement for limited pollen recruitment areas, while Sourhope can be expected to have a relatively high 'regional' (in excess of a few hundred metres distance) component. Nevertheless, in regard to the hypothesis being tested, this lowest site is the least sensitive and least critical.

Temporal concerns

A secure chronology for the deposits is critical, because Parry's most detailed reconstructions of crop failure frequency are calculated on annual instrumental temperature records. Three techniques have been used to construct

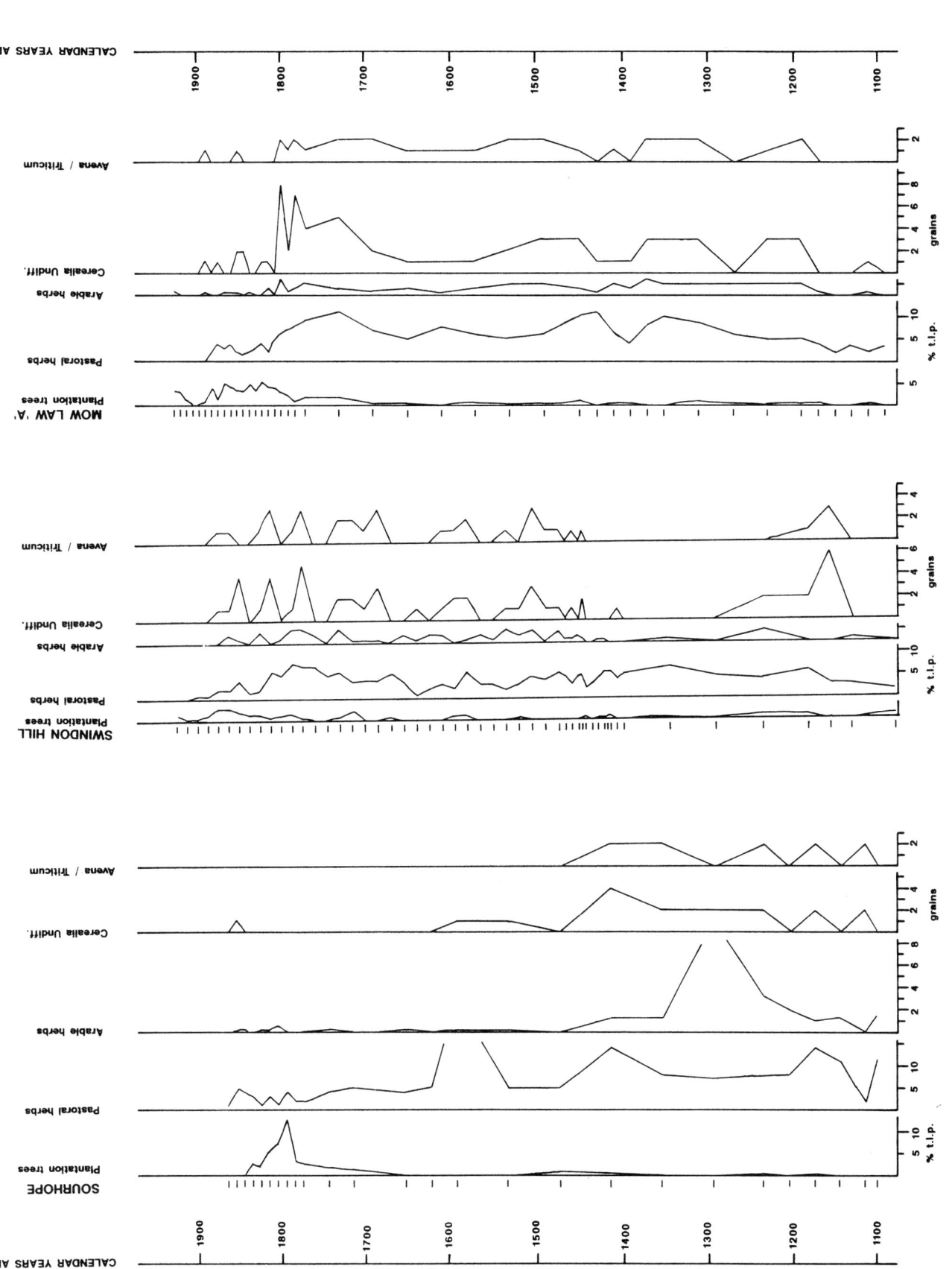

Figure 1.2 Simplified pollen diagrams for the three pollen sites, (a) Sourhope, (b) Swindon Hill and (c) Mow Law 'A' for the period from AD 1100 to the present day, depicting the percentage tlp frequencies for 'Plantation' trees (see text) and pastoral and arable herbs, and grains per 300 count for Cerealia undiff. and Avena/Triticum.

the chronology employed in Figures 1.2a, 1.2b and 1.2c. Radiocarbon dating is of limited value in the late-historic period because it has an error of between 10–20% of the mean value (at one standard deviation), cannot be used to date material younger than approximately 150–200 years old, and calibration can result in a number of intercept dates. Nevertheless, it can be used, as at Sourhope, where thirteen ^{14}C assays within the 315 cm of peat have established a secure chronology for the last 10,000 years, and at Swindon Hill, where five assays show the 152 cm of peat to post-date 4800 BP, to construct a mean sediment accumulation rate that can be extrapolated to the ground surface. At Mow Law 'A' three ^{14}C assays were obtained throughout the 77 cm of peat, but two samples appear to have been affected by rootlet penetration from above, and only the basal date is valid, being in agreement with a similar basal date at Mow Law 'B'.

Dating by ^{210}Pb was used to obtain sediment accumulation rates extending back in time from the present day to about 200 years ago at two sites, Swindon Hill and Mow Law. The results were tested against the third method, pollen-marker dating, at Mow Law, and against the first (^{14}C) at Swindon Hill.

This third method uses the palynological signal of exotic arboreal pollen, from trees planted in the Agricultural Improvements following the AD 1720s, as a marker horizon in the pollen stratigraphy. Figure 1.2 shows that the sum of such tree types (*Pinus, Picea/Abies, Fagus, Ulmus*) has a distinctive peak at Sourhope and Mow Law, though this is not seen at Swindon Hill. Although some plantations for both commercial and aesthetic reasons were established in the Bowmont Valley, the major developments were in lowland areas of the Merse (Anderson 1967). Anderson documented the development of such plantations, and allowing for a delay before they became sufficiently plentiful to be recognized in the regional pollen 'rain', a date of AD 1775 plus or minus 25 years, is ascribed to this peak at the two pollen sites. This agrees closely with the mean sediment accumulation rate of 0.25 cm per year calculated at Mow Law 'A' from ^{210}Pb analyses. It is also in agreement with the extrapolated mean sediment accumulation rate derived at Sourhope from ^{14}C assays. At Swindon Hill the ^{210}Pb derived sediment accumulation rate is in good agreement with that established by ^{14}C assays.

At each site the chronology used in the interpretation has been constructed from more than one technique, and no contradictions are seen between methods. Nevertheless, although on Figure 1.2 dates in calendar years AD are depicted, caution is required in the interpretation of the data, but it is likely that the chronology is accurate to plus or minus 25 years.

Of equal concern in the analyses is the temporal resolution of the analyses, in other words, how frequently pollen analyses are made from the sediment stratigraphy. Figures 1.2a, 1.2b and 1.2c show the approximate positions in time of the analyses, and show that the temporal resolution varies depending on the rate of sediment accumulation and the rate of vegetation change depicted in the analyses. At Sourhope, above AD 1800, samples record changes occurring nearly every decade (9.5 years), and below this an average of one sample per 30 years. At Swindon Hill the stratigraphy is sampled approximately every 14 years from the present day to *circa* AD 1500, every 8 years for a short period just before AD 1500, but before AD 1400 only every 40–45 years. The temporal resolution between AD 1750 and AD 1925 at Mow Law 'A' is around 8 years, about 35 years on average between AD 1200 and AD 1750, and every 20 years before AD 1200.

Throughout the text the term 'Little Ice Age' is used, but the precise chronology of this period of climatic decline is poorly understood (Grove 1988). At a local level there is some palynological evidence for the appearance at one site (Swindon Hill) of plant taxa found today on the summit plateau of The Cheviot, dated to between *circa* AD 1500 and AD 1700, and this period is likely to define the most intense phase of the climatic deterioration.

Cereal pollen identification
The criteria of Andersen (1979) were used to identify large grass pollen grains to cereal type. Two categories are depicted on Figure 1.2. 'Cerealia undiff.' comprises those grains too badly damaged to allow more detailed identification, and may in addition to oats include some grains of barley or wild grasses. *Avena/Triticum* includes both oats and wheat, but it is assumed here that only oats are represented, a view supported by documentary records (Dodgshon pers comm). One weakness in the analyses is that no attempt was made to enhance the representation of cereal pollen either by coarse micro-sieving techniques (Bowler & Hall 1989) or by low-magnification scanning (Edwards & McIntosh 1988). Consideration as to the meaning of the cereal pollen curves will be made following presentation of the results.

RESULTS

The pollen diagrams for each site in Figure 1.2 contain pollen curves for 'Plantation' trees (a composite curve of those tree types planted in the 18th century), pastoral and arable indicator herbs (cf Behre 1981), calculated as % total land pollen (tlp), and Cerealia undiff. and *Avena/ Triticum*, expressed as grains per count: all counts are comparable, being of 300 ± 10 tlp. Complete pollen diagrams covering the full sedimentary sequences will be published elsewhere. The diagrams in Figure 1.2 are sufficient for the present purposes, since the only concern is to test whether cereal pollen continues to be represented through the climatic deterioration of the 'Little Ice Age'.

Given the simplicity of the pollen diagrams, the results can be succinctly summarized:

Sourhope (297 m OD)
(1) cereal and oat pollen present from *circa* AD 1100;

(2) no oat pollen recorded after *circa* AD 1500;

(3) cereal pollen not recorded after *circa* AD 1650.

Swindon Hill (375 OD)

(1) cereal pollen, including oat, present between *circa* AD 1150 and *circa* AD 1300;

(2) both cereal and oat pollen not recorded between *circa* AD 1300 and *circa* AD 1475;

(3) erratic representation above *circa* AD 1475, with the possible absence of oat pollen between *circa* AD 1650 and *circa* AD 1700 (three pollen spectra);

(4) representation of both cereal and oat pollen until *circa* AD 1900.

Mow Law 'A' (420 m OD)

(1) cereal and oat pollen consistently present from *circa* AD 1150;

(2) possible decline in representation of cereal pollen between *circa* AD 1400 and *circa* AD 1450;

(3) sustained presence until *circa* AD 1800.

(4) limited representation until *circa* AD 1900, then absent.

DISCUSSION

Cereal pollen representation during the 'Little Ice Age'

A number of questions still need to be addressed concerning the representation of the cereal pollen before conclusions can be drawn.

Does the cereal pollen represent the local presence of the plant?

Irrespective of the basin size in determining the pollen recruitment area (above) it is clear from the representation of 'Plantation' trees that some pollen is not of local origin. It is possible that the cereal pollen is of a similar long-distance origin.

However, several lines of evidence mitigate against this suggestion. Firstly, there is very good evidence for the representation of cereal pollen to be strongly related to the presence of the plant (Vuorela 1973; Hall 1988), the latter study being most pertinent, deriving from traditionally managed oat fields in an essentially treeless and windy landscape. This is because cereals are self-pollinated, and have very restricted dispersal distances away from the parent plant. Secondly, the differences between sites in the temporal representation of cereal pollen, particularly between Sourhope and the two higher-altitude sites, suggest that no cereal pollen from Swindon Hill or Mow Law 'A' reached Sourhope, at least after *circa* AD 1500, despite the latter site having a considerably larger pollen recruitment area. Price & Moore (1984) and Holland (1975) have described the wind-blown transport of pollen from lowland areas to uplands, but again the differences in cereal

representation between sites suggests that in the Bowmont Valley this is not a serious problem.

Does the cereal pollen represent the plants' local growth?

This question concerns the possibility that cereal pollen could be introduced in fodder for grazing animals, and so need not represent cereals growing at high altitudes.

There is no evidence from the pollen diagrams to disprove this possibility, since the curves for pastoral indicator herbs indicate that grazing was part, and from documentary evidence the main part, of the agricultural activities at each pollen site. Nevertheless, it is not considered likely because the practice then as now was to bring livestock down from the hills in winter to the farm, which we know from Roy's map (1747–55) lay on the valley floor.

Does the cereal pollen represent the cultivation of crops?

At face value this point is not critical in testing the hypothesis that the growth of oats at critical altitudes was prevented by climatic deterioration in the nadir of the 'Little Ice Age'. Providing that the arguments for local presence and growth are accepted, it does not matter whether the crop was planted or self-set, because Parry's suggested controls on crop failure are physiognomic limitations based on the climatic demands of the crop. However, mis-interpretation of sustained records for oat pollen is possible if the plant is capable of surviving for long periods in an uncultivated state, because Parry's model specifically predicts the abandonment of oat cultivation. Uncultivated oats, growing wild, will continue to produce pollen in the 'good' years even where, as at Mow Law 'A', crop failure is predicted from Parry's model to occur once in two years. So it is important in regard to the fuller implications of Parry's thesis to understand whether a cereal crop can survive following a farmer's abandonment of the field.

Once sown as a crop and abandoned, it is possible for cereals to persist within 'natural' plant communities, but almost certainly not for prolonged periods. Cereals are annuals, and would not naturally perennate the following year; they require to be planted. Bare ground is essential to growth, since they do not easily tolerate competition, and the increasing closure of a grassland sward would militate against their survival. In addition, oats in particular can be expected to have undergone selective grazing by herds and flocks since the straw is highly nutritious, and sustained grazing, intensified after the mid-18th century (below) might be expected to have rapidly depleted the crop. In other circumstances it would be useful to refer to the representation of arable indicator herbs (Figure 1.2) as evidence for continued cultivation, but at sites where comparable open ground communities can originate from climatically-induced soil break-up this could be misleading.

It is argued next that high-altitude cereal cultivation in fact had a significant role to play in the agricultural economy, providing good grounds for the sustained pollen records in the northern Cheviots to be meaningful. It is considered that purposeful cultivation of cereals was continued into the late 18th century at least.

The role of high-altitude cereal (oats) cultivation in the agricultural economy

By AD 1100 palynological evidence suggests that the Bowmont Valley was virtually treeless, and the different parts of the valley were in more-or-less intensive agricultural usage. The valley appears to have been densely settled and intensively cultivated, and was a 'highly developed area of land' (Gilbert 1979, 236). Cereal cultivation was clearly of some importance, and the parish contained at least one mill in the late 12th century AD. Crops may have been grown in the uplands, since between AD 1234 and 1249 Kelso Abbey was given a 'toft and croft' (Jeffrey 1855, 274) on the moors near to the outlet at Wytelawe (Whitelaw), on the English Border, and the toft in question must have lain at around 350–400 m OD.

However, if we can accept the cartographic evidence of Pont/Blaeau (1590s) and Roy (1747–55) for the locations within the valley of the main areas of tillage in the early-modern period, it is clear that neither Swindon Hill or Mow Law lay within the cultivated land, which is concentrated on the valley bottoms and lower slopes. However, the maps may not be accurate or comprehensive. There are indications that ploughing was undertaken at considerable altitudes. The First Statistical Account (1795; see Sinclair 1979) for Morebattle parish (which after *circa* AD 1672 included all of the study area) reported that during the Anglo-Scottish wars (periodically from AD 1296 to AD 1603) '... great portion of the lands on the borders were kept under white crops [oats], as it was not so easy for the plundering parties, in these unhappy times, to carry off crops of grain, as it was, had the land been in pasture, to drive away the cattle' (Sinclair 1979: Vol. III, 591). At some time in the reign of Henry VIII, Scots encroached into the College Valley between Whitelaw and the head of Halter Burn, at around 350 m OD, and the warden of the East March was advised to '... see yt no plowing, sowinge or other possession be made' (Bowes 1847). Bowes reported that on the border ridge between Outer Cocklaw and Gamel's Path, above *circa* 500 m OD, there had been houses built (Bowes differentiates these structures from shieling huts) 'in tymes past' (Hodgson 1828, Part III, Volume II, 209), only to have the territorial rights disputed, such that they were then (AD 1542) only shieling grounds. This may imply that shieling did not have a long history (there is no mention of this activity in the 12th–14th century AD land grants; above), and that permanent upland settlement was the norm in more peaceful times.

In the upland part of the valley there was a progressive rather than dramatic decrease in farm populations from the late 17th century until around AD 1750 (Dodgshon 1972 and pers comm). The farms were infield-outfield systems, but the evidence suggests that large areas of the upland were taken up with arable production before *circa* AD 1750, larger than during the subsequent Napoleonic Wars. There was in the period preceding AD 1603 the need to maintain a high tenant population for defence of the territory, but this agricultural economy appears to have been sustained beyond the Act of Union into the 18th century.

Dodgshon (pers comm) sees little change on these upland farms between AD 1750 and the Napoleonic Wars, except in the linking of these to lowland farms for stock fattening and winter feeding of fodder crops (turnips and sown grasses), which did result in reductions in upland arable. By AD 1790–91, the Minister of Hownam parish could comment that 'Several of the farms in the higher part of the parish have scarcely been ploughed in the memory of man' (Sinclair 1979: Volume III, 465), although this may have been an exaggeration, since Hownam Farm in 1769 carried 105 acres of arable (Dodgshon 1976), or 11% of the total acreage.

There are a number of agricultural practices that would support this documentary record for the maintenance of cereal crops at localities distant from the infield. The first of these is tathing (Dodgshon 1975), a form of shifting cultivation involving the seasonal utilization of part of the pasture to increase crop yield, in which the herd or flock is folded in certain areas in one year, moved the next and the manure-enriched land ploughed and sown.

Shielings were the pastoral end of a transhumant system (Bil 1990), and are known to have existed in the Bowmont Valley from place-name, documentary and archaeological evidence. Both Swindon Hill and Mow Law are extensive plateau surfaces and lie at comparable altitudes (around 400 m OD) to localities known to support shielings in the northern Pennines (Ramm *et al* 1970) and Tynedale (Charlton & Day 1979), just over the Border. Shielings could have an arable component, either utilizing quick-ripening crops or, where the shieling lay reasonably close to the farmstead as probably in the Bowmont Valley, planting the crop before 'summering' commenced in May. These systems were maintained in adjacent upland areas until around AD 1660–1700 (Charlton & Day 1979), although there is evidence from the Bowmont Valley that shieling was maintained only until the turn of the 17th century AD (Bain 1896).

Shielings appear to have died out before the beginning of the 18th century AD, but after *circa* AD 1740 specialist sheep units were built in their place, these being the stock or store farms. The change in organisation of pastoral farming, with lowland farms given over to the fattening of stock rather than breeding, meant that upland farms concentrated on the latter (Dodgshon 1976, 1983). This change to increased stock numbers can be discerned in the full pollen diagrams from the three pollen sites (unpublished data) by the dramatic conversion of *Calluna* moorland to species-poor grassland. The timing of this change is different between sites, at Sourhope occurring at *circa* AD 1750, Mow Law 'A' at *circa* 1850 and Swindon Hill at around AD 1890. The strong suggestion is that overgrazing occurred, but the introduction of sown grasses and turnips to upland pasture in the later 18th century, perhaps after AD 1780, eased the pressures on grassland, and it is possible that cereals could have been grown as an additional fodder crop. Nevertheless, some of these farms, generally lying above 220–300 m OD (unpublished data) appear to have maintained a small amount (*circa* 5%) of arable, not for fodder crops but for subsistence. In Tynedale (Charlton & Day 1979) these permanent settlements were often located on shieling sites because of their nutrient-enriched soils and their capacity to support a subsistence cereal crop.

Testing the hypothesis

It is concluded from this detailed discussion that cereals were (i) present close to each pollen site and (ii) purposefully grown as a crop. As such, the results of the analyses can be used to test the hypothesis developed earlier.

The results do not support the hypothesis, and it is rejected. Both high-altitude sites, Swindon Hill and Mow Law, show that cereal cultivation continued through the most intense period of the 'Little Ice Age'. Crop growing was possible and was undertaken prior to, during and after the climatic nadir of the 16th and 17th centuries. The pollen record cannot establish whether crop failure increased at high altitudes after *circa* AD 1500, estimated here to have been the start of the climatic decline, but any increases in crop failure frequency were certainly insufficient to lead to the abandonment of these uplands. The cessation of cereal-growing at these sites, in the early 19th century at Mow Law, later at Swindon Hill, occurred after the end of the 'Little Ice Age'.

However, in one perhaps critical detail the Cheviot Hills are different to Parry's study area of the Lammermuir Hills. It has been stressed above that the localities utilized for high-altitude crop-raising in the Bowmont Valley were distant from the infield surrounding farms. Rig-and-furrow is mapped in the valley and is found to be extensively developed on valley floors and gentler slopes at the foot of hills. The expanses of upland medieval ridged cultivation seen in the Lammermuir Hills (see for instance Plates 7 and 8 of Parry 1978) are not apparent in the Cheviots. The significance of this is unclear, but might indicate that the upland plateaux studied here were not core arable areas as they perhaps were in the Lammermuirs, and so were not regarded with the same intense interest by farmers. This aspect needs additional exploration. However, the observation that the Cheviot uplands could be and were cropped does strongly suggest that the arable land of the Lammermuirs were also capable of maintaining viable crops during the 'Little Ice Age'. It is now left to consider the reasons for the apparent failure of Parry's hypothesis to explain upland de-population in south east Scotland.

The nature of climatic change during the 'Little Ice Age'

There is no evidence from the northern Cheviots for climatically-induced cessation of crop-growing, or for the abandonment of land. Decreases in population numbers which commenced in the 18th century were due to an economic re-alignment of the upland agricultural system.

The maintenance of cereal cultivation at high altitudes means that our understanding of the climatic stresses on oat cultivation in the 'Little Ice Age', the basis of Parry's argument, needs revision. There are several points to be considered here. Firstly, Parry's calculation of the critical altitudinal limits to oat cultivation may be in error, and be too low (Duncan 1992). Duncan suggests Parry's underestimate to amount to around 70 m. If correct, it would imply, for example, that the 1:2 crop failure risk at Mow Law 'A' is an over-estimate of the risk. Clearly, any re-evaluation of the climatic limits to cultivation would require a re-assessment of the spatial association between climatic limit and abandoned land that Parry (1978) found so convincing. Secondly, it is possible that the resistance of oats to climatic stress was under-estimated by Parry. His data on climatic stresses related to 19th century varieties of oat commonly grown in southern Scotland (Parry & Carter 1985), but prior to the 19th century more hardy varieties (grey and black oats) were grown (Symon 1959; Whyte 1979).

Thirdly, and probably most importantly, the nature of climatic change that took place in what has been called the 'Little Ice Age' needs to be re-assessed. This is not an original point, and in Parry's later papers (eg Parry & Carter 1985) this re-assessment had taken place, with the emphasis moving from the effects of long-term (decadal to centennial) declines in mean annual temperatures (eg Lamb 1977) to the identification of short-lived 'risk' periods (runs of three to ten years) in which contiguous crop failures may have forced abandonment of land on farmers.

Such an approach recognises that the 'Little Ice Age' was typified by highly variable weather patterns, with extraordinary extremes of both cold and warmth, often occurring in consecutive years. Probert-Jones' (1984) statistical analysis of Manley's (1974) series of annual temperature records for central England showed that no period of sustained cold longer than a decade can be recognised within the 'Little Ice Age'.

One decade was exceptionally cold (though not outside the expected range); AD 1688–1698. This decade was seen by Parry & Carter (1985) to be one of three clusters of years in Manley's series where cool summers might be expected to have resulted in consecutive crop failures above 300 m OD. This decade was also the focus of Whyte's (1981) discussion of the effect of sustained wet and cold periods on pastoral farming in southern Scotland which although well documented, fails to illustrate the theoretical results of climatic extremes with examples.

Flinn (1977) has indicated that this period of the 1790s, the 'Seven Ill Years', was less severe than popular folklore would maintain, and Probert-Jones (1984) shows that this exceptionally cold decade was promptly followed by one of the warmest decades in the instrumental record.

This pattern of very short-lived climatic excursions cannot be detected in the pollen record from the northern Cheviots. The temporal resolution of samples is not sufficient to be able to detect single or even clustered crop failures, but in its 'smoothed' depiction of cereal production it does allow more sustained monitoring of the course of events than the erratic and inconsistent documentary record, and provides the observation that whatever short-lived setbacks such short-term climatic extremes presented, no long-term disenchantment with the prospects for crop-growing on the plateaux of the Cheviot Hills is seen.

CONCLUSIONS

The analyses presented here suggest that purposeful cultivation of oats was maintained in the northern Cheviot uplands from the high-medieval period until the end of the 18th century. No permanent abandonment of cultivated land due to repeated crop failure occurred. Changes in the agricultural use of the region are thought to have been solely socio-economically induced. This result questions the hypothesis that the 'Little Ice Age' climatic deterioration had a significant role to play in the abandonment of land at high altitudes.

ACKNOWLEDGEMENTS

I would like to thank Historic Scotland for funding of the work in the Bowmont Valley, and Noel Fojut, Patrick Ashmore and Richard Welander of HS for their continued interest and support. Roger Mercer (RCAHMS), then of the Department of Archaeology at Edinburgh University, kindly invited me to participate in the Bowmont Valley Archaeological Survey. For fieldwork assistance in often inclement weather I am grateful to Pete McKeague, Charles Le Quesne, Harry Chrisp, Frank Matsaaert and Jane Webster, and for laboratory assistance to Andrew Akhtar. Grateful thanks are accorded Gordon Cook (^{14}C) and Gus Mackenzie (^{210}Pb) at the Scottish Universities Research and Reactor Centre, East Kilbride, and to HS for funding the radiocarbon assays. My interpretations have benefited greatly from discussions with Robert Dodgshon (Aberystwyth), Strat Halliday (RCAHMS), Sheila Boardman and Alan Fairweather, and discussants at the AEA Edinburgh conference.

REFERENCES

Andersen, S Th 1979 'Identification of wild grasses and cereal pollen', *Danmarks Geologiske Undersogelse* 1978, 69–92.

Anderson, M L 1967 *A history of Scottish forestry*. Edinburgh: Thomas Nelson.

Bain, J 1896 *The Border papers*. Edinburgh: HM General Register House.

Behre, K-E 1981 'The interpretation of anthropogenic indicators in pollen diagrams', *Pollen et Spores,* 23, 225–245.

Bil, A 1990 *The shieling 1600–1840*. Edinburgh: John Donald.

Bowes, Sir R 1847 *The English Border in the days of Henry VIII*. Newcastle: M A Richardson.

Bowler, M & Hall, V A 1989 'The use of sieving during standard pollen pre-treatment of samples of fossil deposits to enhance the concentration of large pollen grains', *New Phytologist*, 11, 511–515.

Charlton, J D B & Day, J J C 1979 'Excavation and field survey in upper Redesdale: Part II', *Arch Ael*, 5th series, VII, 207–233.

Dodgshon, R A 1972 'The removal of runrig in Roxburghshire and Berwickshire 1680–1766', *Scottish Studies*, 16, 121–137.

Dodgshon, R A 1975 'Farming in Roxburghshire and Berwickshire on the eve of improvement', *Scottish Historical Review*, 54, 140–154.

Dodgshon, R A 1976 'The economics of sheep farming in the Southern Uplands during the age of improvement, 1750–1833', *Economic History Review*, 29, 551–569.

Dodgshon, R A 1983 'Agricultural change and its social consequences in the Southern Uplands of Scotland, 1600–1780', *in* Devine, T M & Dickson, D (eds), *Ireland and Scotland 1600–1850*. Edinburgh: John Donald, 46–59.

Duncan, K 1992 *'A climatic record for south east Scotland and implications for agriculture and disease'*. Unpublished Ph.D. thesis, University of Edinburgh.

Edwards, K J & McIntosh, J 1988 'Improving the detection rate of cereal-type pollen grains from *Ulmus* decline and earlier deposits from Scotland', *Pollen et Spores,* 30, 179–188.

Flinn, M 1977 *Scottish population history from the 17th century to the 1930s.* Cambridge: Cambridge University Press.

Gilbert, J M 1979 *Hunting and hunting reserves in medieval Scotland*. Edinburgh: John Donald.

Grove, J 1988 *The Little Ice Age*. London: Methuen.

Hall, V A 1988 'The role of harvesting techniques in the dispersal of pollen grains of Cerealia', *Pollen et Spores*, 30, 265–270.

Hodgson, J 1828 *A history of Northumberland*. Newcastle.

Holland, S M 1975 *A pollen analytical study concerning settlement and early ecology in Co. Down, Northern Ireland*. Unpublished PhD thesis, Queens University Belfast.

Jacobson, G L Jr & Bradshaw, R H W 1981 'The selection of sites for palaeovegetational studies', *Quaternary Research,* 16, 80–96.

Jeffrey, A 1855 *The history and antiquities of Roxburghshire*. Jedburgh: Walter Easton.

Lamb, H H 1977 *Climate: Present, past and future*. London: Methuen.

Manley, G 1974 'Central England temperatures: Monthly means 1659 to 1973', *Quarterly Journal of the Royal Meteorological Society*, 100, 389–405.

Parry, M L 1973 *Changes in the upper limit of cultivation in south east Scotland 1600–1900*. Unpublished PhD thesis, University of Edinburgh.

Parry, M L 1975 'Secular climatic change and marginal agriculture', *Transactions Institute of British Geographers*, 64, 1–17.

Parry, M L 1978 *Climate change, agriculture and settlement*. Folkestone: Dawson & Sons.

Parry, M L 1981a 'Climatic change and the agricultural frontier: a research strategy', *in* Wigley, T M L, Ingram, M J & Farmer, G (eds), *Climate and history*. Cambridge: Cambridge University Press, 319–336.

Parry, M L 1981b 'Evaluating the impact of climatic change', *in* Delano Smith, C & Parry, M L (eds), *Consequences of climatic change*. Nottingham: Department of Geography, University of Nottingham, 3–16.

Parry, M L & Carter, T R 1985 'The effect of climatic variations on agricultural risk', *Climatic Change,* 7, 95–110.

Price, M D R & Moore, P D 1984 'Pollen dispersion in the hills of Wales: A pollen shed hypothesis', *Pollen et Spores,* 26, 127–136.

Probert-Jones, J R 1984 'On the homogeneity of the annual temperature of central England since 1659', *Journal of Climatology,* 4, 241–253.

Ramm, H G, McDowall, R W & Mercer, E 1970 *Shielings and Bastles*. London: HMSO.

Symon, J A 1959 *Scottish farming past and present*. Edinburgh: Oliver & Boyd.

Sinclair, Sir J 1979 *Statistical Account of Scotland 1791–1799 – Vol III The eastern Borders*. Wakefield: EP Publishing.

Vuorela, I 1973 'Relative pollen rain around cultivated fields', *Acta Botanica Fennica,* 102, 1–27.

Whyte, I 1979 *Agriculture and society in 17th century Scotland*. Edinburgh: John Donald.

Whyte, I 1981 'Human response to short- and long-term climatic fluctuations: the example of early Scotland', *in* Delano Smith, C & Parry, M L (eds), *Consequences of climatic change*. Nottingham: Department of Geography, University of Nottingham, 17–29.

2. Bad for trees – Bad for humans?

Michael G L Baillie

Abstract

By definition, archaeologists are interested in what happened to people in the past. Often, however, conventional archaeological evidence gives little indication of the nature of the events which caused change in the archaeological record. When we find that marginal areas were abandoned, or indeed colonized, we can infer that environmental change may have taken place though its nature often remains a mystery. This is where tree-ring evidence can begin to offer a new perspective. Trees represent a parallel biological system and, with the availability of long tree-ring chronologies, we can begin to assess both when trees were affected by environmental factors and, to some extent, the nature of the changes involved.

A growing body of evidence suggests that at times when trees suffered due to environmental downturns people also suffered. We might expect that marginal areas would feel the effects first and that the fate of marginal populations might therefore be most intimately tied up with the well-being of trees.

INTRODUCTION

At first glance, *marginality* looks as if it should be easy to define. We could suggest that marginality involves living in an area which can, with change in just one significant environmental factor, become uninhabitable. Alternatively, we could suggest that marginality involves living in an area which can, with change in just one significant environmental factor, become uninhabitable in the absence of significant social or technological innovation. More generally, we might simply define a marginal situation for a human population as being one which can become untenable due to a change in some parameter outside the control of that population. Again, we might suggest that if a population expands to a point where any reduction in resources inevitably leads to a reduction in population, then that population has placed itself in a marginal situation.

However, all of these 'definitions' have a parochial ring to them. In reality there is a hierarchy of marginality. For a limiting case we could consider trying to maintain any trace of human civilization on Earth in the face of (a) a large asteroid or cometary impact, (b) a Toba class volcanic eruption or (c) a neighbouring supernova. Any one of these 'environmental' changes could (indeed almost certainly would) render civilization extinct and, moreover, would be unstoppable and probably unannounced. By and large, since the return time of such events is regarded as long, human activity takes no account of the 'natural marginality of the planet'. Be that as it may, 'return time' for catastrophic events must be an important factor in assessing marginality.

To take another limiting case, marginality can be the act of living anywhere which, in retrospect, turns out to have been ill-advised; the citizens of Pompeii discovered the marginality of their situation on the 24th and 25th August AD 79. Of course, we assume that they did not know their situation was marginal; but is that really the case? We now know that Vesuvius had erupted fairly spectacularly in the second millennium BC (Vogel *et al* 1990). We assume that this previous eruption had been forgotten by Roman times – but had it? We, and the people of Naples, know all about AD 79 and its effects on the Naples area but the citizens of Naples still live beside not only Vesuvius but the much larger Phlegraean Fields caldera. So, while one would think that human populations would take note of past events and past variations, and, sensibly, should adapt their location, or limit their numbers to a level which can survive the worst conditions within their group memory, in fact there is little evidence for this outside aboriginal populations. Marginality turns out to be a multi-strand issue involving, among others, such factors as time depth, perception of 'normal' conditions, level of resource saturation, buffering, ability to innovate and even human stubbornness if not actual stupidity.

The complexity of the subject in fact encourages a search for a simpler definition and it seems that this is best provided archaeologically. Most of the time, archaeology takes little account of human intentions but can be expected to show the *effects* of marginality. If, in the archaeological record, a population which has been active suddenly disappears or shows clear signs of reduction, if an area which has been occupied is abandoned then at the very least we can suspect that conditions became marginal. Irrespective of how marginality is defined, it seems inherently likely that marginal areas will be among the *first* to suffer in the event of any environmental downturn. This is where dendro-chronology may have a role to play. If studies of tree-growth can define either points (or periods) in time when conditions in general became 'worse', then those should be the times when marginal populations would have been most likely to be affected. This is something which the archaeologist can eventually hope to test.

In 1988 it was demonstrated that oak trees, growing on north Irish peat bogs, had apparently been recording the environmental effects of significant volcanic dust-veil events (Baillie & Munro 1988). The events were marked by the widespread occurrence of narrowest growth rings, ie rings which were, for some trees, the narrowest in their lifetimes. If the underlying hypothesis is true, then these precisely-dated events in the tree-ring record mark the dates of sig-nificant environmental downturns which might be associ-ated, indeed *seem* to be associated, with problems for human populations. So, in the first instance, we should use environ-mental events, such as those in the 1620s BC, the 1150s and 1140s BC, at 208 BC, 44 BC and AD 536–545 to *test* for change in marginal areas. Elsewhere I have suggested the concept of 'marker dates' – dates, like those just mentioned, which might be expected to show up in the archaeological record as periods of change (where change could be related to factors such as population movement, population de-crease, defensive construction, etc) (Baillie 1991). Indeed, as well as dust-veils, there are other dates which fall out of the tree-ring record which hint at environmental change and which may also prove to be marker dates where populations are likely to have responded to environmental pressure.

It is simple to generate a list of some of the more obvious effects which can be deduced from tree-ring evidence. Apart from the narrowest-ring volcanic events, these include the following;

a)   other periods of reduced growth which are 'bad' in the sense of how trees responded to conditions during their lifetimes.

b)   From the dating of *archaeological* timbers we can get

a sense of when people were felling oaks and putting them into wet contexts where they have survived for us to study. Again, this is only one aspect of human activity but it represents a starting point. We can date phases of this particular type of human activity.

c)   From the same, precisely dated, archaeological ring patterns we can assess the *ages* of the trees being used at different times. Similarly, we can identify when the trees were starting to grow; thus we might look for regeneration phases in the past. As will be discussed below, regeneration can, on occasion, be associated with reduction of human 'pressure' on the landscape.

d)   From naturally-occurring bog oaks we can assess when individual bogs were supporting oak trees, when those trees started growing and when the conditions on the bog surfaces finally became unsuitable for supporting oak growth.

e)   From the growth histories associated with oaks from multiple bogs we can begin to get pictures of general, as opposed to local, environmental change. If a lot of trees started growing at around the same time on different bogs we might infer that conditions had changed to favour bogs supporting trees, ie the surfaces may have dried out in a relative sense. Alternately if oaks die out at the same time on different bogs – and we can now look for such synchronous die-off periods across Ireland, England and northern Germany – we might infer a general increase in wetness. If such episodes can be identified, we might expect marginal populations to have been affected.

Overall, dendrochronology in a region can provide a suite of precisely dated strands of environmental evidence as a backdrop to the historical or archaeological story of the area. Examples can be drawn from anywhere that dendro-chronology is established. In the semi-arid southwestern United States, where Douglass first proved the basic principles of dendrochronology by building chronologies for the prehistoric period, it was immediately apparent from the tree-rings that there had been a severe drought episode in the AD 1280s. Not surprisingly, it was subse-quently discovered that this coincided with a major aban-donment of many Amerindian agricultural settlements. The precise dating effectively rules out argument about the connection between the drought and the abandonment – the two coincide.

In this paper I will look at a few such dates which are suggested by tree-ring evidence. These examples are restricted to the last two millennia but serve to illustrate the main types of proxy environmental reconstruction from tree-rings. It is apparent, in at least some of the examples, that there are hints of marginality.

The Irish Potato Famine AD 1845

We do not have to go very far back in time to find a quite surprising tie up between oak trees and people in marginal situations. In the course of identifying the narrowest-ring events in prehistoric trees, an excursion was made into modern oaks in the north of Ireland. If one asks the question 'which decade in the last 200 years was the worst from the point of view of north Irish oak trees?', the answer is unequivocally 'the 1840s', see Figure 2.1a. The same question asked of humans would elicit the same answer; the 1840s undoubtedly represented the worst decade for the human population of Ireland because there was an immediate decline of up to 1 million (or more if some sources are to be believed) and an overall decline of about 3 million over the following few decades, see Figure 2.1b.

> '... when in 1845 potato blight finally caught up with the potato in Ireland, disaster was inevitable. During the famine years (1845–51) about 800,000 people died, and twice as many emigrated. The heart was knocked out of Ireland and the population continued to fall without interruption until 1930...' (Mitchell 1986).

So here we have a case where two quite different biological populations both suffered at essentially the same time. Of course it could be argued that this is no better than a coincidence, and such an argument would appear to be bolstered by the observation that the really narrow tree-rings were in the period 1840–1844 while the famine took place in 1845–51. One could, however, restate the case and ask what the chances are that the four 'worst' years recorded by the north Irish oaks should be the four years immediately prior to 1845? Is this pushing coincidence a little far perhaps? In a sense, arguments about whether or not this coincidence is significant only become important if one is trying to prove a causal relationship between poor growth in trees and human starvation. Such proof is not being sought here. The important thing is that the 1840s offer an example where oak trees and humans both suffered within the same decade. It is also clear that the section of the human population worst affected were those whose lifestyle was most marginal, ie those living on marginal land, dependent on a single (and vulnerable) staple and without resources to purchase alternative nourishment – marginality in depth!

Interestingly, it is now known that another close juxtaposition between severely reduced oak growth, this time across northern Europe, and a notable reduction in human population occurred at AD 1740–42. In this case an earlier Irish famine is reputed to have killed 300,000 people in just a few years (Drake 1968). This population reduction was part of a widespread demographic crisis, believed to have been the last of the pre-industrial era (Post 1985). In this case, the more marginal sections of the Irish population succumbed to famine in the same years recorded by the oaks.

The Black Death

Early on in the construction of the Belfast oak chronology,

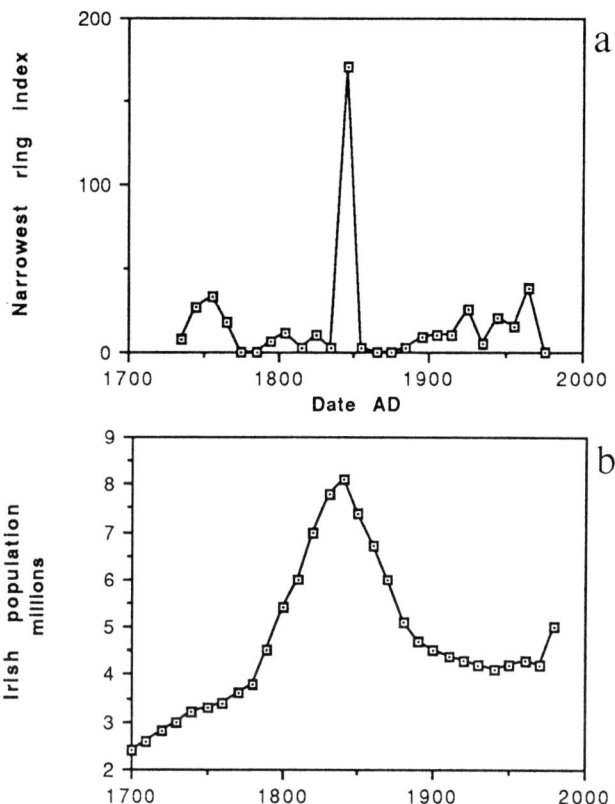

Figure 2.1 (a) Index of percentage of trees with narrowest rings times number of sites plotted against time showing the widespread narrowest ring event in the AD 1840s; (b) Plot of Irish population since AD 1700 (after Mitchell 1986).

it was noted that most trees which had been felled in the 16th and 17th centuries AD had started growth around AD 1360–1400. It proved difficult in both the north of Ireland and in the Dublin area to find oaks whose ring patterns spanned the 14th century. It was not until 1977 that suitable bridging samples were acquired from crannogs in the west of Ulster (Baillie 1977). The problem was not restricted to Ireland. Timbers from Castle of Park in Scotland had started growing in the 1350s and two different English chronologies started within a decade of 1350. This pattern implied that a lot of the oaks used by builders in Britain and Ireland in the 16th and 17th centuries AD had regenerated after the Black Death which arrived in these islands in 1348.

This observation implied that the oak record, which showed an overall depletion/regeneration phase centred on AD 1350, was interrupted just when the human population suffered a major setback. We now know that a building hiatus shows up very clearly in both Germany and Greece, with no buildings producing tree-ring dates between AD 1348 and 1425/1440 (Hollstein 1980; Kuniholm & Striker 1983; 1987). So, although in this case trees do not 'suffer' in terms of narrow growth rings, their survival record documents a major human trauma (Baillie 1995).

The important aspect of the 14th century tree-ring 'problem' (a problem because the building hiatus and the regeneration phase produced difficulties for dendrochronologists trying to build chronologies across the century) is that it gives a clue as to what we should look for in earlier records. With the 14th century experience documented we are sensitized to take note of earlier examples of building hiatus and/or chronology building problems. Thus the Black Death sets the scene for a number of observations associated with a specific tree-ring problem around AD 800.

Something going on around AD 800–1000

In the same way that bridging the period around AD 1350 represented a hurdle to chronology building in Ireland, it became apparent, as the chronology was extended back in time, that there were problems in finding timbers whose ring patterns extended back before the mid-9th century AD. The robust Dublin chronology ran back to AD 855; a chronology for the north of Ireland ran back to AD 919; a Scottish chronology ran back to AD 946 (Baillie 1982). Overall, some six Irish sites produced chronologies which started in the 9th century; the longest extension being one timber from the south of Ireland which ran back to AD 830. There seemed to be a definite limit to the extension of the medieval oak chronology. What was noticeable was the similarity of this situation

to the experience of several workers in Germany. They had run into a similar 'wall' when constructing basic oak chronologies – Huber back to 832 AD (Huber & Giertz 1969); Hollstein to 822 AD (Hollstein 1965); Becker to 820 AD (Becker & Delorme 1978). In addition, it was notable that the German workers had managed to construct chronologies which ran *forward* to around this same period, for example Göttingen 4008 BC to AD 785 (Leuschner & Delorme 1984) and Hohenheim up to AD 755 (Becker & Delorme 1978).

So, although the German workers had eventually bridged the gap around AD 800 (Hollstein 1980; Becker 1981) it was clear that in the chronology building process they had identified a difficult period spanning AD 785 to 820. While in itself this was merely interesting, when it is independently observed that in Ireland no oak trees *started* growing between AD 745 and 830 (Mallory & Baillie 1988) it began to appear that something was going on (see Figure 2.2a). Once this date range was specified, it became apparent that, in Ireland, this short interval included the main period of horizontal mill building. Of twenty seven dated mill sites which fell between AD 630 and AD 1222, no less than fifteen (56%) were constructed with timbers felled between AD 770 and AD 850 (see Figure 2.2b). Thus it seems that the main mill-building phase was associated with a period when almost no oaks were regenerating in two widely separated areas. Since mills by definition suggest cereal production, we are left with a

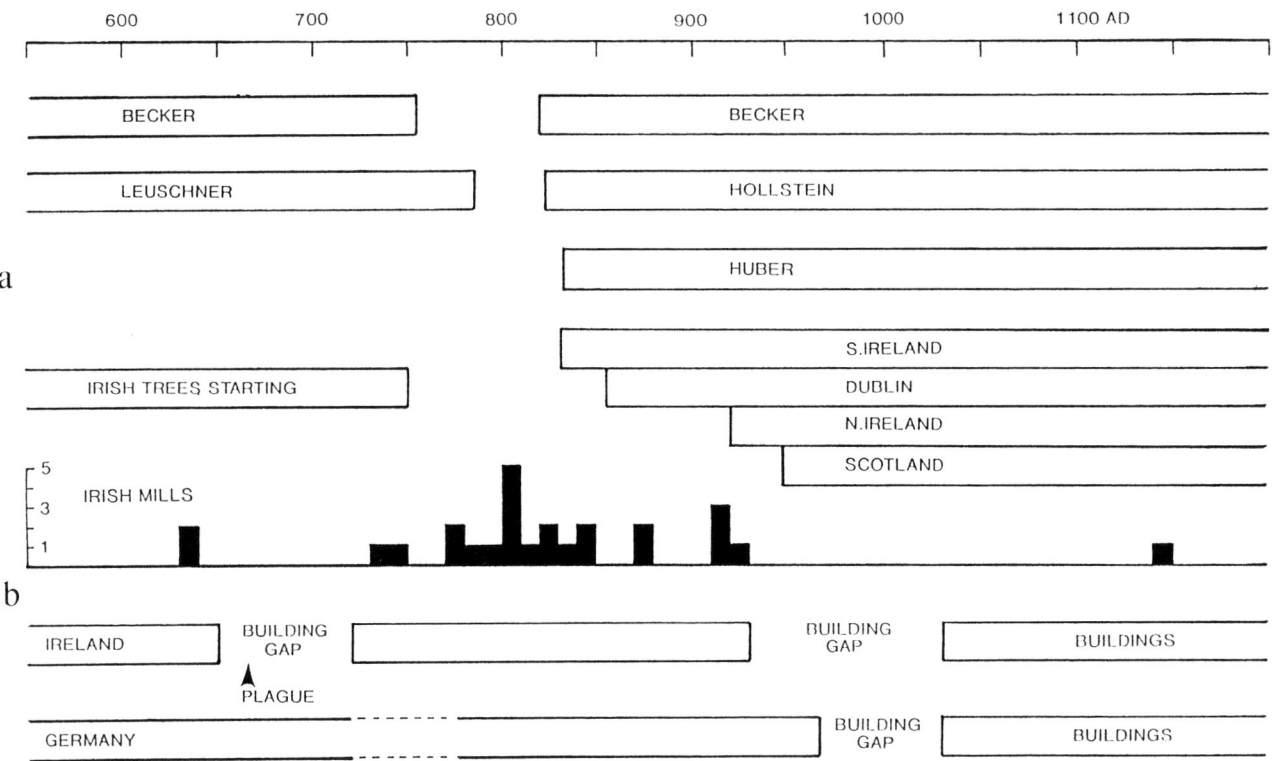

Figure 2.2 (a) Information from chronology building identifies few oaks regenerating in the period around AD 800, with many trees regenerating after AD 820; (b) Precisely dated information from archaeological sites showing periods of construction and gaps in the building record.

suggestion that the *lack* of tree regeneration at this time may be due to human/agricultural pressure.

The impression of something going on around this period is compounded by the additional observation that in Ireland we have no oaks which were *felled* in the century between AD 930 and AD 1030. No mills have been found dating to this century. In Germany, Hollstein (1980) also observed a building hiatus between AD 966 and AD 1030. Reference to this accumulated information in Figure 2.2 shows a picture of human activity around AD 800 and decreased activity after AD 930, while the widespread regeneration of oaks from the ninth century onwards also has to be noted.

It has to be pointed out that this picture of an episode of activity when few trees were regenerating, followed by a period of reduced activity when a lot of trees were regenerating, is based solely on precisely dated tree-ring information. If we were to add in other environmental information, for example Barber's wetness/dryness estimates for Bolton Fell Moss in England (Barber 1981) we might be able to suggest that the *circa* AD 800 activity phase was associated with 'dry' conditions. These might lead to increased cereal production and thus help to explain the proliferation of mills while after AD 930 we might be able to suggest a 'wet' phase leading to decreased cereal production which might explain the mills going out of fashion.

Once we become comfortable with the possibility of a change from 'good' to 'less good' conditions we can begin to look further afield. For example, if we look at the number of dated archaeological sites in the arid, and thus marginal, American southwest (where archaeological tree-ring dates again allow us to look in real time) we see a significant reduction in site construction in the 10th century AD, see Figure 2.3. So perhaps some widespread environmental effects were responsible for the downturn that we seem to be seeing in northern Europe.

Now, obviously, every one of these points could be argued over. The reason for presenting this information is to use it as an example for the discussion of marginality. If the information suggests anything, it appears to suggest an 'upturn' around AD 800 and a 'downturn' in the 10th century. So, yet again, here is a well dated episode where we might look for signs of change in marginal areas.

Plague in Ireland AD 664 to 668

In plotting out the dates of all the 1st millennium Irish archaeological sites and structures, which had been dated by dendrochronology during the 1970s and 80s, it was discovered that there was a very clear 70-year break in the record between AD 648 and 720 (Mallory & Baillie 1988). It was also discovered that in the Irish annals there is a tight cluster of references to plague in four of the five years AD 664 to 668. This plague was well known in medieval history but there is no doubt that the tree-ring evidence lends colour to the possible impact of the disease.

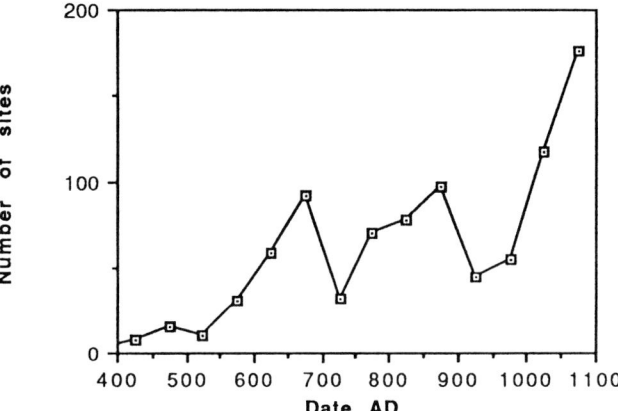

Figure 2.3 Plot of the number of dendrochronologically-dated archaeological sites and phases in the American Southwest showing a significant reduction in site construction in the 10th century AD (data from Robinson & Cameron 1991).

It would seem reasonable, given the example of the Black Death, where the pandemic was associated with a building hiatus, to suggest that this 7th century plague was a factor in the observed building hiatus. The question here is whether the plague might itself be associated with famine conditions. Were that the case we might reasonably ask if more marginal populations or areas were affected.

The Justinian Plague AD 540s

Moving back in time, there appears to be another excellent example of a link between poor conditions for trees and for people in the episode bracketed by AD 536–545. This episode, which has been documented elsewhere, includes historical records of a severe dry fog (dust veil) followed by famines in China, the Mediterranean, England and Ireland (Stothers & Rampino 1983; Weisburd 1985; Baillie 1994; 1995); from AD 542 the Justinian plague sweeps across Europe. These effects are backed up by notable growth reductions in trees from around the northern hemisphere; an independent observation which witnesses some associated environmental downturn.

If we accept that there was a hemispheric downturn in the period AD 536–545, again we might *expect* to see change in the archaeological record of marginal areas. Indeed, we should perhaps have the confidence to realize that previously observed change in the archaeological record 'around AD 600' or 'in the 6th century' or 'circa 550', especially in areas which are inherently marginal, almost certainly relates to this 536–545 event, the deviation from 536–545 being due to the inherent difficulty of producing tight chronologies with conventional archaeological typologies or radiocarbon dating. Overall, the clear evidence for the reality of this event makes it a near certainty that many marginal populations would have been reduced or forced to move.

DISCUSSION

For the sake of discussion, I have identified six tree-ring related phenomena within the last two millennia. In the first three cases, in the 1840s, 1740s and the 1340s, we see different aspects of tree-ring evidence affected at the same time as human populations. In the case of Ireland in the 1840s and 1740s marginal human populations were adversely affected in the years immediately after narrowest ring events. In the 1340s it seems clear that with the radical drop in human population, at the time of the Black Death, there was reduced demand for building timber and marginal land (possibly areas previously in coppice) was allowed to regenerate into forest.

So the equation is not a simple one. However, there is no doubt that the observation of narrowest ring events in the 1840s and 1740s would have been reasonable grounds for asking what happened to marginal populations at those times. With the aid of history we know that the most marginal sections of the Irish population were indeed badly affected. Equally, the observed building gap in the later 14th century, and the associated oak regeneration phase, would itself have raised questions about what happened to human populations. In that case we know that it was not just one section of the human population which was marginal; the whole population of Europe found itself in a marginal situation when confronted with a pandemic against which it had low immunity. This takes us back into the first millennium AD where we have documented tree-ring-related events of various kinds between AD 780 and 1030, between AD 648 and 720 and between AD 536 and 545. Of these, the second appears to be principally an Irish phenomenon where a building hiatus may just possibly be associated with a severe plague episode very much in the manner of the 14th century Black Death. The later 1st millennium package of information implies that we may be seeing the effects of a change in overall environmental conditions with hints of population and agricultural expansion, centred around AD 800, followed by some sort of agricultural and building downturn in the 10th and early 11th centuries. This downturn seems to be associated with a considerable amount of oak regeneration which is again suggestive of reduced human activity. If the Irish landscape was being heavily exploited around AD 800 it seems that it was a lot less exploited between AD 930–1030. This lack of exploitation suggests either a very different form of human activity or that there were less people around. This raises the question of what happened in more marginal areas after AD 930.

Ironically, the earliest of the 1st millennium events is in some ways the best documented. Burgess (1985) had already concluded that there had been a population reduction in Britain, in the 6th century, of similar magnitude to that associated with the later Black Death. We can now see the full extent of the hemispheric downturn with widespread growth reduction in trees, widespread famines and the Justinian plague which we know reached as far as Ireland

(Baillie 1994; 1995). This is the event where we should see the clearest evidence for effects on marginal populations in Britain and Ireland. The only problem is, as noted above, that with conventional archaeological dating the evidence for impact on marginal areas may be couched in terms of 'around AD 600' or 'in the 6th century' or 'circa 550'. The test will be to squeeze the chronology of marginal areas to find out if, in reality, there was a collapse not 'in the 6th century' but between 536 and 545.

It would be possible to proceed back in time picking out and documenting downturns where marginal populations would have been vulnerable; some of those marker dates are alluded to above and it should be noted that there are other episodes in the last two millennia which could be added to the six given here, although a definitive list is still some way off. However, the main point has been made that we are beginning to lay out the dates to which activities in marginal areas are likely to conform.

REFERENCES

Baillie, M G L 1977 'The Belfast oak chronology to AD 1001', *Tree-Ring Bulletin*, 37, 1–12.

Baillie, M G L 1982 *Tree-ring dating and archaeology*. London: Croom-Helm.

Baillie, M G L 1991 'Marking in marker dates; towards an archaeology with historical precision', *World Archaeology*, 23, 233–243.

Baillie, M G L 1994 'Dendrochronology raises questions about the nature of the AD 536 dust-veil event', *The Holocene*, 4 (2), 212–217.

Baillie, M G L 1995 *A slice through time: dendrochronology and precision dating*. London: Batsford.

Baillie, M G L & Munro, M A R 1988 'Irish tree-rings, Santorini and volcanic dust veils', *Nature*, 332, 344–346.

Barber, K E 1981 *Peat stratigraphy and climatic change*. Rotterdam: Balkema.

Becker, B 1981 'A 2350 year South German oak tree-ring chronology', *Fundberichte aus Baden-Wurttemberg*, 6, 369–386.

Becker, B & Delorme, A 1978 *Oak chronologies for central Europe. Their extension from medieval to prehistoric times*, Oxford: BAR International Series, 59–64 (=Brit Archaeol Rep Int Ser, 51).

Burgess, C 1985 *Population, climate and upland settlement*, Oxford: BAR British Series, 195–229 (=Brit Archaeol Rep Brit Ser, 143).

Drake, M 1968 'The Irish demographic crisis of 1740–41', *in* Moody, T (ed), *Historical Studies VI* (Dublin, June 1965). London: Routledge, 101–124.

Hollstein, E 1965 'Jahrringchronologische von Eichenholzern ohne Waldkande', *Bonner Jahrbuch*, 165, 12–27.

Hollstein, E 1980 *MittelEuropaische Eichenchronologie*. Mainz am Rhein: Phillip Von Zabern.

Huber, B & Giertz, V 1969 'Our 1000 year oak chronology', *Conference Report of the Austrian Academy of Science*, 178, 32–42.

Kuniholm, P I & Striker, C L 1983 'Dendrochronological investigations in the Aegean and neighbouring regions, 1977–1982', *Journal of Field Archaeology*, 10, 411–420.

Kuniholm, P I & Striker, C L 1987 'Dendrochronological investigations in the Aegean and neighbouring regions, 1983–1986', *Journal of Field Archaeology*, 14, 385–398.

Leuschner, H H & Delorme, A 1984 'Verlängerung der Göttingen Eichenjahrringchronologien für Nord- und Süddeutschland bis zum Jahr 4008 v. Chr.', *Forstarchiv*, 55, 1–4.

Mallory, J P & Baillie, M G L 1988 'Tech ndaruch: The fall of the House of Oak', *Emania*, 5, 27–33.

Mitchell, F 1986 *The Shell guide to reading the Irish landscape.* Dublin: County House, 228.

Post, J 1985 *Food shortage, climatic variability and epidemic disease in pre-industrial Europe.* Ithica: Cornell University Press.

Robinson, W R & Cameron, C M 1991 *A directory of tree-ring dated prehistoric sites in the American Southwest.* Tucson, Arizona: University of Arizona.

Stothers, R B & Rampino, M R 1983 'Volcanic eruptions in the Mediterranean before AD 630 from written and archaeological sources', *Journal of Geophysical Research*, 88, 6357–6371.

Vogel, J S, Cornell, W, Nelson, D E and Southon, J R 1990 Vesuvius/Avellino, one possible source of 17th Century BC climatic disturbances, *Nature*, 344, 534–537.

Weisburd, S 1985 'Excavating words: A geological tool', *Science News*, 127, 91–96.

3. The response of marginal societies and ecosystems in Britain to Icelandic volcanic eruptions

John Grattan

Abstract

The correlation between ice core acidity peaks from the Greenland ice cap, extremely narrow tree rings from the Irish dendrochronology, and the apparent abandonment of settlements in Northern Scotland has led to speculation that the 1159 BC Hekla 3 eruption was responsible for a severe climatic deterioration.

However, not all volcanic eruptions are climatically effective, and in addition historical documentary evidence suggests that we must not assume that volcanic eruptions will have widespread climatic impacts. Therefore, we must carefully define the impact of the volcano and the nature of the societies and ecosystems affected. Only under specific conditions can we assume that a deleterious effect will occur.

Research suggests that volcanoes may emit millions of tons of volatiles. These can be transported by atmospheric air circulation across thousands of miles and may have a severe impact on vulnerable ecosystems. The impact of these volatiles is described and is shown to be capable of severely acidifying environments which are poorly buffered. The severity and longevity of the impact depends on the ability of the ecosystem to replenish its buffers.

This paper challenges the assumption that volcanic eruptions will have a severe and long lasting climatic effect across wide areas, and instead identifies mechanisms by which the volatiles emitted in an eruption will have a severe impact on a limited range of specifically defined ecosystems.

INTRODUCTION

Archaeological evidence suggests that widespread abandonment of marginal lands, in northern Britain, occurred late in the second millennium BC. Marginal lands are here defined as those which are higher in altitude, or further north or west, than those which we recognise as capable of sustaining high levels of agricultural settlement today. In practise these areas are relatively easy to identify; the Highlands of Scotland (Barclay 1985; Hunt 1987), the uplands of northern England and southern Scotland (Annable 1987; Burgess 1985; Gates 1983; Halliday 1982; Jobey 1980) and the uplands of Wales (Taylor 1980).

Traditional explanations for these changes emphasise the impact of the climatic change then in progress, from Sub-Boreal to Sub-Atlantic climatic conditions, in particular decreasing summer temperatures and increasing precipitation (Frenzel 1966; Godwin 1975). These changes,

coupled with anthropogenic modification of the environment, are part of an evolutionary process, by which the soil cover on the uplands and in the north and west of the British Isles developed from brown forest soils to podzolic soils, gleys and peats (Ball 1975; Bridges 1978, Dimbleby 1965; 1976; Maguire *et al* 1983; Moore 1973; 1975).

The concept of thresholds will be used extensively in this paper, it can be applied equally to societies as to environmental processes. Unless a significant environmental threshold has been crossed, gradual environmental change is likely to be met by changing farming practice and lowered yield expectations (Gates 1983). Neither a gradual climate change, nor a slowly degrading soil resource could account for the apparent suddenness of the settlement abandonment, nor for the hiatus in settlement (Burgess 1985) which is apparent in the late Bronze Age.

In recognition of this problem alternative solutions have been offered, including plague and economic or social crisis (Burgess 1985). The environmental impact of volcanic eruptions, perhaps the Hekla 3 eruption *circa* 1159 BC, provides a further mechanism to explain these events (Baillie 1989; Burgess 1989; Grattan & Gilbertson 1994).

ICE CORE ACIDITY, TREE-RINGS AND VOLCANIC ERUPTIONS

The association between volcanic activity and settlement abandonment in northern Scotland was made possible by the development of a long tree-ring chronology for western Europe (Pilcher *et al* 1984), and a chronology of volcanic eruptions, based on the presence of volcanically induced acid peaks in the Greenland ice cores (Hammer 1977; Hammer *et al* 1980; 1981). Research prompted by the Greenland ice-core acidity work showed that decadal length bands of extremely narrow tree-rings in Irish bog oaks were correlated with volcanic events recorded in ice cores (Baillie & Munro 1988). The long term environmental stress indicated by the Irish tree-ring record led to speculation that volcanically-induced climatic change may persist over decades rather than the few years that most research suggests (Baillie 1989; Burgess 1989). This persistent environmental stress may be the mechanism which triggered settlement abandonment in northern Scotland.

The Irish chronology is unique in the persistence of the trees' response to the volcanic events. All other work which correlates volcanic eruptions and narrow or damaged tree-rings suggests that the stress is largely confined to a single year. All the trees which appear to experience environmental stress in response to volcanic eruptions are those growing at or near the limits of their environmental range (Jacoby & Ulan 1982; Jacoby *et al* 1988; LaMarche & Hirschboek 1984; Scuderi 1990). In these locations, even slight changes to temperature have the potential to affect the trees. The trees' ability to tolerate even slight climatic deterioration is severely limited. Hence these trees experience environmental stress, recorded by narrow or damaged tree-rings. Narrow or damaged tree-rings, therefore, need not indicate great climatic fluctuation. Rather they demonstrate that slight climatic or environmental change may have a severe impact in marginal locations.

The single exception to these short term responses to volcanic events are the oaks described by Baillie & Munro (1988), growing on mires in Ireland.

Volcanoes and climate change

Are there any mechanisms by which a volcanic eruption may severely affect climate for decades and thus account for the response in the Irish oaks?
It is the latitude of a volcano and the ability of an eruption to inject sulphur gases and other volatiles into the stratosphere, which determines its effectiveness as an engine of climatic change (Pollack *et al* 1976; Rampino & Self 1984; Lough & Fritts 1987).

The volatile output of the Hekla 3 eruption was considerable with a yield of 1.58 X 10^{11} g Sulphur, 2.48 X 10^{11} g Chlorine, 4.84 X 10^5 tonnes H_2SO_4 and 2.55 X 10^5 tonnes HCl (Devine *et al* 1984), and a total acids output of 7.39 X 10^5 metric tons (Palais & Sigurdsson 1989). However, in terms of sulphuric acid output and total erupted mass, the Hekla 3 eruption cannot be considered exceptional. There have been many eruptions on a similar scale and several of a far greater order of magnitude.

Across what scale are volcanic eruptions capable of affecting climate, and what degree of impact can we assume the Hekla 3 eruption to have had?
Extensive research in the past decades has identified relatively minor surface cooling after individual eruptions (Kelly & Sear 1984; Mass & Portman 1989). Mt Agung (1972), an eruption with a similar sulphur output to Hekla 3, caused a maximum temperature decrease of 0.2 – 0.5 °C (Hansen *et al* 1978; Kerr 1981; Newell 1981). The massive eruption of Tambora, in 1815, may have caused a cooling of up to 0.7 °C in northern hemisphere temperatures (Lamb 1970; Harington 1992). Taylor *et al* (1980) identified weak summer cooling, up to 0.5 °C, in the wake of high latitude eruptions. A recent study (Mass & Portman 1989) of nine volcanic events between 1883 and 1982 concluded that for the largest eruptions there was post-eruptive cooling of 0.3 °C in the composite temperature records. However, conflicting evidence was produced of temperature declines of a similar scale to post-eruptive cooling in the ten months preceding the eruptions and temperature rises were noted in the 2–3 months immediately preceding the eruptions.

Given a standard deviation for northern hemisphere temperatures of 0.7 °C in January and 0.2 °C in July (Self *et al* 1981; Kelly & Sear 1984) it is difficult to accept that short lived volcanically induced temperature fluctuations, on the scale observed and modelled, are capable of a serious impact on the ecosystems and societies of temperate latitudes.

Volcanic eruptions and circulation response

Are slight temperature differences capable of altering global circulation patterns?
Handler (1986) and Handler & O'Neill (1987) noted an association between low latitude volcanic aerosol input into the stratosphere and Southern Hemisphere seasonal rainfall. The phenomenon of the El Niño/Southern Oscillation (ENSO) was used as a proxy for assessing the activity of low latitude volcanic aerosols. In addition Handler (1989) constructed a model, based on the subsequent steepening of the thermal gradient between the sea and land, which would assign climatic significance to the relatively minor temperature fluctuations caused by volcanic activity. Linkage was also observed between the presence of stratospheric aerosols and the intensity of the Indian and Sri Lankan Monsoon (Mukherjee *et al* 1987). As with all models of global weather and circulation the

inherent problem is one of assigning cause to effect. Nicholls (1988) noted that the trend in pressure anomalies, indicating that the ENSO phenomenon is imminent, begins well before the volcanic events held responsible.

The above discussion emphasises the slight nature of the climatic change instigated by volcanic eruptions in the past hundred years. Yet the tree-ring record, archaeological evidence and contemporary accounts indicate that the aftermath of some eruptions yield significant environmental consequences.

Do eruptions which are famed for their impact on climate stand up to close scrutiny of the available records?

The eruption of the Icelandic volcano, Laki, in 1783, was one of the largest of the Holocene. This eruption is widely held to have generated serious climatic change, largely as a result of the observations made by Benjamin Franklin, and the disaster, amongst the people and livestock of Iceland, which followed.

The Laki fissure eruption
8 June 1783 – 7 February 1784

Benjamin Franklin (1784) held that an eruption in Iceland was the most likely culprit for the cool summer he observed in Paris in 1783. The impact of the eruption, he believed, resulted in an early and harsh winter:

'During several of the summer months of the year 1783, when the effect of the sun's rays to heat the earth in these northern regions should have been greatest, there existed a constant fog over all Europe and great part of North America. This fog was of a permanent nature; it was dry, and the rays of the sun seemed to have little effect towards dissipating it, as they easily do a moist fog arising from water ... Of course, their summer effect in heating the earth was exceedingly diminished. Hence the surface was early frozen. Hence the first snows remained on it unmelted, and received continual additions. Hence the air was more chilled, and the winds more severely cold. Hence perhaps the winter of 1783–4 was more severe, than any that had happened for many years. The cause of this universal fog is not yet ascertained. Whether it was adventitious to this earth ... Or whether it was the vast quantity of smoke, long continuing to issue during the summer from Hecla in Iceland, and that other volcano which arose from the sea near that island, which smoke might be spread by various winds over the northern part of the world, is yet uncertain. It seems however worth the enquiry, whether other hard winters, recorded in history, were preceded by similar permanent and extended summer fogs.' (Franklin 1784).

The clear conclusion to be drawn from Franklin's observations is that the conditions which he observed in Paris were typical of a far wider area stretching from Europe to North America. It should then be possible to support Franklin's conclusions and observations from contemporary European records.

Gilbert White's 'Natural History Of Selbourne' contains another contemporary description of the strange atmospheric phenomena of 1783. While similar in many respects to Franklin's observations, his description of the climate is strikingly different.

'The summer of 1783 was an amazing and portentous one, and full of horrible phenomena; for besides the alarming meteors and thunder-storms that affrighted many counties of this kingdom, the peculiar haze or smokey fog, that prevailed for many weeks in this island and in every part of Europe, and even beyond its limits, was a most extraordinary appearance, unlike anything known within the memory of man. By my journal I find that I had noticed this strange occurrence from June 23 to July 20 inclusive, during which period the wind varied to every quarter without making any alteration in the air. The sun, at noon, looked as blank as a clouded moon, and shed a rust coloured ferruginous light on the ground, and floors of rooms; but was particularly lurid and blood coloured at rising and setting. *All the time the heat was so intense that butchers meat could hardly be eaten on the day after it was killed; and the flies swarmed so in the lanes and hedges that they rendered the horses half frantic, and riding irksome.*' White (1789): my italics.

White's descriptions of the smokey fog, the blank sun and the lurid colours of the rising and setting sun, would be recognised by any current volcanologist as the result of stratospheric veiling, caused by a volcanic eruption. Where it differs from Franklin's work is in his account of the extreme summer heat. On this count both records are mutually incompatible. We can test the validity of these statements by studying the temperature records which are available for both England and Scotland.

White's observations are supported by temperature records collected by Manley (1974), for central England. July 1783 was the warmest recorded in the Central England record until 1983 (Parker *et al* 1992). The warmest July in the period 1771–1791, in central England, is that of 1783 (Figure 3.1). There is no suggestion in the data that the summer of 1783 was cooler than usual.

What of Franklin's other statement that: ' ... the winter of 1783–84 was more severe than any that happened for many years.' Central England temperatures do indicate that the winter of 1783–84 was cooler than any which had occurred for many years, but only marginally so (Figure 3.2). The data also records a similarly cold winter in 1779 and three far more extreme events in the ninety two year period presented in Figure 3.2, none of which are related to a major volcanic eruption. Data recorded in Edinburgh (Mossmann 1896) present a less ambiguous picture (Figure

Figure 3.1 Mean July temperatures for Central England, 1771–90.

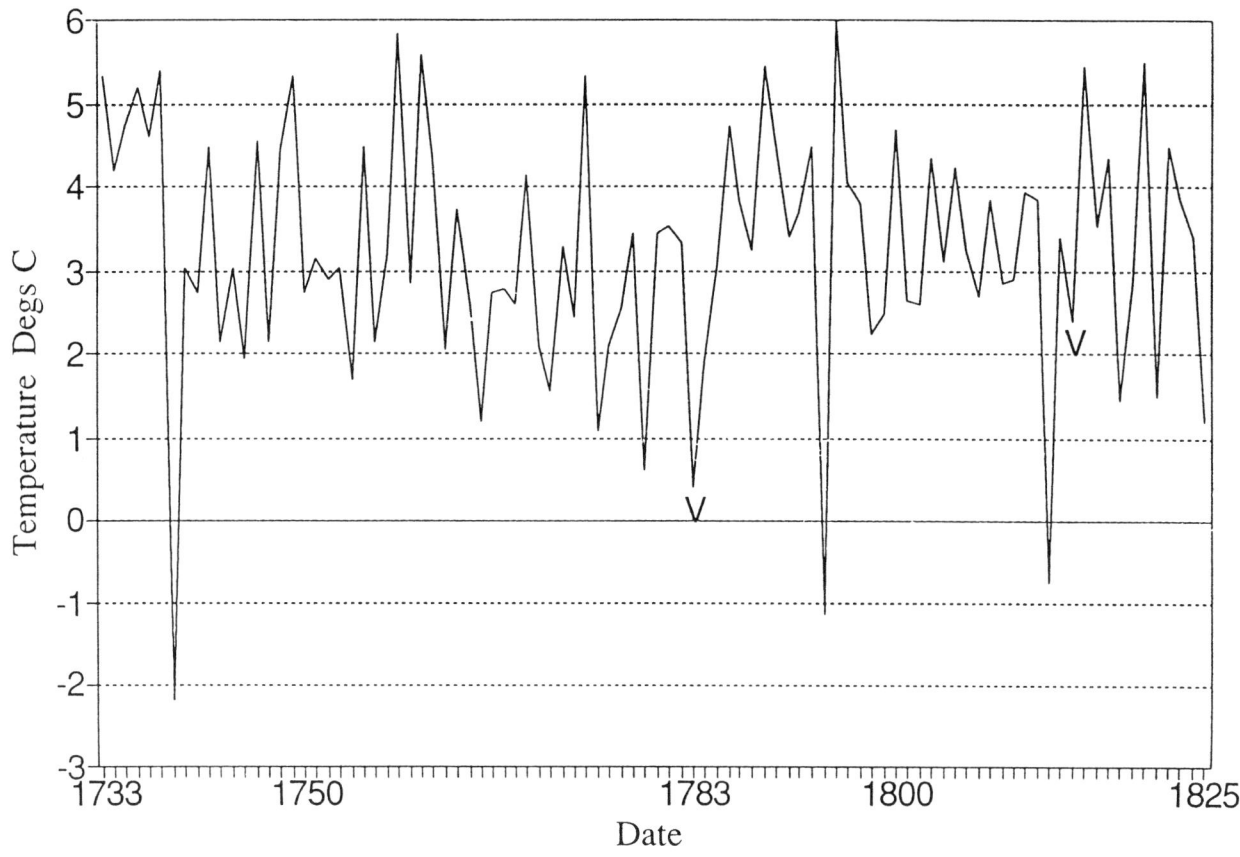

Figure 3.2 Mean winter (January/February) temperatures for Central England, 1733–1825.

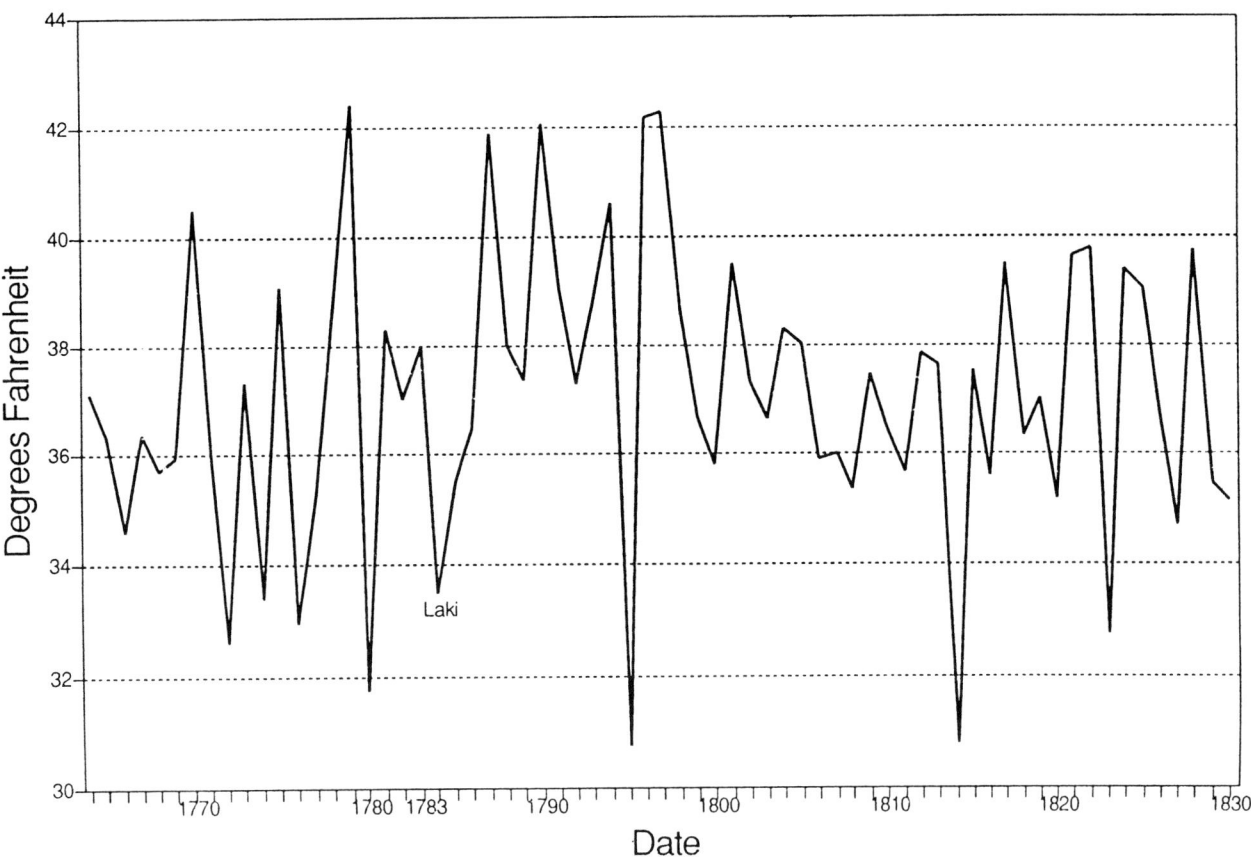

Figure 3.3 Mean winter (January/February) temperatures for Edinburgh, 1764–1830.

3.3). There were four winters harsher than 1783–84 in the preceding 12 years; none of these were associated with a major eruption.

It is not the purpose of the above discussion to cast doubt on the veracity of Benjamin Franklin's statements. What it attempts to demonstrate is that observations of poor climate, even when positively correlated with an eruption cannot be held as proof of a causal link between the two events. Nor can such observations be held to be true for a wider geographical area, unless corroborating temperature records or reliable observations exist, no matter how attractive the correlation.

If an eruption on the scale of Laki, 1783, failed to trigger pronounced wide scale cooling, how may we account for the environmental stress evidenced in the Irish tree-ring record?

The ability of tephra fall to kill or damage plants and insects at a great distance from the eruption is amply demonstrated by the aftermath of the 1783 eruption of Laki at such diverse locations as Bergen, Caithness and Kent (Lamb 1970; Thórarinsson 1981). Lamb (1970) reported the death of plants and insects in southern England. In late June 1783 a sulphurous fog which blighted a large number of plants was noted in Holland (Brugmans 1787; Symons 1888). Ash fall in Caithness was so heavy that the fields were choked; one hundred years later people

in Caithness still remembered 1783 as 'the Year of the Ashie' (Geikie 1893).

The transport of volcanic gases to Bergen in Norway led to the deaths of plants and animals:

'Here we could only smell the smoke and yet we got sick. It was the fog which approached the country, which must have diluted the poison, it fell on the leaves of various vegetation which then withered. It is, therefore no wonder that the valleys and fields of Iceland were quite destroyed and became useless as pastures for the cattle.' (Brun 1786 cited in Thórarinsson 1981).

In Iceland the effects of the eruption were more pronounced:

'The horses lost all flesh, on some the hide rotted all along the back ... The head swelled inordinately, whereupon followed a paralysis of the jaw, so that the beast could not graze or feed ... The entrails corrupted, the bones withered, quite drained of marrow ... Sheep suffered yet more grievous harm ... What passed for meat was both rank and bitter, and thereto full of strong poison, wherefore the eating of it proved the death of many a man.' (From Steingrimsson 1915–1917, translated by Hannesson, in Thórarinsson 1979).

In the aftermath of this event 24% of people, 50% of cattle, 79% of sheep and 76% of horses in Iceland perished (Thórarinsson 1969).

The symptoms described above are largely those of fluorosis. Fluorine is not the only noxious volcanic emission; ammonia, carbon monoxide, hydrochloric acid, hydrofluoric acid, hydrogen sulphide, sulphur dioxide and sulphuric acid are also emitted (Herron 1982). All are potentially damaging to either plants or animals (Le Guern *et al* 1988). The thresholds at which damage occurs can vary between as little as 0.1–500 ppm in air (Garrec *et al* 1977; Mandl *et al* 1975).

The suspended load within the eruption cloud contains the finest particles which have the longest residence time in the stratosphere (Mackinnon *et al* 1984). Thus the fraction of finer particles is expected to increase towards the margins of the tephra deposit (Oskarsson 1980; Rietmejer 1988). The finer particles, which travel farthest, and have the longest residence time within the gaseous eruption cloud, may entrain volatile acid droplets and gels (Rose 1977). The phenomenon was described in Alaska in 1913, following the eruption of Katmai: 'Leaves of the currants, Salmon berries and many other of the shrubs and herbs ... were blighted by the dust or acid rain which fell there. This effect, curiously enough, did not occur in the district of thicker ash.' (Martin 1913).

The nature of the vegetation may modify and exaggerate the impact of acidity. Trees can scavenge sulphur from the atmosphere, both from rainfall and occult deposition from mists and low cloud (Lindberg & Garten 1988). Forest micro climates may also enhance the impact of acidity; evaporation from the trees concentrates dissolved volatiles in cloud or mist droplets which have a greater impact than rain (Unsworth 1984), and canopy will concentrate acidity around the trunk (Lindberg & Garten 1988).

Indirect consequences of acid precipitation include reduced decomposition rates, alterations to soil microbial populations and soil chemistry and reduced root activity (Aber *et al* 1982; Mason 1990). The direct consequences of increasing acidity will mobilise aluminium and other heavy metals in the soil (Krug & Frick 1983; Fernandez 1989; Mulder *et al* 1989; Mason 1990). In high concentration these may be phytotoxic (Cresser *et al* 1988). Leached heavy metals finding their way into streams may further damage aquatic biota (Morrison & Battarbee 1988).

The impact of acid deposition will be magnified if contained within snow cover and then released in the spring melt (Leivistad & Muniz 1976; Koerner & Fisher 1982). This is graphically demonstrated by the changes in fluorine concentrations in the river Ytri-raga, Iceland, during the 1991 eruption of Hekla, when following snow melt fluorine concentrations rose from 0.7 ppm to 4.1 ppm (Gudmundsson *et al* 1992).

Environments vulnerable to acid impacts have a low Critical Load, defined as 'the damage threshold for the response of ecosystems to acid deposition' (Bull 1991).

The potential of Icelandic volcanoes as producers of acid volatiles is important to any assessment of the impact of the eruptions on the downwind ecosystems. Distance is no guarantee that the degree of impact will be lessened. The deposition of acid volatiles either as an occult dust, or as acid rain, or both, will have serious consequences in ecosystems which have a limited capacity to neutralise acid inputs.

The environmental and social impact of acid volatiles emitted by an Icelandic volcanic eruption is graphically demonstrated in 1783 AD when an acid fog was experienced across Europe (Camuffo & Enzi 1995; Grattan & Brayshay 1995; Grattan & Charman 1994; Grattan & Pyatt 1994; Grattan *et al* 1996) often with dramatic consequences. The following description from Holland is typical of the experience of many regions of Europe:

On many days after the 24th June, in both the town and countryside there was a strong, persistent fog which attracted attention both because of the extraordinary phenomenon and because of its bad effects. ... On the 24th the fog was very dense and accompanied by a very strong smell of sulphur, especially in the morning ... it was noticeable not only because of the smell but also because of the taste.

After the 24th, many people in the open air experienced an uncomfortable pressure, headaches and experienced a difficulty breathing exactly like that encountered when the air is full of burning sulphur, asthmatics suffered to an even greater degree: ... the fog brought about a great extermination of insects, particularly amongst leaf aphids, but only those on the leaves which were affected.

On the morning of the 25th the land offered an aspect of severe desolation, the green colour of the plants had disappeared and everywhere the leaves were dry ... This affected a wide variety of plants: some were covered in spots, others changed gradually while some leaves dried up completely. Some leaves did not entirely deteriorate and these continued to grow, but their leaf tips were decayed. Another noticeable change was that in a moment the colour could change from green to brown, black, grey or white.' (Brugmans 1787).

The role of acid aerosol deposition as a powerful agent of environmental change across Europe is now widely recognised (Blackford *et al* 1992; Camuffo & Enzi 1995).

ACID IMPACTS IN SCOTLAND

While acid input is at levels for which there is an adequate supply of bases there will be no appreciable acidification, new base cations weathering to replace those lost through leaching. In a soil where there is already an inadequate supply of bases, increased soil acidity will further lower the buffering capacity, reducing the ability of the soil to respond to acid inputs. Skiba *et al* (1989) found that areas in Scotland experiencing acid deposition levels as low as 0.8 kg H^+ ha^{-1} yr^{-1} produced peats with the lowest pH, 3.0, and lowest base saturation, 10%. Recent estimates of tephra deposition in Ireland after the Hekla 4 eruption are as high as 1 $tonne^{km2}$ (Pilcher & Hall 1992).

Such levels of deposition represent extreme acid impacts. While in the short term acid sensitive plants would suffer stress, in the longer term the removal of buffers from the soil would lower the pH. In the case of a raised bog, where buffers are largely replaced by rainfall, the ability of the ecosystem to recover from such an event would be limited (Lee *et al* 1988). Recovery has been estimated as taking decades or centuries for some sandy soils (Mulder *et al* 1989).

DISCUSSION: VOLCANIC EMISSIONS, ECOLOGICAL RESPONSE AND THE SETTLEMENT RECORD

Was the Hekla 3 eruption responsible for the abandonment of settlement in marginal zones in North Britain in the Late Second Millennium? The discussion above, while questioning the scale and longevity of climatic response to volcanic eruptions, offers a mechanism by which this abandonment could have been caused.

The modelling of the response of ecosystems with a low critical threshold to volcanically generated acid rain is important in understanding the environmental impact of volcanic eruptions. The deposition levels for potentially toxic Icelandic volcanic material in northern Britain are several orders of magnitude higher than the levels of acid deposition which today will critically lower the pH of a poorly buffered soil.

The narrow tree-rings recorded in the Irish dendrochronology (Baillie & Munro 1988; Baillie 1989) have been taken to indicate climatic impact or long term waterlogging. These may instead record changes such as the removal of buffers and the lowering of the bog pH, in response to the deposition of acids on a scale beyond the buffering capacity of the bog, followed by the slow replacement of base cations by inwash or by atmospheric deposition.

Pedological processes had already led to the formation of a range of podzols and gleys in Northern Scotland (Futty & Towers 1982). These processes involve increasing acidification, in which the buffering ability of the soil is reduced. All these soil types may be vulnerable to events of extreme acid deposition, from which recovery would be slow. Soils evolved on base rich parent material, and on alluvial material would be less vulnerable to this process.

In the late second millennium BC, settlement was widespread on soils and at altitudes which today are considered marginal. The long, slow process of acidification and climate change had not led to land abandonment, but the resultant falling crop yields and increased

grazing pressures may have encouraged the exploitation of further marginal lands in order to maintain production. With anthropogenic exploitation of a degraded soil resource an extreme acidification event may have intensified the marginality of the society. In the short term, foliar leaching could damage or destroy crops (Parnell & Burke 1990), and fluorosis may have decimated the sheep and cattle stocks (Thórarinsson 1979; 1981).

Is there any evidence of such a direct impact?
Are volcanic eruptions in Iceland capable
of a direct environmental impact
on the mainland of Britain and Ireland?
Evidence now suggests that the Hekla 4 eruption, dated to 3700 ± 70 BP, may have been responsible for a sharp decline in *Pinus* pollen (Blackford *et al* 1992). The presence of volcanic ash shards, in the pollen core, at the same level makes a direct environmental impact a distinct possibility.

Documentary evidence also makes some tantalising suggestions. The life of St Columba, written in the seventh century AD, appears to describe such an event.

> 'While the saint was living in the island of Io ... he saw a heavy rain cloud that had risen from the sea in the north, on a clear day. Watching it as it rose, the saint said to one of his monks ... "This cloud will be very hurtful to men and beasts; and on this day it will quickly move across and in the evening drop pestiferous rain upon Ireland ... from the stream that is called Ailbine to Ath-clíath (Dublin) and it will cause severe and festering sores to form on human bodies and the udders of animals. Men and cattle who suffer from them, afflicted with that poisonous disease will be sick even to death". Following the saints' instruction ... Silnan arrived, with the Lords' help, at the place aforesaid; and found the people of that district ... devastated by the pestiferous rain falling upon them from the cloud'. (*Adomnan's Life of Columba*. From the translation by Anderson & Anderson 1961).

This account appears to describe an Icelandic volcanic eruption and the deposition of noxious volatiles including fluorine carried in the cloud and rained on the people of Ireland. The description of the sores and suffering would be very familiar to any Icelander suffering fluorosis in the aftermath of the 1783 eruption of Laki.

Such an event would not necessarily cause settlement abandonment, but the deposition of sufficient acids to deplete the base reservoir of soils with a low critical load to the point from which recovery would be measured in months if not years is at least a possibility. The impact of a volcanic eruption alone is unlikely to cause land abandonment. Only if the societies settled in marginal zones were themselves already vulnerable would the deposition of acid volatiles be sufficient to trigger abandonment.

This model suggests a mechanism by which the Hekla

3 eruption may have caused settlement abandonment in Northern Britain which is based on an understanding of both the nature of volatile output and transport from Icelandic volcanoes, and the response to acid inputs by soils with a low critical threshold. This model accounts for the abandonment of marginal zones without attributing this phenomenon to an extreme and indiscriminate weather fluctuation; rather it relies on the mechanisms of soil acidification and the concept of thresholds, and the mechanisms by which these are lowered and then crossed.

ACKNOWLEDGEMENTS

I would like to acknowledge the help and support of a CASE award from SERC and the Scottish Development Department.

REFERENCES

Aber, J D, Hendry, G R, Francis, A J, Botkin, D B & Mellio, J M 1982 'Potential effects of acid precipitation on soil nitrogen and productivity of forest ecosystems', *in* D'Itri, F M (ed), *Acid precipitation effects on ecological systems*. Ann Arbor: Springer-Verlag.

Anderson, A O & Anderson, M O 1961 *Adomnan's Life of Columba*. London: Thomas Nelson.

Annable, R 1987 *The later prehistory of Northern England*. Oxford: BAR British Series (= Brit Archaeol Rep Brit Ser, 160).

Baillie, M G L 1989 'Hekla 3: how big was it?', *Endeavour, New Series*, 13, 78–81.

Baillie, M G L & Munro, M A R 1988 'Irish tree rings, Santorini and volcanic dust veils', *Nature*, 322, 344–346.

Ball, D F 1975 'Processes of soil degradation', *in* Evans, J G & Limbrey, S (eds), *The effect of man on the highland zone*. London: Council for British Archaeology.

Barclay, G J 1985 'Excavations at Upper Suisgill, Sutherland, *Proceedings of the Society of Antiquaries of Scotland*, 115, 159–198.

Blackford, J J, Edwards, K J, Buckland, P C; Dumgore, A J & Cook, G T 1992 'Icelandic volcanic ash and the mid-Holocene Scots pine (*Pinus sylvestris*) pollen decline in northern Scotland', *The Holocene*, 2, 260–265.

Bridges, E M 1978 'Interaction of soil and mankind in Britain', *Journal of Soil Science*, 29, 125–139.

Brugmans, S J 1787 *Naturkundige verhandeling over een zwavelatigen, nevel den 24 Juni 1783 inde provincie van stad en lande en naburige landen waargenomen*. (A physical treatise on a sulphuric smog as observed on the 24th of July 1783 in the province of Groningen and neighbouring countries). Leyden.

Bull, K 1991 'Critical load maps for the UK', *NERC News, July 1991*, 31–32.

Burgess, C 1985 'Population, climate and upland settlement', *in* Spratt, D & Burgess, C (eds), *Upland settlement in Britain: The second millennium BC and after*. Oxford: BAR British Series, 195–229. (=Brit Archaeol Rep Brit Ser, 143).

Burgess, C 1989 'Volcanoes, catastrophe and the global crisis of the late second millennium BC', *Current Archaeology*, 117, 325–329.

Camuffo, D & Enzi, S 1995 'Impacts of clouds of volcanic aerosols in Italy during the last 7 centuries', *Natural Hazards*, 11, 135–161.

Cresser, M S, Harriman, R & Pugh, K 1988 'Processes of acidification in soils and freshwater', *in* Ahmore, M, Bell, N & Garrety, C (eds), *Acid rain and Britain's natural ecosystems*. ICCET.

Devine, J D, Sigurdsson, H, Davis, A N & Self, S 1984 'Estimates of Sulfur and Chlorine yield to the atmosphere from volcanic eruptions and potential climatic effects', *Journal of Geophysical Research*, 89, 6309–6325.

Dimbleby, G W 1965 'Post-glacial changes in soil profiles', *Proceedings of the Royal Society*, B161, 355–362.

Dimbleby, G W 1976 'Climate, soil and man', *Philosophical Transactions of the Royal Society*, B275, 197–208.

Fernandez, I J 1989 'Effects of acid precipitation on soil productivity', *in* Adriano, D C & Johnson, A H (eds), *Acidic precipitation. Volume 2: Biological and ecological effects*. Ann Arbor: Springer-Verlag.

Franklin, B 1784 'Meteorological imaginations and conjectures', *Memoires of the Literary and Philosophical Society of Manchester*, 2, 373–377.

Frenzel, B 1966 'Climatic change in the Atlantic/Sub Boreal transition on the Northern Hemisphere: botanical evidence', *in* Sawyer, J S (ed), *World climate from 8000 to 0 BC*. London: Royal Meteorological Society, 99–123.

Futty, D W & Towers, W 1982 *Soil and land capability for Agriculture. Northern Scotland*. Soil Survey of Scotland: Macaulay Institute.

Garrec, J P, Lounowski, A & Plebin, R 1977 'The influence of volcanic fluoride emissions on the surrounding vegetation', *Fluoride*, 10, 153–156.

Gates, T 1983 'Upland agriculture in Northumberland', *in* Chapman, J C & Mytum, H C (eds), *Settlement in North Britain 1000 B.C. – 1000 A.D.* Oxford: BAR British Series, 103–148. (=Brit Archaeol Rep Brit Ser, 118).

Geikie, Sir A 1893 *Text book of geology*. London: Macmillan.

Godwin, Sir H 1975 *The history of the British flora*. Cambridge: Cambridge University Press.

Grattan, J P & Brayshay, M B 1995 'An amazing and portentous summer: environmental and social responses in Britain to the 1783 eruption of an Iceland volcano', *The Geographical Journal*, 161, 125–134.

Grattan, J P & Charman, D J 1994 'Non-climatic factors and the environmental impact of volcanic volatiles: implications of the Laki Fissure eruption of AD 1783', *The Holocene*, 4, 101–106.

Grattan, J P & Gilbertson, D D 1994 'Acid-loading from Icelandic tephra falling on acidified ecosystems as a key to understanding archaeological and environmental stress in northern and western Britain', *The Journal of Archaeological Science*, 21, 851–859.

Grattan, J P & Pyatt, F B 1994 'Acid damage in Europe caused by the Laki Fissure eruption – an historical review', *The Science of the Total Environment*, 151, 241–247.

Grattan, J P, Charman, D & Gilbertson, D 1996 'The environmental impact of Icelandic volcanic eruptions: a Hebridean perspective', *in* Gilbertson, D D, Kent, M & Grattan, J P (eds), *Search: the environment of the Outer Hebrides, the last 10,000 years*. Sheffield: Sheffield Academic Press, 51–58.

Gudmundsson, A, Oskarsson, N, Gronvold, K, Saemundsson, K, Sigurdsson, O, Stafansson, R, Gislason, S R, Einarsson, P, Brandsdottir, B, Larsen, G, Johannesson, H & Thordarson, T 1992 'The 1991 eruption of Hekla, Iceland', *Bulletin Volcanologique*, 54, 238–246.

Halliday, S P 1982 'Late prehistoric farming in SE Scotland', *in* Harding, D W (ed), *Late prehistoric settlement in South East Scotland*. Edinburgh: University of Edinburgh Department of Archaeology Occasional Paper 8, 74–91.

Hammer, C U 1977 'Past volcanism revealed by Greenland ice sheet impurities', *Nature*, 270, 482–486.

Hammer, C U, Clausen, H B & Dansgaard W 1980 'Greenland ice

sheet evidence of post glacial volcanism and its climatic impact', *Nature*, 288, 230–235.

Hammer, C U, Clausen, H B & Dansgaard W 1981 'Past volcanism and climate revealed by Greenland ice cores', *Journal of Volcanology and Geothermal Research*, 11, 3–10.

Handler, P 1986 'Stratospheric aerosols and the Indian monsoon', *Journal of Geophysical Research*, 91, 14475–14490.

Handler, P 1989 'The effect of volcanic aerosols on the global climate', *Journal of Volcanology and Geothermal Research*, 37, 233–249.

Handler, P & O'Neill, B 1987 'Simultaneity of response of Atlantic Ocean tropical cyclones and Indian monsoons', *Journal of Geophysical Research*, 92, 14621–14630.

Hansen, J E, Wang, W & Lacis, A A 1978 'Mt. Agung eruption provides test of a global climatic perturbation', *Science*, 199, 1065–1068.

Harrington, C R (ed) 1992 *The year without a summer?* Ottawa: Canadian Museum of Nature.

Herron, M M 1982 'Impurity sources of F, Cl, NO and SO in Greenland and Antarctic precipitation', *Journal of Geophysical Research*, 87, 3052–3060.

Hunt, D 1987 *Early farming communities in Scotland: aspects of economy and settlement 4500–1250 BC*, Oxford: BAR British series (=Brit Archaeol Rep Brit Ser, 159).

Jacoby, G C & Ulan, L D 1982 'Reconstructing past ice conditions in a Hudson Bay estuary using tree rings', *Nature*, 298, 637–639.

Jacoby, G C, Ivanciu, L S & Ulan, L D 1988 'A 263-year record of summer temperature for northern Quebec reconstructed from tree-ring data and evidence of a major climatic shift in the early 1800's', *Palaeogeography, Palaeoclimatology, Palaeoecology*, 64, 69–78.

Jobey, G 1980 'Unenclosed platforms and settlements of the later 2nd Millennium BC in northern Britain', *Scottish Archaeological Forum*, 10, 12–26.

Kelly, P M & Sear, C B 1984 'Climatic impact of explosive volcanic eruptions', *Nature*, 311, 740–743.

Kerr, R A 1981 'Mt. St. Helens and a climatic quandary', *Science*, 211, 371–374.

Koerner, R M & Fisher, D 1982 'Acid snow in the Canadian high arctic', *Nature*, 295, 137.

Krug, E C & Frick, C R 1983 'Acid rain on acid soil: a new perspective', *Science*, 221, 521–525.

LaMarche, V C & Hirschboek, K 1984 'Frost rings in trees as records of major volcanic eruptions', *Nature*, 307, 121–126.

Lamb, H H 1970 'Volcanic dust in the atmosphere; with a chronology and assessment of its meteorological significance', *Philosophical Transactions of the Royal Society*, Series A, 266, 425–533.

Le Guern, F, Faivre-Pierret, R X & Garrec J P 1988 'Atmospheric contribution of volcanic sulphur vapour and its influence on the surrounding vegetation', *Journal of Volcanology and Geothermal Research*, 35, 173–178.

Lee, J A, Press, M C, Studholme, C & Woodin, S J 1988 'Responses to acid deposition in ombrotrophic mires in the UK', *in* Hutchinson, T C & Meema, K M (eds), *Effects of atmospheric pollutants on forests, wetlands and agricultural ecosystems.* London, Paris, Tokyo: Springer-Verlag.

Leivistad, H & Muniz, I P 1976 'Fish kill at low pH in a Norwegian river', *Nature*, 251, 391–392.

Lindberg, S E & Garten, C T 1988 'Sources of sulphur in forest canopy throughfall', *Nature*, 336, 146–151.

Lough, J M & Fritts, H C 1987 'An assessment of the possible effects of volcanic eruptions on North American climate', *Climatic Change*, 10, 219–237.

Mackinnon, I D R, Gooding, J L, McKay, D S & Clanton, U S 1984 'The El Chichón stratospheric cloud: solid particulates and

settling rates', *Journal of Volcanology and Geothermal Research*, 23, 125–146.

Maguire, D, Ralph, N & Fleming A 1983 'Early land use on Dartmoor: palaeobotanical and pedological investigations on Holne Moor', *in* Jones, M (ed), *Integrating the subsistence economy.* Oxford: BAR International Series, 57–105. (=Brit Archaeol Rep Int Ser, 181).

Mandl, R H, Weinstein, L H & Keveny, M 1975 'Effects of hydrogen fluoride and sulphur dioxide alone and in combination on several species of plants', *Environmental Pollution*, 9, 133–143.

Manley, G 1974 'Central England temperatures: monthly means 1659 to 1973', *Quarterly Journal of the Royal Meteorological Society*, 100, 389–405.

Martin, G C 1913 'The recent eruption of Katmai Volcano in Alaska', *National Geographic*, 24, 131–181.

Mason, B J 1990 'Acid rain – cause and consequence', *Weather*, 45, 70–79.

Mass, C F & Portman, D A 1989 'Major volcanic eruptions and climate: a critical evaluation', *Journal of Climate*, 2, 566–593.

Moore, P D 1973 'The influence of prehistoric cultures upon the initiation and spread of blanket bog in upland Wales', *Nature*, 241, 350–353.

Moore, P D 1975 'Origin of blanket mires', *Nature*, 256, 267–269.

Morrison, W R J & Battarbee, R W 1988 'Effects on freshwater flora and fauna (excluding fish)', *in* Ashmore, M, Bell, N & Garretty, C (eds), *Acid rain and Britain's natural ecosystems.* ICCET.

Mossman, R C 1896 'The meteorology of Edinburgh', *Transactions of the Royal Society of Edinburgh*, 39, 63–207.

Mukherjee, B K, Indira, K & Dani, K K 1987 'Low latitude volcanic eruptions and their effects on Sri Lankan rainfall during the North-East Monsoon', *Journal of Climatology*, 7, 145–155.

Mulder, J, Van Breemen, N & Eijck, H C 1989 'Depletion of soil aluminium by acid deposition and implications for acid neutralization', *Nature*, 337, 247–249.

Newell, R E 1981 'Further studies of the atmospheric temperature change produced by the Mt. Agung eruption in 1963', *Journal of Volcanology and Geothermal Research*, 11, 61–66.

Nicholls, N 1988 'Low latitude volcanic eruptions and the El Nino Southern oscillation', *Journal of Climatology*, 8, 91–95.

Oskarsson, N 1980 'The interaction between volcanic gases and tephra: Fluorine adhering to tephra of the 1970 Hekla eruption', *Journal of Volcanology and Geothermal Research*, 8, 251–266.

Palais, J M & Sigurdsson, H 1989 'Petrologic evidence of volatile emissions from major historic and pre-historic volcanic eruptions', *in* Berger, A, Dickinson, R E, & Kidson, J W (eds), *Understanding climate change.* Geophysical Monograph 52. IUGG Volume 7, 31–53.

Parker, D E, Legg, T P & Folland, C K 1992 'A new central England temperature series', *International Journal of Climatology*, 12, 317–342.

Pilcher, J R, Baillie, M G L, Schmidt, B & Becker, B 1984 'A 7272-year tree ring chronology for Western Europe', *Nature*, 312, 150–152.

Pilcher, J R & Hall, V A 1992 'Towards a tephrochronology for the Holocene of the north of Ireland', *The Holocene*, 2, 255–299.

Pollack, J B, Toon, O B, Sagan, C, Summers, A, Baldwin, B & Van Camp, W 1976 'Volcanic explosions and climatic change: A theoretical assessment', *Journal of Geophysical Research*, 81, No 6, 1071–1083.

Rampino, M R & Self, S 1984 'Sulphur rich volcanic eruptions and stratospheric aerosols', *Nature*, 310, 677–679.

Rietmejer, F J M 1988 'Enhanced residence of submicron Si rich particles in the lower stratosphere', *Journal of Volcanology and Geothermal Research*, 34, 173–184.

Rose, W I 1977 'Scavenging of volcanic aerosol by ash: atmospheric and volcanological implications', *Geology,* 5, 621–624.

Scuderi, L A 1990 'Tree ring evidence for climatically effective volcanoes', *Quaternary Research,* 34, 67–85.

Self, S, Rampino, M R & Barbera, J J 1981 'The possible effects of large 19th and 20th century volcanic eruptions on zonal and hemispheric surface temperatures', *Journal of Volcanology and Geothermal Research*, 11, 41–60.

Skiba, U, Cresser, M S, Derwent, R G, & Futty, D W 1989 'Peat acidification in Scotland', *Nature*, 337, 68–69.

Symons, G J 1888 *The eruption of Krakatau and subsequent phenomena: Report of the Krakatau Committee of the Royal Society of London.* London: Trubner.

Taylor, J A 1980 *Culture and environment in prehistoric Wales.* Oxford: BAR British Series. (= Brit Archaeol Rep Brit Ser, 76).

Taylor, B L, Gal-Chen, T & Schneider, S 1980 'Volcanic eruptions and long term temperature records: an empirical search for cause and effect', *Quarterly Journal of the Meterological Society,* 106, 175–199.

Thórarinsson, S 1969 'The Lakagigar eruption of 1783', *Bulletin Volcanologique,* 33(3), 910–929.

Thórarinsson, S 1979 'On the damage caused by volcanic eruptions with special reference to tephra and gases', *in* Sheets, P D & Grayson, D K (eds), *Volcanic Activity and Human Ecology.* New York: Academic Press, 125–159.

Thórarinsson, S 1981 'Greetings from Iceland. Ash falls and volcanic aerosols in Scandinavia', *Geografiska Annaler,* 63 A, 109–118.

Unsworth, M H 1984 'Evaporation from forests in cloud enhances the effects of cloud deposition', *Nature*, 312, 262–264.

White, G 1789 *The natural history of Selbourne.* Reprinted 1977. London: Penguin.

4. Human responses to marginality

Ian Armit

Abstract

The development of settlement patterns in the Western Isles from the Neolithic to the Early Historic period demonstrates the unpredictability of social responses to an increasingly marginal environment. Hints to this unpredictability are to be found in the history of the area. The islands may have been economically marginal in the Norse period, but this was not reflected by political marginality until the Late Middle Ages.

Economic marginality in the Western Isles seems to have developed after the early part of the Neolithic. Peat growth and machair development combined gradually to force human settlement onto the coastal belt, with an abandonment of the interior of the islands by the last centuries BC or possibly earlier. Nonetheless, the flowering of monumental architecture in this very period seems to negate the marginal character of the area. This paper argues that while some important, long term settlement trends may be attributable to environmental forces, we have to take an altogether different approach to short to medium term change in societies which we perceive as marginal.

INTRODUCTION

General

The aim of this paper is to examine the relationship between two ideas: environmental stress and marginality. Environmental stress can have direct economic effects leading to the economic marginalisation of an area. This can take the form of vulnerability to resource failure, inability to generate surplus production, difficulty in meeting subsistence needs over extended periods, etc. The term economic marginality will be used throughout this paper as a shorthand to describe such direct effects of environmental stress. The term will be applied to the case study area as a unit although clearly the effects of environmental stress will not be uniform across any landscape.

The concept of marginality has wider implications than simply the economic restrictions outlined above. It implies a cultural, social and political component. Clearly, environmental deterioration will limit the economic options open to human societies and constrain their economic development. This paper seeks to examine how such a deterioration might affect other aspects of such societies and to discuss how far the concept of marginality and 'marginal' societies

is useful in explaining social, cultural and political development. It will be argued that assumptions of marginality based on environmental circumstances run the risk of projecting modern political and cultural relationships onto past societies and can be deleterious to modern populations of 'marginal' areas by naturalising their peripheral relationship to the wider socio-political units of which they form a part.

The paper deals specifically with human responses to the deterioration of the natural environment in the Western Isles of Scotland (Figure 4.1), in particular the island of North Uist, from the Earlier Neolithic to the post-medieval period.

The Western Isles

The Western Isles encompass the 'long island' from Lewis down through the Uists, to the small islands south of Barra. The quantity of archaeological research in this area has increased dramatically in recent years. Edinburgh and Sheffield Universities have been involved in long term research projects incorporating both field survey and excavation (cf Harding & Armit 1990, SEARCH 1993).

The Centre for Field Archaeology, University of Edinburgh, has also become involved in the intensive survey of machair regions (Armit 1994) in collaboration with palaeoenvironmental research by Sheffield and Birmingham Universities. This initiative should provide a highly detailed picture of human interaction with the environment in the Bhaltos peninsula, one of the key areas for archaeological research in the islands.

The reconstruction of the changing environment of the region is developing quickly now after a long period of stagnation. Much of the detail of what is said in this paper regarding the current palaeoenvironmental picture for the Western Isles will soon be out of date. The basic processes of environmental and settlement pattern change in the islands are, however, becoming apparent, even if their rates and chronologies of development remain to be integrated.

Assumptions and implications of marginality

The Western Isles are fashionably marginal. The current campaign by Sheffield University took the marginality of the islands as a central premise (cf Research Seminar on the SEARCH Project 1987). Dun Carloway, the celebrated Lewis broch tower, adorned the poster for the conference from which this volume derives. The Western Isles, then, provide a useful area for the discussion of the term 'marginality' and for a consideration of its utility in describing and explaining human cultural behaviour.

It is easy to see why the Western Isles have acquired this label. Their language is different from that of the Scottish mainland, and is spoken by few archaeologists. Their religion is distinctly out of step with an increasingly atheistic mainland society. The islands are clearly a distinct cultural zone, as any exposure to a Lewis Sabbath will quickly establish. Whether they are seen as a medieval backwater or a haven of sanity, they are culturally different from mainstream British society, peripheral to the Scottish economy and heavily dependent on outside economic support. Add to this their geographical position and a miserable climate and it easy to assume that the islands were laid out for archaeologists as a definitional aid for the term marginality.

This presents problems. The archaeological record for the Western Isles from the Neolithic to the Early Historic period demonstrates the unpredictability of social responses to an increasingly unfriendly environment. Hints at this unpredictability are to be found in the more recent history of the area. The islands may have been economically marginal in the Norse period, but this was not reflected by political marginality until the Late Middle Ages. The present position, outlined above, of cultural, political and religious alienation from Lowland Scotland results from a complex of factors most immediately based on political events. The Western Isles were only ceded to Scotland in 1266 by which time the dominant powers in Scotland had already become heavily anglicised and culturally trans-

formed from their Celtic origins. The cultural clash between the islands and the Scottish kingdom was a severe restriction to any kind of prosperity or integration in the Middle Ages. Prior to this period, the islands had occupied a pivotal position in the Norse Kingdom of the Isles. Cultural and political marginality, then, may not have been part of the same package as economic marginality.

THE MARGINAL ENVIRONMENT

The modern environment

Although they supported essentially subsistence economies for millennia, the Western Isles have not been integrated into the modern Scottish economy. How far this is a factor of political and cultural factors (eg the collapse of the kelp industry, the failure of schemes to introduce commercial fish-processing, etc), rather than environmental ones, is a matter of debate.

It is clear that the environment of the Western Isles has become less conducive to human settlement since the first settlement of the area. Several processes have combined to reduce the economic value and stability of the islands both in absolute terms and relative to other areas: these are principally deforestation, peat growth, coastal erosion and machair change. The evidence for each of these processes as they affect the archaeology of the islands has recently been reviewed in detail (Armit 1992a, Chapter 2) so only a brief summary will be presented here.

Deforestation

The present treeless state of the islands does not appear to reflect their earlier prehistoric condition. Work by Wilkins (1984) on sub-peat arboreal remains from 40 sites in Lewis and Harris has demonstrated the former existence of birch, willow and pine forest in areas now used for peat cutting. Pine forest appears to have survived on the sample sites into the 2nd millennium BC, and may have survived longer in areas less exposed to human exploitation. Palynological work by Bohncke on the Leobag peninsula near Callanish has demonstrated the former presence of birch forest in sheltered west coast areas of Lewis (Bohncke 1988). Most recently, work by Bennett *et al* has indicated varied woodland composition throughout the long island with South Uist supporting oak and elm as well as birch and hazel until gradual loss began around 4000 BP (Bennett *et al* 1990)

The results of these studies conflict with earlier work at Little Loch Roag, which suggested that the islands had never had substantial forest (Birks & Madsen 1979). Interestingly, however, the proportions of pollen recorded by Wilkins in association with preserved sub-peat timber were similar to those recorded by Birks and Madsen, although the absolute quantities of pollen at Little Loch Roag were very low (Birks & Madsen 1979). The Little

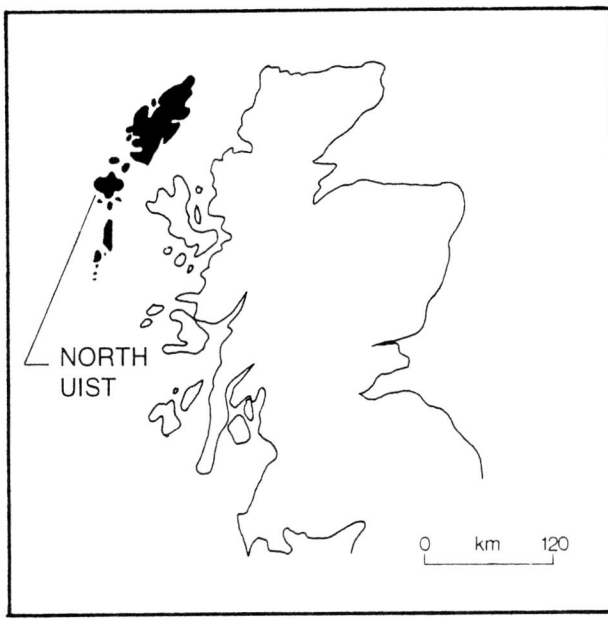

Figure 4.1 Location map.

Loch Roag site might be unrepresentative of the wider situation on Lewis, partly due to its exposed west coast location. That there were timber supplies present in the Neolithic and later is clear from the work of Wilkins and others. Evans' work on the molluscs from Northton suggests that woodland managed to regenerate in south Harris between the Later Neolithic and Iron Age occupations of the area (Evans 1971).

The evidence from Leobag and the work on sub-peat deposits have begun to construct a picture of the decline of woodland in the Western Isles, but there are still many large areas which have not been subject to detailed environmental work. The lack of published environmental work on areas away from human settlement is a problem for any attempt to reconstruct the off-site environment. The east coast of Lewis, for example, away from the Stornoway / Eye peninsula area (which has been examined by Newell 1988), or parts of the Harris Hills, are areas which may have been wooded much later than more intensively occupied areas such as North Uist and the west coast of Lewis. The current Sheffield work in the Uists and Barra may help address this problem, although it is probably the less intensively occupied parts of Lewis and Harris, outside the area of the SEARCH Project, which are more likely to have remained wooded later.

It is worth mentioning in this context that at Eilean Domhnuill, Loch Olabhat, North Uist, timber appears to have been a scarce resource, comprising principally birch and willow, with no sign of substantial timbers and clear evidence for the use of larch *(Larix decidua)*, presumably obtained as driftwood (Skinner pers comm). Peat was apparently the main fuel here even in the Neolithic. At Eilean an Tighe, another Neolithic settlement on North

Uist, birch, willow and hazel were also recovered, while a somewhat different picture derives from the presence of oak and pine at Unival chambered tomb (Scott 1948, 1).

Timber was, however, used in quantity on certain sites of probable early prehistoric date. The North Tolsta crannog in east Lewis appears to have comprised a timber-built artificial islet (Blundell 1913, 298), and a similar construction is recorded for the crannog in Loch Airidh na Lic, Lewis (RCAHMS 1928, no. 51).

The roofing and flooring requirements of broch towers and related roundhouses in the later prehistoric period must have pushed timber supplies to the limit. It is not clear where this timber would have been obtained or what kinds of timber were used. By the post-medieval period, the islands would have been entirely incapable of supplying timber in the quantities required for the earlier profusion of atlantic roundhouses.

Peat growth

The chief characteristic of much of the Western Isles land surface is its peat cover. This is seen to most dramatic effect in the northern part of Lewis which lies under a blanket of peat covering *circa* 230 square miles. Peat has also been removed to reclaim large areas of land around many of the settlements antecedent to the modern crofting townships.

Peat was apparently absent from soil profiles found under the chambered tombs at Unival and Clettraval, both on North Uist (Scott 1948, 1), now in the midst of the peat-covered interior of North Uist, although it was present in Neolithic North Uist in quantities sufficient to make it the principal fuel in at least the later phases of the Neolithic settlement of Eilean Domhnuill.

The formation of the peatland landscapes of the islands has yet to be properly calibrated and assessed against contemporary human settlement. The role of anthropogenic factors in enabling the expansion of peatland has also yet to be fully explored in the area. There can be little doubt, however, that the extent of peat cover has extended dramatically over the period of human settlement in the islands and has had a marked effect on human settlement and economy.

Coastal erosion and machair change

Relative sea level rise has caused the progressive loss of coastal areas, particularly along the western part of the islands, although its rate and chronology have yet to be fully established. The effects of this process on human settlement are similarly unclear; it is perhaps relevant that the present and indeed post-medieval exclusivity of coastal settlement was not a feature of prehistoric settlement.

The effects of coastal change, certainly since the Neolithic, have been most directly manifested through machair movement and redeposition. This has led to the

loss of rich areas of land such as the now inter-tidal
Valley Strand, on which numerous later prehistoric settlements focused (Armit 1992a, Chapter 2). A similar, more
gradual process has taken place at Bhaltos in Lewis
(Armit 1994). Sudden loss of machair lands through
coastal erosion is recorded from the post-medieval
period. The best example is perhaps the settlement of
'Hussaboste' recorded in 1389 but absent from later
records. This is now represented by a reef, Sgeir Husa-
bost, off the west coast of Baleshare, itself a tidal island
off the west of North Uist. Local tradition also asserts
that this was the location of a settlement and the name
of the surviving settlement Baleshare, meaning 'east
town', further suggests the presence of a former 'west
town' (Angus & Elliot 1992, 5). Numerous other losses
of farming land on the machair are recorded from the
15th through to the 19th century.

Although the machair sand is redeposited further inland
and the overall Western Isles machair system does not
suffer loss, key areas of human settlement have been lost.
At Bhaltos the hills behind the machair have prevented
the development of a wide machair plain and the surface
area of the Bhaltos machair has greatly reduced as the
sand has accumulated in depth against the foot of the hills.
At Valley much former machair was submerged and little
redeposition has taken place further inland.

The processes of coastal change have affected human
settlement at two scales. The long term relative rise of sea
levels has removed substantial areas of the islands, and
has fragmented the island chain. Individual episodes of
machair movement, sometimes catastrophic in scale, as
possibly at Valley and as documented at Baleshare, have
caused localised economic and social dislocation. This
latter process is likely to have become more intense in its
direct effects on human communities as settlement focused

more and more on the machair from the later prehistoric
period onwards.

Summary: environmental marginality

The environmental marginality of the Western Isles seems
to have intensified after the early part of the Neolithic.
Although we can now understand the basic processes of
environmental worsening we are a long way from establishing their full interrelationships and chronologies of
change. The human settlement evidence provides another
dimension to this picture of environmental decline.

SETTLEMENT PATTERN CHANGE

General

Settlement pattern change is easiest to analyse for North
Uist where the most intensive survey has been carried out
and where the records for post-medieval settlement have
been collated (Crawford 1965).

Post-medieval settlement

Despite the social upheavals of the past two hundred years,
the lotting of crofts, the clearances, and subsequent
resettlement, the areas occupied by present human settlement are broadly similar to those of the post-medieval
period. Crawford's study of the baile settlements of North
Uist from the early 17th to the 19th century demonstrates
that post-medieval settlement concentrated almost exclusively on the machair fringes of the north and west
coasts (Figure 4.2). This enabled these communities to
exploit a wide range of resources, with agriculture focused

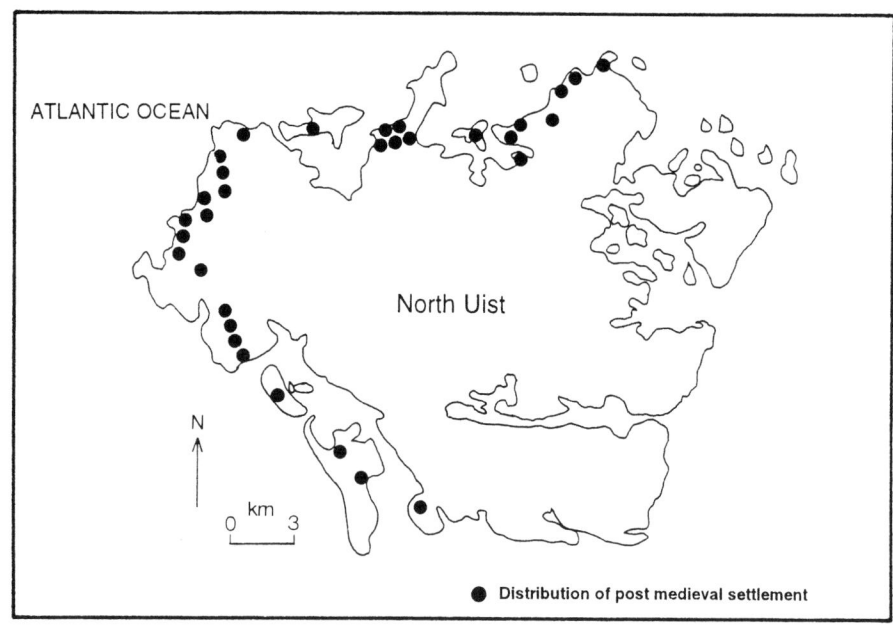

Figure 4.2 Distribution of post-medieval settlement in North Uist.

on the machair and reclaimed peatland. The inland parts of the island were avoided for permanent settlement, as were the south and east coasts until the population expansion of the mid to late 18th century (Crawford 1965, 44–47). Settlement in this period was thus very localised and heavily dependent on the resources provided by the machair.

Later prehistoric settlement

The distribution of later prehistoric atlantic roundhouses on North Uist shows the occupation of similar areas to that of post-medieval settlement, but with less complete avoidance of the southern and eastern parts of the island (Figure 4.3). More of the land area of North Uist, therefore, appears to have been viable for settlement in the later 1st millennium BC. We have less evidence for settlement distributions between the later 1st millennium BC and the post-medieval period due to the disappearance of monumental, above-ground structures in the early 1st millennium AD. The limited evidence available suggests that the wheelhouses and cellular structures of the 1st millennium AD generally, with a few important exceptions, occupied the machair. There appears to have been a gradual concentration of settlement on the machair on the north and west coasts at the expense of settlement in the interior, and on the south and east of the island.

Earlier prehistoric settlement

A number of settlements of non-monumental character, the walled islets and causewayed islets, occupy the interior and eastern part of North Uist. Their distribution overlaps with that of the atlantic roundhouses but also covers the eastern interior of the island, areas now covered by expanses of peat bog and virtually unused in the island's subsistence economy in later periods (Figure 4.4). From their distributions and their lack of characteristics typical of atlantic roundhouses these sites appear to represent earlier settlements abandoned prior to the period of monumental architectural construction in the 1st millennium BC (Armit 1992a).

Without more extensive field evaluation it is impossible to assign these islet sites to particular archaeological periods. Their chief common attributes are negative traits, for example the lack of collapsed massive stone structures characteristic of atlantic roundhouse sites. Excavations at Eilean Domhnuill (Armit 1992b) and Eilean an Tighe (Scott 1951) have revealed that these were both Neolithic islet settlements. While not all of these sites are likely to be Neolithic, it appears that they represent a distribution of settlement sites which had been abandoned prior to the later 1st millennium. Contemporary sites of earlier prehistoric date on the north and west coast areas would often, by contrast, have continued as settlement foci and many would have been the sites of later atlantic roundhouses.

Long term settlement change

It is possible to construct a model for settlement change in which the earlier prehistoric site distributions cover the majority of the North Uist land mass. With a worsening environment, the expansion of peat, and other environmental factors, the inland and eastern sites were abandoned and settlement contracted to the coasts. By the later 1st millennium BC this process was well established and the settlement distribution had begun to resemble that of the post-medieval period. Machair may have become increasingly important as a resource as this process progressed, making settlement more vulnerable to the effects of coastal change. The adoption of monumental architecture

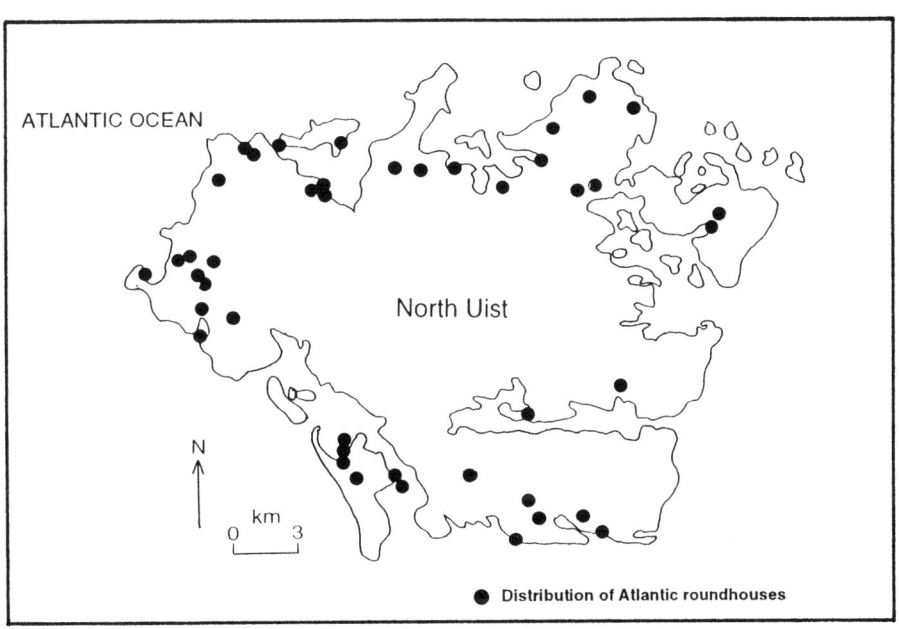

Figure 4.3 Distribution of atlantic roundhouses in North Uist.

led to the fossilisation of one period of settlement in the archaeological record. Together with the 'tide-mark' of probable earlier prehistoric settlement and the post-medieval settlement pattern reconstructed from documentary sources, we have three snapshots of the process of settlement contraction in action.

MONUMENTAL ARCHITECTURE

In the long term, over a period of *circa* 4000 years, settlement pattern change developed along lines which could be predicted from a simple processual model based on environmental decline. This long term picture of settlement contraction can be interpreted as the result of the increasing economic marginalisation of the Western Isles. One period which appears, however, to defy the label of marginality in its wider sense is the later 1st millennium BC when the monumental atlantic round-houses, including the broch towers, were constructed (Armit 1990b).

The flowering of monumental architecture which produced impressive structures like Dun Carloway and Loch na Berie broch seems to negate the marginal character of the area. Atlantic roundhouses represent a form of monumentality which, in environmental terms, was highly ill-adaptive. Their above ground, multistorey construction exposed them to the full blast of the Hebridean gales, which structures of most other periods in the islands have avoided by being low and squat, or semi-subterranean. Their circular, open interiors would have required substantial timbers for flooring and roofing. It is not always appreciated that these monumental stone buildings may have made their most excessive demands on their inhabitants' resources in their requirement for timber.

The adoption of broch architecture and the atlantic roundhouse form as the standard settlement type of the later 1st millennium BC (cf Armit 1988) shows that the islanders were part of the Atlantic cultural continuum which included the Northern Isles, Caithness and Sutherland, as well as the north-west mainland and Inner Isles. They were engaged in complex social interaction with each other and with a wide network of contacts throughout this Atlantic region, as is shown by the parallel development of architectural styles including, at various times, broch and wheelhouse architecture and cellular buildings. The decoration of pottery and bone artefacts (as well as organic materials, such as the decorated woodwork from Dun Bharabhat, Lewis; N Dixon pers comm) in combination with the elaboration of their architecture does not suggest a society in constant threat of basic economic collapse.

This Early Iron Age period of Western Isles history, like the Norse period, shows the islands as part of a wider social and cultural province. The complexity of social interaction within the island and between the islanders and other groups cannot be explained adequately by reference to the concept of marginality.

Yet we have seen that this period lies within a long-term process of environmental decline and settlement contraction. What we appear to be seeing is short-term social and cultural behaviour cross-cutting the long-term, environmentally-constrained economic marginalisation of the islands.

The communities who built the Western Isles atlantic roundhouses do not appear to have reached the same levels of internal stratification as those of the Orkneys. The nucleated settlements focused on broch towers as at Gurness in Orkney appear to be absent from the Western Isles, at least on the scale and complexity of the Orcadian examples (Armit 1990a). Instead we see the development of a new form of monumental architecture of a profoundly different character. The development of wheelhouse architecture, apparently somewhat later than the atlantic roundhouses (though the chronological relationships of the two requires considerably more attention), represents the emergence of what may be termed an 'adaptive monumentality'. Wheelhouses were generally semi-subterranean, revetted into sand hills or pre-existing structures, so were well insulated and stable. They required no large timbers for roofing so were more suited to the environment of the Western Isles which was presumably, by then, largely treeless. By the end of the 1st century AD monumental architecture appears to have been replaced by essentially functional cellular building forms. Possibly the Western Isles were by then part of a more extensive political structure, perhaps centred on the Pictish sub-kingdom of the Orkneys (Armit 1990a; 1996).

It is possible to argue that the failure of the Western Isles to develop power structures sufficiently strong to figure independently in the political world of the Early Historic period was due to their increasing economic marginalisation. Nonetheless, different approaches to interpretation have to be adopted to deal with the initial emergence and development of monumental architecture.

REFLECTIONS ON MARGINALITY

The present palaeoenvironmental record for the Western Isles suggests a gradual worsening of the regional environment from the earliest human occupation of the islands. Aside from notable local catastrophes, such as the sudden loss of machair recorded sporadically in the post-medieval period and inferred from archaeological evidence at Vallay, this worsening would have been too slow to be detected by individual generations.

The economic marginality of the islands would have increasingly affected settlement patterns and restricted economic options. Direct economic pressures could have profoundly affected social stability in the short term by causing social dislocation as resources failed, and in the long term by preventing the establishment and maintenance of social structures based on the reliable production of surpluses. Alternatively, the contraction of settlement, which appears to have happened over several thousands

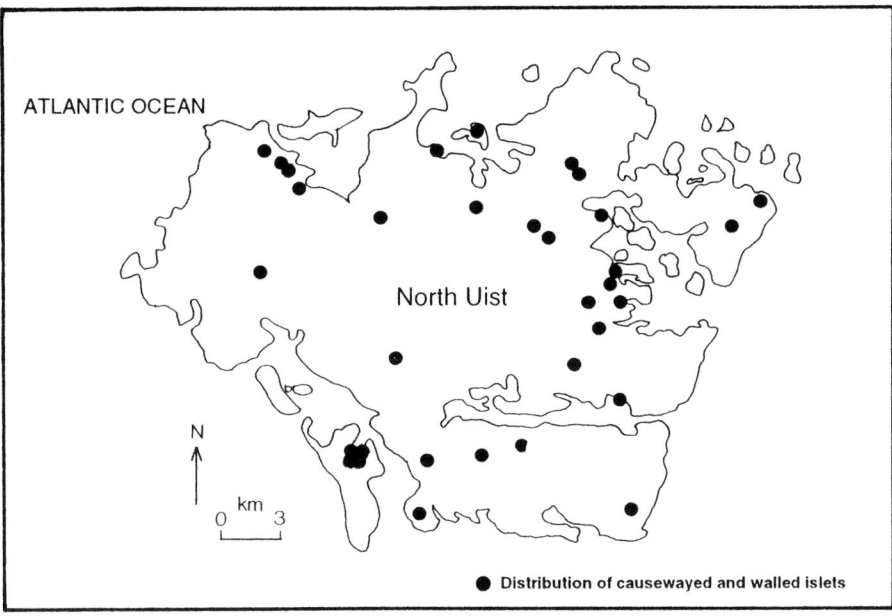

Figure 4.4 Distribution of causewayed and walled islets in North Uist.

of years, may have been entirely accommodated by social reorganisation, natural fluctuations in population numbers through disease, crop failure, and small-scale population movement. If human populations were able to survive with reasonable security, and to engage in social interaction with their neighbours and with wider social networks outwith the islands, then it hard to see what relevance the wider concept of marginality had to these people.

The Western Isles case study suggests that the constraint of economic marginality is a useful factor in the analysis of settlement, economic and social change in the long term. The gradually deteriorating environment can be read as providing a fatalistic backdrop to the whole history of the Western Isles. The present woes of the Western Isles, however, are not the product of economic marginality alone. Political marginalisation brought about by fluctuations in the power of societies who controlled or influenced the Western Isles have been at least as decisive. The later 1st millennium BC is not the only period when the communities of the Western Isles cannot easily be interpreted as culturally marginal. The Norse period is another example, when the islands formed part of an important medieval kingdom. One could argue, however, that fluctuations over a few hundred years are nonetheless constrained within an overall downward spiral conditioned by economic marginality.

This highlights one of the central points in the relationship between long term environmental decline and the wider concept of marginality: the changing scale of analysis. While from the long term perspective marginalisation may form an explanatory framework, in the shorter term, within and between conventional archaeological periods, it need not determine human action. Fluctuations over a few centuries, such as the period of monumental atlantic

roundhouse construction, cannot be adequately explained or predicted by models based on marginalisation. The variety and complexity of human behaviour which produces material culture, whether it be broch towers or decorated pottery, requires modes of explanation based on a more detailed level of analysis. The difference between these two levels equates broadly with the difference between processual and post processual forms of analysis. Post processual critiques of processual archaeology have often focused on their inability to explain specific human behaviour on this shorter timescale, eg production and use of specific items of material culture. Such critiques do not, however, generally allow for the explanatory value of processual and environmental models at other levels of analysis.

We can begin to trace the development of economic marginality in the Western Isles from the palaeoenvironmental record and from the analysis of settlement pattern change. We can identify long term economic constraints imposed by environmental decline and locally catastrophic decline in the case of machair loss. For the long term historical development of economy and settlement in the Western Isles economic marginalisation is a useful model. When we attempt a closer analysis of human behaviour over shorter periods within that time span, the explanatory value of the marginalisation hypothesis is reduced. Social and cultural behaviour does not mirror the increasing economic marginalisation of the islands. As the islands enter the historical period, their political position can be seen to fluctuate and their culture continues to develop and change.

We can only begin to describe the Western Isles as marginal, in the sense set out in the introduction to this paper, when they were ceded to Scotland. This marginality, with its cultural and social implications, was the outcome of a series of political events. Its relationship to the environmental decline and economic marginality which has developed over the past few thousand years cannot be assumed. As the scale of the socio-political unit changes, so too do perceptions of the marginality of its peripheral components: from being a part of an independent Scotland, to part of the United Kingdom and perhaps to a part of a united Europe, the islands have seemed less and less relevant to the political centre. The assumption of perennial marginality projects a modern political situation onto prehistory. The people of the Western Isles and other Gaelic communities are presently trying to demonstrate the validity of their culture and the importance of their heritage, to help define their role in the present. If there is any message from the history of human settlement in the islands it should be that while the environment has constrained and shaped its economic and settlement potential, human culture and society has certainly not been tied to continual decline.

ACKNOWLEDGEMENTS

Preliminary drafts of this paper were read by Dr Bill Finlayson and Diane Nelson and their comments are gratefully acknowledged.

REFERENCES

Angus, S, & Elliot M M 1992 'Erosion in Scottish machair with particular reference to the Outer Hebrides', *Proceedings of the 3rd European Dune Congress*, Galway.

Armit, I 1988 'Broch landscapes in the Western Isles', *Scottish Archaeological Review*, 5, 78–84.

Armit, I 1990a 'Epilogue', *in* Armit, I (ed), *Beyond the brochs*. Edinburgh University Press, 194–210.

Armit, I 1990b 'Broch-building in northern Scotland; the context of innovation', *World Archaeology*, 21.3, 435–445.

Armit, I 1992a *The later prehistory of the Western Isles of Scotland*. Oxford: BAR British Series (= Brit Archaeol Rep Brit Ser, 221).

Armit, I 1992b 'The Hebridean Neolithic', *in* Sharples, N & Sheridan, A (eds), *Vessels for the ancestors*. Edinburgh: Edinburgh University Press, 307–321.

Armit, I 1994 'Archaeological field survey in the Bhaltos (Valtos) peninsula, Lewis', *Proceedings of the Society of Antiquaries of Scotland*, 124, 67–93.

Armit, I 1996 *The archaeology of Skye and the Western Isles*. Edinburgh: Edinburgh University Press.

Bennett, K D, Fossitt, J A, Sharp, M J & Switsur, V R 1990 'Holocene vegetational and environmental history at Loch Lang, South Uist, Western Isles, Scotland', *New Phytologist*, 114, 281–298.

Birks, H J B & Madsen, B J 1979 'Flandrian vegetational history of Little Loch Roag, Isle of Lewis, Scotland', *Journal of Ecology*, 67, 825–842.

Blundell, F O 1913 'Further notes on the artificial islands in the Highland area', *Proceedings of the Society of Antiquaries of Scotland*, 47, 267–302.

Bohncke, S J P 1988 'Vegetation and habitation history of the Callanish area, Isle of Lewis, Scotland', *in* Birks, H H, Birks, H J B, Kaland, P E & Moe, D (eds), *The cultural landscape – Past, present, future*. Cambridge: Cambridge University Press, 445–461.

Crawford, I A 1965 'Contributions to a history of domestic settlement in North Uist', *Scottish Studies*, 10:2, 34–63.

Evans, J G 1971 'Habitat change on the calcareous soils of Britain; the impact of Neolithic man', *in* Simpson, D (ed), *Economy and settlement*. Leicester: Leicester University Press, 11–26.

Harding, D W & Armit, I 1990 'Survey and excavation in west Lewis', *in* Armit, I (ed), *Beyond the brochs*. Edinburgh: Edinburgh University Press, 71–107.

Newell, P J 1988 'A buried wall in peatland by Sheshader, Isle of Lewis', *Proceedings of the Society of Antiquaries of Scotland*, 118, 79–93.

RCAHMS 1928 *The Outer Hebrides, Skye and the Small Isles*. Edinburgh: HMSO.

Scott, W L 1948 'The chambered tomb of Unival, North Uist', *Proceedings of the Society of Antiquaries of Scotland*, 82, (1947–8), 1–49.

Scott, W L 1951 'Eilean an Tighe: a pottery workshop of the second millennium BC', *Proceedings of the Society of Antiquaries of Scotland*, 85 (1950–1), 1–37.

SEARCH 1993 *The Western Isles project: 6th interim report*. Sheffield: University of Sheffield.

Wilkins, D A 1984 'The Flandrian woods of Lewis (Scotland)', *Journal of Ecology*, 72, 251–258.

5. The spread of cultivation into the marginal land in Ireland during the 18th and early 19th centuries

Jonathan Bell

Abstract

The population of Ireland increased rapidly from the mid-eighteenth century until the Great Famine of the 1840s. During this period the need for land on which to grow subsistence crops of potatoes drove millions of the rural poor to settle on previously unreclaimed mountain and bogland. This paper examines some of the techniques which were used to bring this marginal land into cultivation, including drainage, ridgemaking, paring and burning, liming, and the use of organic manures. All of Ireland, including marginal areas, was divided into large estates. The paper outlines some of the complex relationships between Irish landlords and their poorest tenants, who were engaged in land reclamation. Discussion of the technical and socio-economic evidence presented leads to the pessimistic conclusion that very little of this complexity could be deduced from the material remains alone.

INTRODUCTION

Ireland has a cool temperate west coast climate, with mild temperatures and high rainfall. Very few parts of the country rise above 900 m, but climatic conditions mean that marsh and blanket bog are extensive in lowland areas, and heath and bog common over hills (Aalen 1978, 208). During the second half of the eighteenth century cultivation increasingly encroached on these waste areas, and contemporary writers and modern historians have worked at length to identify factors leading to this. These factors are generally agreed to have included commercialisation of farming, an improved transport system, the encouragement of tillage by the state, and above all, a large population increase. Between 1735 and 1841 the population of Ireland rose from 3 million to 8.2 million (Cullen 1972, 118). Urbanisation and industrialisation did not absorb this increase, and contemporary observers concluded that the salvation of the Irish poor lay in farming or emigration. In 1847, the state body known as the Devon Commission identified three possible remedies for '... the overstocked labour market of Ireland.' These were:

* Cultivation of waste.
* Draining and subsoiling the lands already productive which required this operation.

* Emigration (Devon Commission 1847, 565).

Some contemporaries saw the creation of tiny farms on reclaimed waste land as being not only of benefit to both tenants and landlords but also as an effective means of restraining social discontent (Young 1780, Vol 2, Part 1, 173).

Contemporaries and historians have debated whether the amount of reclamation of waste should have been greater, and what its profitability might have been. In the 1840s it was claimed that 6,290,000 acres of land capable of improvement still lay waste (Devon Commission 1847, 607). By this time, however, cultivation was reaching an altitude of 210 metres, the limit claimed by some to be the present upper level for profitable farming in Ireland (Solar 1983, 77).

The process of land reclamation was well under way by the 1770s, but accelerated greatly in the early nineteenth century. In 1810, one commentator described the situation in County Cork:

'Some thirty years ago, population and culture were almost confined to the sea coast. A few scattered hamlets excepted, the inner parts presented nothing to the view but bogs and wilds, overgrown with heath and furze, with some woodlands. Within that period, population has advanced with such rapidity, that both furze and heath, as well as wood, have almost disappeared, and hills, once deemed unfit for anything but coarse summer feeding, are cultivated to their very tops' (Townsend 1810, 310–311).

Cultivation was also encroaching on to lowland bogs. In 1832, in County Roscommon, for example, 'In almost every part of the county, wherever a road is carried over a bog, and wherever drains are made for keeping the road dry, which drains also serve to dry the bog, the peasantry, if left to themselves, will immediately fix, and build huts' (Weld 1832, 82–83).

The most important source of capital available for reclamation projects was that controlled by the individual Irish landlords. The Devon Commission identified three main approaches taken by landlords to the reclamation of waste land on their estates. These were as follows:

* The proprietor could leave the whole operation to the unassisted exertion of his tenants.
* The proprietor could reclaim at his own cost, and then let the land at its full improved value.
* Tenants could be located on unimproved land and assisted '... in those operations which ... [were] beyond [their] ... own means, or which tend[ed] to bring the land more rapidly forward to a perfect state of tillage' (Devon Commission 1847, 571).

The approaches adopted by landlords towards the reclamation of waste had direct implications for the scale at which the work was carried out, the techniques used, and to some extent, the permanence of the improvement. Most reclamation schemes involved the same range of operations but these were often carried out in very different ways. The main operations were:

* Road-building
* Draining
* Ploughing and/or digging
* Paring and burning
* Liming and/or manuring with dung, clay, etc.
* Planting

ROADS, DRAINS AND RIDGES

In 1802, one observer argued that in reclaiming bog and mountain, road making and drainage should go together. Effective drainage could not be achieved when access was not possible, and road-building should not be carried out on undrained, very wet land (McEvoy 1802, 190). Some large-scale drainage was carried out during the later eighteenth century. Arthur Young recorded several examples of rivers being widened, or navigable channels being cut through bogs (Young 1780, vol 1, 146–150). By the beginning of the nineteenth century, agriculturalists were stressing that drainage should be scientific. The

methods developed by the Englishman, Joseph Elkington of Warwickshire were generally favoured (Archer 1801, 73; Coote 1802, 120). However, in most parts of Ireland observers lamented the lack of application of the new principles. McEvoy's comment on County Tyrone in 1802 describes what was probably common practice: 'Open drains are in common use, only temporary to save crops in moist situations, when the latter end of the spring happens to be wet. The secret of hollow draining is very little understood in any part of the county much less the intercepting, or cutting off springs. Sod drains are not known in the county' (McEvoy 1802, 115).

By the 1840s, the drainage system advocated by the Scot, James Smith of Deanston in Perthshire (Smith 1843), was the one preferred by most agriculturalists. Smith's system had drains laid down at regular intervals; twenty-one feet apart buried at a depth of two and a half feet. The angle of drains to the slope, the distance between them, and methods of making drainage tiles, were all topics discussed at length in nineteenth century farming texts. However, throughout the period, there was widespread agreement that on uneven, stony and hilly ground, spades, shovels, picks and crowbars were the most effective implements for improving land. From the eighteenth century the cultivation ridges made by small farmers were included as part of the large scale drainage schemes, or more commonly, as alternatives to them. High sided ridges could be made by ploughs, plough and spade used in conjunction, or more often, by spades alone. The narrow spade ridges, which determined so many aspects of Irish small farmers' cultivation techniques, were often condemned by contemporary writers. Increasingly, however, their use was also defended. Arthur Young, for example, found that at Killarney 'Mr Herbert has cultivated potatoes in the common lazy-bed method [where ridges were built up on strips of untilled land] ... and he is convinced, from repeated experience that there is no way in the world of managing the root that equals it, especially for bringing in waste land' (Young 1780, Vol 2, 117). When associated with potato cultivation, it became widely accepted that Irish spades (Lambert 1845, 23), and the ridges constructed using them '... supplied the most minute conceivable system of artificial drainage' (Devon Commission 1847, 108–109). Spade ridges prevented water-logging, and were also deep enough to achieve subsoiling, breaking through a hard stratum of clay known in Ireland as *lacklea*.

We are still finding unrecorded techniques of making ridges, during fieldwork. Within any area, the technique and size of the resulting ridges might be varied to take account of soil, slope, aspect, and time of year crops are planted, the crop grown and its place in a rotation (Bell 1982). The possible range in technologies used in ridge making can be seen nowadays in Counties Cavan and Leitrim. At a ploughing match in Carrigallen, County Leitrim, ridges are made with one sided spades or loys, using a horse drawn swing plough, or even using a similar plough pulled by a tractor. The ploughs used are fitted with socks, or shares, which are up to 30 cm in width. Ridge making or 'coping' in these counties is still used by farmers to provide drainage and to build up enough depth of soil to grow potatoes on land which is underlain by a hard pan known locally as 'channel' (unpublished field-work observations).

PARING AND BURNING

One practice widely reported throughout the eighteenth and nineteenth centuries was the burning of sods and top soil. Paring off the land surface was usually a preliminary to this operation. 'Paring and burning was ... carried out prior to tillage on lea on previously uncultivated ground and consisted in stripping off the surface sod, allowing it to dry, burning it, and then spreading the ash on the soil as a fertiliser' (Lucas 1970, 99). In the early eighteenth century the practice was widely condemned by landlords, and in 1743 the Irish Parliament passed an act forbidding it (Lucas 1970, 101). This act was not very effective, however, and by the early nineteenth century the condemnations of the practice by contemporaries were not so sweeping. This was the period when the reclamation of land was increasing rapidly, and it was recognised that if properly controlled, paring and burning could speed up the process signifi-cantly. Arthur Young, though often critical of the practice, made use of it in the reclamation scheme he himself organised on the south side of the Galtee mountains (Young 1780, vol 2, part 2, 69). The change in attitude is very apparent in the county statistical surveys published in the early years of the nineteenth century. Coote, for example, writing in 1802 stated that, '... the effects of burning land were not well understood, when the legislature imposed the heavy penalty against this process' (Coote 1802, 114). Thompson, also writing in 1802, recorded a case which does not appear to have been unusual of a gentleman farmer systematically reclaiming bog land by paring with modified ploughs, and then having the sods gathered into heaps and burnt (Thompson 1802, 275–276). Modified forms of the process were practised even in some of the most systematic reclamation schemes. At Tullychar, County Tyrone in 1835, for example, paring was performed using the local type of breast-plough, known as a *flachter* (Watson 1979). When burning was found to produce only small quantities of ash however, layers of lime were added to the sod heaps to encourage 'fermentation' (Devon Commission 1847, 610–611). By the mid-nineteenth century, the situation was as described by Lucas: 'Most writers ... were entirely in favour of permitting [paring and burning] ... for the purpose of reclaiming waste land, often alleging that there was no other way of reducing land of this nature to cultivation' (Lucas 1970, 136).

FERTILISERS

The manures added to reclaimed land seem to have been largely similar to those used on ground in which tillage was well-established. Earth and clay, dug elsewhere, were sometimes spread. In extreme cases, such as areas of limestone pavement on the Aran Islands, ridges were made entirely of materials collected elsewhere. Arthur Young (1780) recorded frequent instances of earth being taken from ditches to form composts, by both landlords and small farmers. The removal of sods and top soil from mountains and bogs to spread over land elsewhere was more contro-versial. The practice was widely reported, however. In Queen's County in the 1840s, for instance, '... in all parts [of the county] in which it can be obtained, even at a distance of several miles, bog soil, commonly called "bog stuff" or "manure", is extensively used by large and small farmers, rich and poor ... I know not what would have become of the population, or how they would have subsisted without its aid' (Devon Commission 1847, 66–67).

In coastal areas, and in some cases up to thirty miles inland (Lucas 1970, 113) sea-weed, sea-sand, and sea-shells, were effectively used as fertilisers. Some observers claimed, on the other hand, that animal manure was not efficiently used. Arthur Young was emphatic: 'In the catalogue of manures, I wish I could add the composts formed in well littered farm yards, but there is not any part of husbandry in the kingdom more neglected than this; indeed I have scarce anywhere seen the least vestige of such a convenience as a yard surrounded with offices for the winter shelter, and feeding of cattle' (Young 1780, Vol 2, Part 2, 68).

How far there was neglect of animal waste as a manure is uncertain, however. When Lord George Hill took over his estate in north-west Donegal, he found that in the byre-dwellings of the poor, ten or even fifteen tons of animal manure might be collected (Evans 1957, 34).

Limestone is a common rock in Ireland, as are limestone gravels and marl. By the late eighteenth century, the use of lime in agriculture was widespread. Arthur Young found that:

> 'The manure commonly used in Ireland is lime, inexhaustible quarries of the finest limestone are found in most parts of the island, with either turf or culm at a moderate price to burn it. To do the gentlemen of that country justice, they understand this branch of husbandry very well and practice it with uncommon spirit. Their kilns are the best I have seen and great numbers are kept burning the whole year through' (Young 1780, Vol 2, Part 2, 67).

Limestone gravel was sometimes spread thickly enough on wet bogs to compress these, so assisting drainage (Young 1780, Vol 1, 376), and on some heaths the spreading of lime and marl was found to be the only operation necessary before ploughing to bring the land into cultivation (Young 1780, Vol 1, 291–292). Young claimed that he had not heard of any land being over-limed, but it has been suggested that in fact this may have been a widespread problem (Solar 1983, 79).

The first crops to be taken off newly reclaimed mountain bog varied considerably. Most major grain crops have been recorded; rye, wheat, oats and bere. Rape and cabbage were also used. Arthur Young was convinced that turnips could play a vital role in breaking in ground, although his experiment with this crop failed (Young 1780, Vol 1, 77 and Vol 2, Part 2, 70). In 1812, Dubordieu argued that the final aim in the reclamation of upland should be the production of fine grass land. This has also been the conclusion of some modern historians who argue that this end result was imposed by physical conditions as well as social and economic forces (Dubordieu 1812, 55).

Despite the recommendations of contemporary agriculturalists, and the conclusions of modern historians, from the late eighteenth century onwards the potato was the most important crop on newly reclaimed waste land. The Devon Commission summarised the advantages of the potato in reclaiming land, for a poor farmer: 'In cultivating the potato he was able to divide the labour of completely breaking down the soil over a period of three years, taking a crop each year, which repaid him as he went on. It is doubtful whether any other crop can be substituted, which will admit of this advantageous mode as regards of the poor man' (Devon Commission 1847, 570).

It has been pointed out that Ireland never became a potato mono-culture (Mokyr 1983, 92). It was however the basic food for most of the poor who were the main agents of reclaiming waste (Mitchell 1976, 205). Potato blight (*Phytophthora infestans*) hit the Irish potato crop in 1845, and for several years afterwards. This brought on a famine, known as the Great Hunger, in which more than one million people died. In some areas of marginal land, where there were alternative sources of income or food, population did not fall immediately after the famine. Overall crop production throughout Ireland did not fall immediately either. One feature of Irish life which the famine did end, however, was what had been the most common system of land reclamation; the cultivation of small patches of potatoes by the poor.

DISCUSSION

Most contemporary and historical discussion on land reclamation in Ireland has centred on its relative profitability, or usefulness in quelling social discontent. Some observers, however, also point out that reclamation could lead to detrimental results for the landscape. Cultivation techniques which led to the eventual exhaustion of the soil were often described and condemned. Over-cropping was alleged to be widespread, as was the abuse of paring and burning ground. This latter practice was sometimes claimed to have reduced soil to sand or gravel (Devon Commission 1847, 67–68). The removal of 'bog-stuff', from mountains and bogs for spreading on cultivated land, was alleged to be carried on to such an extent that

in some areas land already under cultivation had 'actually become moory' (Devon Commission 1847, 67–68).

These abuses of reclamation techniques are easily understood when we remember the desperate poverty of most of the reclaimers. For people whose main aim was simply to survive for another year, there was no question of carefully husbanding resources for the long term. Their overwhelming need was to provide food for the present.

Even where the productivity of land was increased by reclamation, the improvement was often only short term. We have seen that the techniques used in small scale reclamation centred on steep sided cultivation ridges, and potatoes. Almost all contemporary accounts are emphatic that land reclaimed for pasture by these techniques reverted to its original state within a few years if the ground was not periodically broken by tillage. Tighe's comment on County Kilkenny in 1802 is typical: 'All the hills ... revert easily into heath when neglected' (Tighe 1802, 65). Townsend, writing about County Cork at the same period, argued that bog also required tillage every few years (Townsend 1810, 261–262). In 1812, Dubordieu pointed out that: 'The most difficult task to accomplish, in draining, is the banishment of the different species of *Juncus*: nothing less than the complete drying of the soil, and repeated tillage, can subdue them: and often, when the latter is given up for some time, they return, though in diminished numbers; for the seed will continue fresh for an indefinite time' (Dubordieu 1812, 319).

In the 1840s, a proprietor from County Galway suggested that land could be prevented from reverting to bog by planting it with potatoes '... every fifth or sixth year'. Where land was not systematically worked, but cropped for '... three or four seasons ... , the improvement is almost entirely effaced; the croft is sometimes a little greener, but in many cases is scarcely distinguishable from the neighbouring bog' (Devon Commission 1847, 631).

The great reclamation movement of the eighteenth and early nineteenth centuries did dramatically change the Irish landscape, and traces of it are still common on marginal land all over Ireland. These traces are the sort of evidence which archaeologists rely on, and this raises the question of how much of the social and economic change which produced them could be deduced by examination of the material remains. The answer to this question must be to a large extent pessimistic. Settlement and field patterns would probably allow a tentative distinction to be made between planned and unplanned reclamation schemes, and traces of cultivation ridges, drains, fertilisers, etc, would permit some intelligent speculation about techniques used. However, the documentary evidence available for the period shows that a diversity of techniques could leave very similar traces, while diversity of social relationships under which reclamation was carried out could not be deduced using only material evidence. An examination of the use of marginal land in Ireland during the eighteenth and nineteenth centuries can be used in speculation on the significance of similar remains from other periods but its main impact should be to make such speculation very cautious.

REFERENCES

Aalen, F H A 1978 *Man and the landscape in Ireland*. London: Academic Press.

Archer, J 1801 *Statistical survey of the County Dublin*. Dublin: Dublin Society.

Bell, J 1982 'A contribution to the study of cultivation ridges in Ireland', *Journal of the Royal Society of Antiquaries of Ireland*, 114, 80–97.

Coote, C 1802 *Statistical survey of the County Cavan*. Dublin: Dublin Society.

Cullen, L 1972 *An economic history of Ireland since 1600*. London: Batsford.

Devon Commission 1847 *Digest of evidence taken before Her Majesty's Commissioners of inquiry into the state of the law and practice in respect to the occupation of land in Ireland Part 1*. Dublin: HMSO.

Dubordieu, J 1812 *Statistical survey of the County of Antrim*. Dublin: Dublin Society.

Evans, E E 1957 (2nd edition 1967) *Irish folkways*. London: Routledge and Kegan Paul.

Lambert, J 1845 *Agricultural suggestions to the proprietors and peasantry of Ireland*. Dublin: Farmer's Gazette.

Lucas, A T 1970 'Paring and burning in Ireland: A preliminary survey', *in* Gailey, A & Fenton, A (eds), *The spade in Northern and Atlantic Europe*. Belfast: Ulster Folk Museum.

McEvoy, J 1802 *Statistical survey of the County of Tyrone*. Dublin: Dublin Society.

Mitchell, F 1976 *The Irish landscape*. London: Collins.

Mokyr, J 1983 'Uncertainty and pre-famine Irish agriculture', *in* Devine, T M & Dickson, D (eds), *Ireland and Scotland, 1600–1850*. Edinburgh: John Donald.

Smith, J 1843 *Remarks on thorough draining and deep ploughing*. Stirling: Drummond.

Solar, P 1983 'Agricultural productivity and economic development in Ireland and Scotland in the early nineteenth century', *in* Devine, T M & Dickson, D (eds), *Ireland and Scotland, 1600–1800*. Edinburgh: John Donald.

Thompson, R 1802 *Statistical survey of the County of Meath*. Dublin: Dublin Society.

Tighe, W 1802 *Statistical observations relating to the County of Kilkenny*. Dublin: Dublin Society.

Townsend, H 1810 *Statistical survey of the County of Cork*. Dublin: Dublin Society.

Watson, M 1979 'Flachters; their construction and use in an Ulster peat bog', *Ulster Folklife*, 25, 61–66.

Weld, I 1832 *Statistical survey of the County of Roscommon*. Dublin: Royal Dublin Society.

Young, A 1780 *A tour in Ireland*. Dublin.

6. The facts don't speak for themselves

Mervyn Watson

Abstract

Drawing on structural, documentary and oral evidence this paper discusses livestock management in marginal areas in Ireland during the historic period. In particular it focuses on the management of cattle and the form of transhumance in Ireland known as booleying. It also examines the inter-relationship between booleying, fodder and the rundale open field system.

The paper uses data from the more recent past and assesses the value of such studies for archaeologists. Looking at the complexities of the practice of booleying, the structural remains and changing role of booley houses in the Gweedore area of County Donegal, the paper highlights problems of interpretation for archaeologists and the dangers of imposing too rigid a model of agricultural practices on past societies.

INTRODUCTION

Although not an archaeologist, the writer's work at the Ulster Folk and Transport Museum requires the study of the past, and in particular agriculture in the past in Ireland. This paper, using structural, documentary and oral sources, discusses fodder and related livestock management in marginal areas in Ireland during the historic period. As regards livestock management it deals mainly with the management of cattle. Also, drawing on the social science maxim 'The facts don't speak for themselves', it attempts to assess the value of such studies, particularly those related to the recent past, for archaeologists.

BOOLEYING

One livestock management practice in marginal areas in Ireland, which has attracted a lot of attention amongst writers and researchers, is a form of transhumance known as *búailteachas* in Gaelic or booleying in its anglicised form. The origins of booleying are obscure and it is usually associated with the movement of cattle. One early reference to the practice was by Edmund Spenser in 1595 who wrote of booleying: 'Have you ever heard what was the occasion and first beginning of this custom ... There is

one use amongst them to keep their cattle and to live themselves the most part of the year in boolies pasturing upon the mountain and waste wild places' (Morley 1890, 67).

Early documentary evidence suggests that the movement of humans and livestock followed a general pattern based on a seasonal rotation. This involved the movement of livestock and people from the lowlands during the summer to the distant upland pastures or booley grounds. O'Flaherty in his account of west Connacht, *H'Iar Connaught*, in 1684 noted that '... in summertime they drive their cattle to the mountains, where such as look to the cattle live in small cabins for that season' (O'Flaherty 1846, 17). West Connacht is on the Atlantic seaboard of Ireland.

In 1744 Walter Harris in *The Ancient and Present State of the County Down* claimed: 'In the bosom of the Mourne Mountains there is a place called the 'Deers' meadow to which great numbers of our people resort in the summer months to graze their cattle. They bring with them their wives, children and little wretched furniture and there live' (Harris 1744, 125). The Mourne Mountains are situated on Ireland's north eastern seaboard.

Later documentary and oral sources suggest a more complex pattern of movement involved in booleying than the general picture of summer migration painted in the earlier evidence. Sir William Wilde in 1836, describing booleying on Achill Island, situated on Ireland's Atlantic seaboard, noted:

> 'During the spring the entire population of several of the villages in Achill close their winter dwellings, tie their infant children on their backs, carry with them their loys [one-sided spades] and some carry potatoes, with a few pots and cooking utensils drive their cattle before them, and migrate into the hills, where they find fresh pastures ... There they build rude huts and summer-houses of sods and wattles, called booleys and then cultivate and sow with corn a few fertile spots in the neighbouring valleys. They thus remain for about two months of the spring and early summer till the corn is sown. Their stock of provisions being exhausted, and the pasture consumed by their cattle, they return to the shore and eke out a miserable precarious existence by fishing. No further care is ever taken of the crops, indeed they seldom ever visit them, but return in autumn, in a manner similar to the spring migration, to reap the corn, and afford sustenance to their half-starving cattle' (Wood-Martin 1888, 238).

Wilde's description of booleying differs from earlier sources in the number of migrations during the year and the time when the migrations took place. Instead of one migration to the booleying grounds Wilde's description has two. One takes place during the spring and early summer and the other during the autumn harvest.

Another complex form of booleying took place in County Donegal around the Gweedore district and is described in Lord George Hill's *Facts from Gweedore* in 1887, in which it states that '... it often happens that a man has three dwellings – one in the mountains, another upon the shore, and the third upon an island' (Hill 1971, 24). Estyn Evans in the preface to the fifth edition of *Facts from Gweedore* claims that livestock were taken from the shore to mountain grazing during the summer and that it was mainly in the autumn that they were taken to the islands.

Later oral testimony adds a further twist to transhumance in the Gweedore area in more recent times. The shore area in Gweedore is called Magheragallan and local man Fred Coll claimed that the landlord moved the people from Magheragallan to the nearby hills to reclaim the land and settled them in the townlands of Stranacorcagh and Seskin Beg (Coll & Bell 1990, 83). As a result of the movement of the permanent dwelling house to the hills there was a reversal in the booleying procedure. Instead of the animals being moved from shore to mountain pasture in the summer, they were moved from mountain pastures to the shore.

The oral evidence indicates that the movement of cattle from the mountains to the shore at Magheragallan persisted well into this century. Dr Desmond McCourt recollected

that until around 1940 during early summer young people from adjacent townlands would bring cattle to Magheragallan and spend a few weeks tending them. The oral evidence also indicates that in its final stages the movement of cattle from the hills to Magheragallan was on a daily basis. Fred Coll claimed:

> 'They kept going back and forth summer and winter until around the mid 1950's ... there was a lot of emigration. Around that time there were those who let their houses fall down and become a ruin. What they would do was ... the women and children would drive cows down the couple of miles to Magheragallan, and then take them back to Seskin Beg at night. At the end there were only two houses belonging to the Ferry's and Doherty's kept so that they would be used for a period during the summertime' (Coll & Bell 1990, 83).

The Ferry and Doherty houses were byre dwellings with the animals living at one end of the house and the people at the other. The Ferry house has since been dismantled and re-erected at the Ulster Folk and Transport Museum, Figure 6.1.

As well as bringing about a change in the movement of cattle, the movement of people from Magheragallan to the hills, by the landlord, changed the role of the houses on the shore from being permanent dwelling houses to booley houses. The changing role of the Magheragallan houses is a cautionary reminder for archaeologists when interpreting archaeological remains. If anything, the physical remains of the houses would suggest that they were permanent dwellings. Their role as booley houses is less obvious. Consequently any assessment of the Magheragallan houses based on structural evidence must be at best conjecture; the facts do not speak for themselves. Booley houses varied in structure and in the materials used to construct them. Some, such as the Magheragallan houses, were substantial structures made of stone whilst others were very temporary structures made of sods or merely holes cut in the side of the bog. These latter types were known in the Gweedore area as bothógs.

RUNDALE

Hay was not widely used in Ireland as a fodder crop until the latter half of the nineteenth century. Consequently, some authorities have argued that the development of booleying was made possible by the adoption of oats. Furthermore, booleying coupled with the growing of oats is perceived by some authorities as having been central to the rundale system as practised in Ireland. Estyn Evans outlining the yearly pattern of feeding livestock in the rundale open field system argued: 'The livestock would be fed in winter mainly on oatstraw, and grazed on the infield stubble, supplemented by the outfield grasses which had been allowed to mature while the bulk of the stock

Figure 6.1 Interior of Magheragallan byre dwelling belonging to the Ferry family re-erected at the Ulster Folk and Transport Museum, Cultra, Northern Ireland.

Figure 6.2 Earthen banks or mearings at Magheragallan, County Donegal.

was at the summer booley grounds' (Evans 1967, 152).

Grazing on commonage was calculated in units equivalent to the amount of pasturage necessary to support a cow. These units were sometimes referred to as sums or collops or in gaelic *áit bó* [cow's grass]. The number of units of grazing which a farmer held was related to the size of their farm.

As with hill pasture the sandy coastal plain known as the *machaire* was also common grazing for cattle. The commonage on the *machaire* at Magheragallan was divided into strips by earthen banks known as mearings, Figure 6.2. The oral record states that grazing on the *machaire* was tightly controlled as over-grazing on the light sandy soil would expose the sand and cause erosion of the *machaire*. It is claimed that sheep were only allowed on the *machaire* at Magheragallan for three weeks as the people believed if the sheep were kept on the *machaire* too long they would fill up with sand. One local informant said that he knew of a butcher in recent times who had killed sheep which had been kept on the *machaire* for a prolonged period and they were found to have half a stone of sand in their gut.

One aspect of pasture management frequently commented on by observers was the length of time Irish farmers in marginal areas kept cattle indoors before releasing them onto summer pastures. McParlan in his survey of County Donegal in 1802 gave two reasons for the practice. He observed: 'All through this county the cattle are housed during the winter months; in the mountain region not only during the winter, but in summer too very much, for the double purpose of collecting manure and avoiding the cruppan which the people fancy the cattle are more subject to by feeding abroad at large than confined in the house' (McParlan 1802, 54).

Fear of the 'cruppan' is a recurring explanation in the evidence for the practice of prolonged confinement. 'Cruppan' is given in *Dinneen's Irish – English Dictionary* as meaning a 'shrinking, a wasting disease' (Dinneen 1927, 276). The writer has been informed that hill pasture is deficient in minerals, notably cobalt, and that this can cause wasting in cattle. Research to define the specific type of disease represented by the 'cruppan' is on-going.

CONCLUSION

In assessing the value of the above material for archaeologists in understanding agricultural practices in marginal areas, it is argued that the use of such material, particularly that related to the more recent past reminds us not to impose too rigid a model of agricultural practices on past societies. This is especially applicable to prehistoric societies where structural remains are the only evidence with which the researcher has to build a model of a society.

This point is reinforced not only by the complexity of the booleying patterns outlined in the later sources but also by the changing role of the houses at Magheragallan. Whilst the buildings themselves remained more or less the same their functions as dwelling houses changed.

Also central to any analysis of farming in marginal areas is the social structure within which farming practices take place. Whilst the environment is a powerful constraint, the determining forces which define the margins are social. In the case of Gweedore it was the landlord who moved the people from Magheragallan to the hill slopes for social and economic reasons not environmental pressures.

REFERENCES

Coll, F & Bell, J 1990 'An account of life in Machaire Gathlán (Magheragallan), early this century', *Ulster Folklife,* 36, 80–85.

Dinneen, P (ed) 1927 *Focloir Gaedhilge agus Béarla.* Dublin: The Educational Company of Ireland.

Evans, E E 1967 (4th edition) *Irish folk ways.* London: Routledge and Kegan Paul.

Harris, W 1744 *The ancient and present state of the County of Down.* Dublin: A Reilly.

Hill, G 1971 (5th edition, ed E Estyn Evans) *Facts from Gweedore.* Belfast: Queens University Press.

McParlan, J 1802 *Statistical survey of the County of Donegal.* Dublin: Dublin Society.

Morley, H 1890 *Ireland under Elizabeth and James I.* London: G Routledge & Sons.

O'Flaherty, R 1846 (1846 edition, ed J Hardiman) *H'Iar Connaught: A chorographical description of west or H'Iar Connaught 1684.* Dublin: The Irish Archaeological Society.

Wood-Martin, W G 1888 *The rude stone monuments of Ireland.* Dublin: Hodges, Figgis & Co.

7. Calf slaughter as a response to marginality

Finbar McCormick

Abstract

High incidences of very young calves are a feature of bone assemblages from many Atlantic Scottish archaeological sites. Evidence for fodder shortage, particularly in the case of the Hebrides, is here considered as an explanation for this phenomenon.

INTRODUCTION

The study of cattle bone assemblages from the Western Isles has generally produced rather unusual age-slaughter patterns. Although many reports are still unpublished, the work of Paul Halstead, Dale Serjeantson, Judith Finley and Barbara Noddle has consistently shown large numbers of juvenile cattle present, and particularly high numbers of neo-natal cattle. At the Late Bronze Age – Early Iron Age site at Baleshare, North Uist, for instance, some 36% of the cattle present were neo-natal while at the Udal, North Uist about 30–50% were less than two or three weeks of age at time of death. Dale Serjeantson noted that the cattle age slaughter pattern at the Udal differed little between about AD 300 and AD 1700. The Iron Age settlement at Cnip, on Lewis, suggested a high incidence of calf mortality, with about 50% of the cattle being less than one year old at time of death while at the wheelhouse at Sollas, in North Uist, about 33% of the cattle found in the so-called ritual pits in the floor bottoms were neonatal individuals (Finley 1991, 144). Again at the broch at Dun Mor Vaul, Tiree, a very high incidence of juvenile animals was evident (Noddle 1974, 190–196). On the basis of the data available, therefore, it can be assumed that the presence of large numbers of neo-natal and juvenile cattle was the normal situation in sites on the Western Isles between the Bronze Age and probably the post-medieval period. If one considers the evidence from the Orkney Islands this pattern can be observed as far back as the early Neolithic (Noddle 1979; 1983). At the Knap of

Howar over 50% of the calves died during their first year, mostly shortly after birth. Noddle's survey of the Orcadian data showed that this pattern of high juvenile cattle mortality continued until the late medieval period.

Generally, pronounced age-slaughter patterns are interpreted as representing a specific livestock economy, ie, either dairying of beef production. Contrasting interpretations of the presence or absence of high incidences of juvenile cattle have been published (Legge 1992; McCormick 1992). It can be suggested, however, that the high incidence of juvenile cattle on the Western Isles sites are not a reflection of economic strategies but are simply a product of the area's marginality in terms of climate and availability of fodder for livestock.

FODDER AND HAY MAKING

The Western Isles, or Outer Hebrides, is today a disadvantaged area in terms of agriculture and it seems that the situation was little different in the past. A range of ecological factors contribute to this situation. Because of its coastal setting the climate of the Hebrides is essentially mild but the weather is anything but so. Active fronts accompanied by heavy rain and driving wind are a common feature of the weather of the Western Isles, especially during winter. While the rainfall may not be significantly greater than much of Highland Scotland the continual wind blasts extenuate its effect. As Boyd & Boyd (1990, 46) state, 'Although the climate is generally mild, the combined effects of the elements makes it severe on plant growth, livestock and wild animal'. Livestock lose heat in high winds, especially when combined with driving rain. A wet, windy and cloudy summer will not only retard the growth of crops and pasture but also leave the livestock entering winter in a relatively poor condition so that they might easily succumb to severe weather during winter. It has been estimated, for instance, that the supplementary feeding of livestock necessary to compensate the energy deficit caused by a normal Hebridean summer near sea level is similar to that prescribed at an altitude of 300 m on the hills of Perthshire (Boyd & Boyd 1990).

Therein lies the problem. Until fairly recently there was little, or no, supplementary feeding available for cattle during the winter in the Western Isles, or indeed elsewhere in the Highlands of Scotland. One tends to think that since hay has definitely been saved in Britain since at least the Anglo-Saxon period, and probably from the Roman period, that it has been universally in use since then. This, however, is not the case.

In the past the saving of hay tended to be restricted to certain economic situations. It was often saved, for instance, when dung was needed for tillage. By keeping cattle housed over winter and feeding them on hay one could much more efficiently collect the dung for manuring. In other instances, the sheer coldness of the winter meant that the little vegetation available would have been inaccessible to cattle as it was buried under snow. This was the case in Norway where at the beginning of the present century it was noted that in some extreme cases cattle could only be left outside to forage for four months of the year (Rasmussen 1989, 63). In such cases hay saving would be necessary and it is not surprising the documentary evidence indicates that the Vikings saved hay (Magnusson & Palsson 1960, 258). The antiquity of hay saving has not yet been established. Swiss Neolithic sites have produced byres that appear to be for the keeping of cattle during the winter but the evidence there indicates that they were fed on leaves, which can easily be dried and preserved (Rasmussen 1989). This type of fodder would have had little potential in the Northern and Western Isles where tree cover has always been extremely limited. The earliest probable evidence for the saving of hay in Britain is provided by the scythe during the Roman period, with several examples from Newstead providing evidence for the practice in southern Scotland (Curle 1911, 284). The fact the Bede notes that hay was not saved in Ireland implies the practice had become well established in England during the eighth century (Sherley-Price 1965, 39).

HAY MAKING IN THE WESTERN ISLES

The earliest evidence for the saving of hay on the Western Isles is in the eighteenth century when the Rev Dr John Walker travelled there in 1764 and 1771 (McKay 1980).

Walker noted that hay was only occasionally cut in the Western Isles and where it was, the practice had been only recently introduced. In Lewis the inhabitants had only begun to save hay less than nine years before his visit and the practice was only introduced to South Uist in 1756 (McKay 1980, 43, 78). Walker (McKay 1980, 208) also notes that cattle were generally let '... run abroad the whole winter, without receiving a mouthful of dry forage.' (McKay 1980, 208). He adds, however, a very important observation about the use of hay when describing South Uist and Isle of Skye (McKay 1980, 78 & 208). He states that the hay that was saved was reserved exclusively for young stirks, ie, calves that have just been weaned. The implication of this is that young calves found it particularly difficult to get through the first winter.

As in the Hebrides, the practice of saving hay was uncommon, or absent, throughout most of Scotland before the late eighteenth century and as a consequence of this widespread mortality of cattle during winter was a common occurrence (Handley 1953, 69). Where it was necessary to house animals during winter, they were fed extremely poor supplements such as straw, boiled chaff and pounded whins/furze. It is estimated that in Scotland as a whole one in five livestock died of starvation during the winter in the eighteenth century while in some instances, such as in Ayr in 1800, as much as one in three of the cows and horses perished (Handley 1953, 69). Handley (1953, 69) vividly describes the effect of poor winter fodder on livestock when he states that

> '... those that survived came forth in the spring bleary eyed and dizzy with weakness, and staggered drunkenly to the pasture grounds. Often they were so weak that they had to be carried. For the first few days much time was spent in dragging them forth from the bogs and marshes into which they had been tempted by the sprouting vegetation on the surface and from which they were too weak to extricate their clogged feet'.

One Perthshire farmer noted in 1808 that he remembered the

> '... poor wives during the nipping cold north-east winds in May, provincially termed the Cowquake, tending their cows, reduced to a skeleton and covered with a blanket, while they picked up spires of grass which had begun to rise in the kailyard or at the bottom of walls and banks. And to such extremities were they reduced at times that I have heard of their taking the half rotten thatch from the roofs of their houses and giving it to the half dead animal as a means of prolonging its miserable existence' (quoted in Handley 1953).

In fact the period between February and May became known as the 'Lifting' because it was an annual operation when all the neighbours were summoned to carry and support the poor beasts at the new grazing (Fenton 1978, 428).

The fact that hay was kept specifically for the stirks in the Western Isles suggests that calves were especially vulnerable during winter. Walker (McKay 1980) notes some of the preoccupations of Hebridean farmers with calf rearing so that they would have a good chance of surviving the first winter. The main objective was to get the new-born calf to arrive at the optimum time, ie, to coincide with the new grass crop. One concern of the farmers was the danger that the calf would arrive too late, and would therefore not have had sufficient time to build up its strength before the onset of winter. The avoidance of late fertilisation of the cow by the bull was therefore of great importance (McKay 1980, 75). Early calving produced another problem. If they arrived too early there would not be enough fodder for the calf, and more importantly, the shortage of fodder for the cow would not enable it to produce sufficient milk for its calf. This problem was noted by Walker in Skye where he states that if a cow calves before the first of March it is sometimes a month or six weeks before the cow has sufficient milk to feed its calf (McKay 1980, 209). Presumably, in many such instances the calf would die.

DELIBERATE CALF SLAUGHTER

These eighteenth century sources clearly show that a high level of natural mortality would have been expected in the Western Isles and this might partly account for the high incidence of calves on archaeological sites. In addition, natural mortality would have been extenuated by the deliberate killing of young calves. The shortage of fodder at its simplest meant the production rate of cattle outstripped the ability of the land to sustain the animals and it is clear that attempting to get all the calves through the first winter was impractical. The remedy was for excess calves to be killed at, or soon after, birth. Thus in the nineteenth century Hebrides it was noticed that '... it was common to raise only one calf upon the milk of two cows', implying that the excess calf was killed at the beginning of the lactation period (McKay 1980, 65). Handley (1953, 69) states that calf slaughter was widely practised at the time. The Statistical Account of Sutherland written in 1820 stated that this practice was '... universally adapted ... on account of the want of winter keep' (Handley 1953, 69). The economic necessity of the practice is succinctly expressed in the Uist proverb '*Is fearr aon laogh na da chraicionn*' – one calf is better than two skins (Carmichael 1916, 256).

CONCLUSIONS AND ARCHAEOZOOLOGICAL IMPLICATIONS

The documentary evidence clearly indicates that the shortage of winter fodder necessitated the killing of large numbers of calves at, or soon after, birth. In addition to this a large number of calves probably died of natural causes during their first winter. This probably accounts for the high incidence of neo-nates and juvenile cattle noted on Hebridean and Orcadian sites dating between the early Neolithic and late Medieval period. The type of age pattern noted in these areas probably does not reflect any particular specialised livestock economy, such as beef production or dairying. Cattle were deliberately killed simply to ensure that others would survive. The limited fodder available, coupled with the winter's acute weather, essentially meant that environmental determinism was the cause of the age slaughter pattern rather than any higher economic strategy.

ACKNOWLEDGEMENTS

I would like to thank all those archaeozoologists and excavators whose reports and sites I have referred to prior to publication, especially Dale Serjeantson, Paul Halstead, Ian Armit, John Barber and Ian Crawford.

REFERENCES

Boyd, J M & Boyd, I L 1990 *The Hebrides*. London:Collins.
Carmichael, A 1916 'Grazing and agrestic customs of the Outer Hebrides', *The Celtic Review*, 10 (1914–1916), 40–54, 144–148, 254–262, 358–375.
Curle, J 1911 *A Roman frontier fort and its people: The fort at Newstead in the parish of Melrose*. Glasgow: James Maclehose & Sons.
Fenton, A 1978 *The Northern Isles: Orkney and Shetland*. Edinburgh: John Donald.
Finley, J 1991 'Ritual Pit deposits from period B1', *in* Campbell, E, 'Excavations of a wheelhouse and other Iron Age structures at Sollas, North Uist, by R J C Atkinson in 1957', *Proceedings Society Antiquaries Scotland*, 121 (1991), 141–148.
Handley, J E 1953 *Scottish farming in the eighteenth century*. London: Faber and Faber.
Legge, A J 1991 *Animals, environment and Bronze Age economy*. (=Excavations at Grimes Graves Norfolk, 1972–1976, Fascicule 4), British Museum Press.
Magnusson, M & H Palsson 1960 *Njal's saga*. Harmondsworth.
McCormick, F 1992 'Early faunal evidence for dairying', *Oxford Journal of Archaeology*, 11, 201–209.
McKay, M M (ed) 1980 *The Rev. Dr. John Walker's Report on the Hebrides of 1764 and 1771*. Edinburgh: John Donald.
Noddle, B 1974 'Report on the animal bones', *in* MacKie, E, *Dun Mor Vaul*. Glasgow: Glasgow University Press, 187–198.
Noddle, B., 1979, 'A brief history of domestic animals in the Orkney Islands, Scotland, from the 4th millennium B.C. to the 18th century', *Archaeozoology*, 1, 226–303.
Noddle, B 1983 'Animal bone from Knap of Howar', *in* Ritchie, A, 'Excavation of a Neolithic farmstead at Knap of Howar, Papa Westray, Orkney', *Proceedings Society Antiquaries Scotland*, 113 (1983), 92–100.
Rasmussen, P 1989 'Leaf-foddering of livestock in the Neolithic: Archaeobotanical evidence from Weier, Switzerland', *Journal of Danish Archaeology*, 8 (1989), 51–71.
Sherley-Price, L L 1965 *Bede's history of the English church and people*. Harmondsworth.

8. On the outside looking in: a view of animal bones in Roman Britain from the North West Frontier

Sue Stallibrass

Abstract

An ongoing study of the sizes and morphology of cattle bones from Roman sites in northern England appears to demonstrate that differences existed, not only between livestock in the north and the south of England, but also either side of the Pennines. The north-west may have had more in common with Wales and the south-west of England than with eastern and southern areas. The differences may be related to differing degrees of Romanisation, in turn influenced by the political and geographical marginality of this region (the north-west frontier of the Roman Empire). Alternatively, the differences could have developed from pre-existing variability established during the Iron Age, but a paucity of relevant data precludes a study of this possibility at this stage.

INTRODUCTION

This extended synopsis introduces a long-term study that was begun in 1992 and which is still ongoing in 1997. The overall aims and methodology are presented in some detail, together with some preliminary results and hypotheses. Tables and figures are not given since the study is incomplete, but a selection of supporting data is presented (most of which are otherwise unpublished).

The marginality considered in this paper is geographical and political, rather than agricultural. Northern England was occupied by the Roman army from the 1st – 4th centuries AD and was, in effect, the north-west frontier of the Roman Empire. Although there is plenty of evidence for contact between northern England and areas further south (such as the finds in northern England of pottery that was manufactured in southern England or on the Continent), there are several lines of evidence that suggest that the degree of Romanisation enjoyed or undergone by the inhabitants of the north was considerably less than that apparent in the south. This evidence includes types and distributions of settlements, artefacts, exotic (ie non-indigenous) foodstuffs, etc.

This study seeks to address two questions:

1) Were livestock types and patterns of animal husbandry different in different regions of the country during the Roman occupation?

2) Can these differences be traced back to pre-existing differences during the later Iron Age, or do they reflect varying degrees of 'Roman' influence?

THE MATERIAL AVAILABLE FOR STUDY

Compared to southern England, few Iron Age sites in northern England have been excavated, and even fewer have produced animal bone assemblages of any great size. Most of the known Iron Age sites lie in rural areas, where the typically shallow stratigraphy (often combined with acidic soil conditions) has produced aggressive leaching conditions that are unconducive to the preservation of unburnt bone. This scarcity of large, well preserved assemblages of Iron Age animal bones in the region precludes any comprehensive comparisons of livestock types or husbandry patterns in the Iron Age and Roman periods. Iron Age sites in northern England with good preservation of organic materials need to be targeted for excavation in order to address several research questions (Huntley & Stallibrass 1995).

In contrast, many of the military and urban Roman sites have deep stratigraphy and good conditions for organic preservation (including waterlogged deposits in several cases). Large collections of well-preserved animal bones have been recovered from several sites and are suitable for study. Rural sites of the Roman period in this region, however, suffer from the same problems as those of the preceding Iron Age. There are almost no Roman villas in England north of North Yorkshire (Scott 1993).

The material available, therefore, consists mainly of collections retrieved from military and urban sites. Most of these (hand-recovered) collections are dominated by the remains of cattle, as are those from similar sites further south. Sites in the north that have been particularly productive include Carlisle (Stallibrass 1991a; 1991b; 1993a; forthcoming a), Ribchester (Stallibrass 1993b; forthcoming b) and Papcastle (Mainland & Stallibrass 1990; forthcoming) in the west, and Catterick (Meddens 1990a; 1990b; Payne 1990) and Piercebridge (Rackham & Gidney unpub.) in the east of the region. For various reasons including preservation conditions, depths of excavations, and historical and military events, much of the material from the western part of the region derives from 1st and 2nd century deposits, whilst much of that from the eastern part derives from the 3rd and 4th centuries. Further data should be available within the next two years from Carlisle (Stallibrass in prep.) and Binchester (O'Connor *et al* in prep.), dating from the 1st to the 4th centuries AD in both instances. In addition, Iron Age material is available from Thorpe Thewles (Rackham 1987) and Catcote (Hodgson 1967), and a large body of data is currently being analysed from Stanwick (Rackham in prep.). These three sites all lie in a small area of County Durham/North Yorkshire within the basin of the River Tees. There are no large Iron Age collections from the western half of the region, and there are no good collections of either Iron Age or Roman period animal bones from the central part of northern England (notwithstanding various excavations at forts along Hadrian's Wall that have produced small assemblages or collections of confused derivation).

This study, therefore, has concentrated on comparisons within the Romano-British period. Any comprehensive understanding of potential Iron Age variability will have to await the availability of large data sets pertaining to the Iron Age in northern England, particularly in the western part of the region.

THE DETAILS OF THE STUDY

There are several aspects of the animal bone assemblages that need to be compared between the sites in the north and south of England that might relate to Roman influence, either socially (regarding fashion or style) or economically (regarding trade and supply). These include: the relative importance of wild species of mammals, fish and birds; the relative importance of the three main domestic species (cattle, sheep and pigs) as contributors of meat to the diet; the types of animal husbandry represented by age and sex distributions for each of the three major domestic species (eg dairying, meat, wool economy or use as traction animals); patterns of butchery and meat distribution for

each of the three major domestic species (eg whole carcases or selected joints of meat); evidence of ritual use or of taboos (eg foundation sacrifices, non-consumption of horse-flesh); and the size and conformation of the animals themselves. It is this latter aspect that is addressed by this study, concentrating on one 'species': domestic cattle.

Cattle bones have been chosen for detailed study and comparison between sites since they are particularly common. Also, being generally quite large, they are less susceptible to recovery biases than are the bones and teeth of smaller species, such as sheep. This means that comparisons between sites are less likely to be affected by conditions and methods of excavation.

Many specialists working with animal bones in continental Europe have noted that, at sites within the Roman Empire, there is an increase in the average sizes of cattle bones due to the presence of larger bones in addition to those of the 'normal' type that continues straight through from the Iron Age (Bökönyi 1974; Teichert 1984). The bones of these so-called 'Celtic short-horn' cattle are typically small and gracile, and the skulls tend to have small horncores. They appear to have been ubiquitous throughout western Europe (including Britain) during the Iron Age and Roman periods. Outside of the Roman Empire, these cattle persisted right through to the Migration period. Within the occupied territories, they also provided the majority of the animal bones found on archaeological sites. The presence of bones from larger animals leads to an increase in the average sizes of the bones together with an increase in the variability of the measurements. The proportions of larger to 'normal' bones tend to be highest towards the centre of the empire and to fall off noticeably towards the peripheries. Since Britain lay on the northwestern periphery of the Roman Empire, it might be expected that it lay at or beyond the limit of the area in which these larger cattle were kept.

Some specialists working with animal bones from Roman sites in southern England have found bones from larger cattle, indicating that the influence of the Roman Empire did extend that far with regard to livestock husbandry (Grant 1989).

Did larger cattle appear in the north-west frontier zone (northern England) during the Roman occupation?

A further question regarding any larger animal bones is: what do they signify? Several researchers (eg Noddle 1983) have suggested that larger animals might simply reflect better husbandry practice. Poor nutrition can restrict growth whilst good husbandry can permit livestock to attain (more nearly) their full genetic potential. If the differences in size are related to nutrition rather than to breed types, then the conformation of the skeletons and the shape of the horncores should be similar regardless of absolute size. 'Breed' (ie genetically-based) differences should be detectable in terms of morphological traits such as the presence, size and shape of horncores and, possibly, the incidence of congenital traits such as missing teeth, or teeth with differing morphologies.

Another factor that is bound to affect the relative sizes of domestic cattle bones is sexual trimorphism. Any consideration of the sizes of a group of animal bones needs to ascertain the sex ratios of females: castrates: entire males before that group can be compared to any others (eg Hodgson 1969).

Much work has already been undertaken regarding the question of Roman livestock conformation. Thomas's (1989) doctoral thesis compared material from different types of Roman site and from different geographical zones. His results concerning different types of sites are interesting, and he was able to demonstrate that producer sites (such as villas and rural settlements) tended to have slightly larger cattle than the consumer sites (such as towns and military establishments) that form the basis for the current study. But Thomas can be criticised for using arbitrary geographical zones that were imposed on the map (using a pair of compasses centred on Dover) without regard to topography, lines of communication, etc. His comparisons also suffered from an unavoidable paucity of data for the northern zone. The current study has access to data for the northern areas that have become available since Thomas undertook his study, and seeks to define the zones by the data themselves, not to impose artificial boundaries.

for splaying (using a comparison of the width across the condyles with that at the fusion line cf Higham 1969). Although it is not always possible to ascertain whether or not this has been done for published data, the incidence of splayed metacarpals in any assemblage is usually very low, and is unlikely to affect comparative studies to any major extent.

Although there is general agreement that metacarpals from cows tend to be smaller (ie narrower) than those from bulls, and that metacarpals from castrated males tend to be longer than those from entire males, the precise ratios of width to length for each 'sex' and the degree of overlap of such ratios are still only poorly known. In the absence of larger data sets of measurements from modern, known sex animals, the ratios used by Howard (1962) are used here, despite their limitations.

For the calculation of withers (shoulder) heights, the factors given by Fock 1966 (in Driesch & Boessneck 1974) have been used for the datasets used in this study. Wijngaarden-Bakker & Bergström (1988) have criticised the use of metapodials for wither height calculations and Bartosiewicz (1987) has concluded that the radius gives the most reliable estimate, but complete cattle radii are extremely rare in archaeological assemblages, and this study uses the metacarpals in order to make use of the same bones as those used for estimates of sex ratios.

METHODOLOGY

This study takes a single element (the metacarpal) for comparison between assemblages. The metacarpal was chosen because:

a) it is robust and survives well;
b) it is one of the first elements in the skeleton to reach its mature size;
c) it is often preserved relatively intact due to low rates of butchery;
d) it is easy to measure;
e) most animal bone reports include raw data or summary statistics of its measurements (due to a, b, c & d, above);
f) it shows sexual di- or tri-morphism and
g) factors have been calculated for its use in estimates of withers height.

This study makes use of distal width and total length measurements, since these can be used in studies of overall size, sexual dimorphism and withers heights. The use of distal width measurements can have a drawback which is that some bones may have become splayed during the animal's life, thus enlarging the distal width measurements (Driesch 1975). This can be a consequence of heavy exertion (such as that caused by pulling a plough through clayey or stony soil, or pulling heavy cartloads). It is possible to eliminate such bones from any statistical analyses, either by initial selection procedures or by testing

RESULTS

The earliest sizeable collection comes from an early timber fort dated to the late 1st century AD at Annetwell Street, Carlisle (Stallibrass 1991a; 1991b). The mean distal breadth of the cattle metacarpals is 52.7 mm, with a range from about 46 – 64 mm (N=33). A collection from the second timber fort at the same site, dated to the early-mid 2nd century has an almost identical distribution: mean = 52.0 mm, range = 45 – 64 mm (N=38). Both distributions are skewed towards the smaller end of the range, and Howard's shape indices suggests that the bones derive from ratios of about 5:3 and 3:2 females:males, respectively.

The next large assemblage from Carlisle dates to the late 1st to early 3rd centuries, and comes from the civilian site at The Lanes (Stallibrass 1993a; forthcoming a). The sample of 41 metacarpals has a mean distal breadth of 52.0 mm, with a range from about 47 – 64 mm. This is statistically indistinguishable from the military collections from Annetwell Street, and indicates that there was no increase in size between the late 1st and early 3rd centuries and also, interestingly, suggests that there was no difference in size between the cattle bones deposited at the civilian and military sites. Other lines of evidence, including the incidence of various congenital traits and the morphology of the horncores, also suggest that the civilian and military cattle bone assemblages derive from a common stock. Once again, the shape indices of the complete metacarpals indicate a ratio of 3:2 females to males.

The latest collection from a fort at Annetwell Street relates to a stone fort of the early 4th century. The preservation of bones from this period is not as good as that for the earlier, waterlogged, timber forts, and the sample is small. Also, there are no complete metacarpals from which shape indices can be calculated. The thirteen metacarpals have distal breadth measurements with a mean of 49.5 mm, ranging from about 44 – 57 mm. In this small sample from animals of unknown sex, there are no particularly large bones, but the lower end of the range is similar to that for all of the other three collections and there is certainly no indication of any increase in size.

Further bones from Roman civilian deposits at The Lanes are due to be studied (by Stallibrass) in 1996–7, and should increase the sample sizes, particularly for the later Roman period (3rd/4th centuries). Also, measurement data were recorded by Rackham for material from the military annexe at Castle Street but were not included in the published report (Rackham *et al* 1991). The deposits date from the late 1st – early 3rd centuries and it is planned that these will be analysed and included in the final study.

Other sites west of the Pennines that have produced Roman period animal bones include the forts at Papcastle and Ribchester. Material from both of these tends to date from the 1st/2nd centuries and confirms the generally small size of the cattle in this part of the region, although the sample sizes tend to be rather small. At Ribchester, Lancashire, cattle metacarpals from the late 1st – mid 2nd centuries AD have a mean distal breadth of 51.1 mm, ranging from 46.2 – 58.1 mm (N=17). Again, the sex ratio is approximately 3:2 females:males. At Papcastle, Cumbria, a collection of late 1st – mid 2nd century metacarpals have a mean distal breadth that is slightly larger: 53.4 mm (range 45.6 – 60.9 mm, N=16) but the sex ratio of this group is completely unknown since none of the bones retained their complete length. A collection from 3rd century deposits had a mean distal breadth very similar to those of the collections from Carlisle and Ribchester (mean = 52.2 mm, range = 43.9 – 64.8 mm, N=15). Again, the sex ratio of this collection is completely unknown.

The overall picture emerging from these analyses is one of stability and uniformity. The cattle metacarpal breadths are small (compared to those of early domestic cattle from Neolithic deposits, or more recent, 'improved' breeds of the post-Medieval and recent periods) and show little variation through time or space within north-western England during the 1st-3rd centuries AD. Similar material has been obtained from other sites in the north-west 'highland zone' at Wroxeter, near Shrewsbury, Shropshire (Armour-Chelu, unpublished data) and Segontium, which lies beneath Caernarfon, Gwynedd (Noddle 1993). At Wroxeter, a very substantial collection (N=148) of late 2nd – early 3rd century cattle metacarpals has a mean distal breadth measurement of 52.3 mm (range = 45.6 – 71.5 mm). The sex ratio is about 4:1 females:males. At Segontium, even a 4th century collection has a mean distal breadth of 50 mm (range = 41 – 62 mm, N=69; sex ratio unknown).

These data contrast with those from the eastern part of northern England. Here, the comparison is hampered by a current paucity of material from earlier Roman deposits, or the failure of the excavator/faunal analyst to separate material from different periods, but the general trend appears to be one of slightly larger measurements (except at the Iron Age site of Thorpe Thewles). At Corstopitum, near Corbridge on Hadrian's Wall, a large collection of material has a mean distal breadth of 54 mm (range = 45 – 73 mm, N=132), but this includes examples from all periods between the 1st – 4th centuries, and it is not possible to ascertain whether there were any changes through time (Hodgson 1967). Nor are shape indices available for judging the ratio of females:males represented. Suggestions that the military bases on Hadrian's Wall were supplied with large cattle imported from southern England cannot be substantiated unless local sites can be found that prove that local cattle were smaller than those at the forts. Hodgson's (1967) own work at the rural late Iron Age/1st century AD site of Catcote (recently returned to County Durham from Cleveland) showed that the small collection of 10 distal metacarpals had an even higher mean distal breadth of 55 mm (range 51 – 61 mm) but, again, the sex ratio is unknown (due, in this case, to the absence of any complete metacarpals).

At Piercebridge, which is also in the lowlands of the Tees Basin on the Durham/North Yorkshire border, and which is close to the major late Iron Age site of Stanwick, a mainly 4th century collection from the fort and vicus has a mean distal breadth of 55 mm (range = 48 – 70 mm, N=70) (Rackham & Gidney unpub.). As at Carlisle, Papcastle and Ribchester, the sex ratio is 3:2 females:males. Statistical comparisons of the means do not give numerically significant results, but the differences between the collections from east and west of the Pennines are consistently biased towards larger mean distal breadths for the eastern collections. At Bainesse Farm, Catterick, North Yorkshire, a roadside settlement along Dere Street (the modern Great North Road / A1) produced a mainly late 2nd – late 4th century collection (Meddens 1990a). The mean distal metacarpal breadth is 54.0 mm (range = 47.1 – 67.8 mm, N=42), and the sex ratio is 5:3 females:males.

The material from Stanwick will be extremely important for indicating the nature of some of the late Iron Age livestock in the area (although the site itself may not be typical of rural Iron Age sites, since it was a very large, high status site: possibly an oppidum). At the moment, the only relatively large collection from an Iron Age site in the north of England comes from Thorpe Thewles, which is in the same general area as Catcote, Piercebridge and Stanwick (Rackham 1987). Occupation at this rural site dates to the mid/late Iron Age – 1st century AD. The cattle metacarpals have a mean distal breadth of 52.4 mm (range = 47.1 – 59.2 mm, N=20), which is similar to all of the Roman material from the west of the region.

The Thorpe Thewles material is difficult to assess using Howard's shape indices. Although six bones have 'female' indices, the other six are all indeterminate, falling in the gaps between Howard's data for females and castrates (although biased towards the female ends of the gaps), and it may be that the ranges that Howard obtained are restricted in their use. A similar problem may occur with the Roman material from the General Accident site, York (O'Connor 1988). Here, some of the shape indices fall below the lower end of Howard's range for females.

Regional differences in livestock need not simply involve absolute size ranges. At both York and Exeter (Maltby 1979), the mean distal breadths are particularly low (about 49 / 50 mm), and both collections are associated with slightly greater than usual proportions of females (approximately 4:1 rather than the more common ratio of 3:2 females:males). At Exeter, pooled collections from various excavations show no differences between material from the mid 1st – late 3rd century and that from 4th century deposits. The York material falls in the middle of this time range, dating to the late 2nd-early/mid 3rd centuries, but the animals would have looked rather different, despite their similar distal metacarpal breadth measurements. The withers heights of the Exeter cattle confirm their overall small size, but those for the York cattle suggest that these animals were as tall as those from sites where the distal breadths were considerably wider. The lack of comprehensive Iron Age animal bone data for the region is frustrating, since cultural evidence demonstrates that part of eastern Yorkshire was occupied in the late Iron Age by people with continental (La Tène) affinities, and it is possible that they brought in, or preferred, a slightly different type of livestock to that established in the rest of the region. This could have become the base stock from which the slender, leggy cattle of the Roman period derived.

Comparisons of distal metacarpal breadths and withers heights of Iron Age and Roman period material, made between those from the north of the country and those from the south, suggest that the cattle whose remains were deposited at the north-western Roman sites were quite similar to those from Iron Age sites in various parts of the country, whereas those from the north-east Roman sites may have been somewhat larger. Those from York may have been a regional type. It is interesting that those from Exeter, in the south-west of the country, follow the pattern for the north-west rather than the rest of the south. It is tempting to suggest that the north and the west of England were 'marginal' during the Roman occupation, compared to the lower lands of the east and south.

Since the weight and height of an animal can be affected by nutrition and husbandry methods as well as by genetic predisposition, a study has been started of the size and morphology of the cattle horncores at these sites. Horncore morphology and size are related to sex as well as to hereditary factors, but the introduction of new genotypes might be recognisable by the appearance of new phenotypes. At York, a small group of late 2nd – early/mid 3rd century horncores is very distinctive in size and shape from the normal distribution, and may indicate the presence of a new type of cattle at the site. At Chester-le-Street in County Durham, a tiny collection includes one 1st/2nd century skull fragment with 'normal' Celtic short-horn type horncores, and one 4th century horncore that falls well within the distribution of the larger type of horncore seen at York (Stallibrass 1991c; 1993c). Since every one of the 40 measurable late 1st/early 2nd century horncores from the fort at Annetwell Street and the 53 measurable 1st/3rd century horncores from the civilian site at The Lanes, Carlisle was of the 'Celtic shorthorn' type, with no evidence for the presence any other types of horncore, the presence of one out of two at Chester-le-Street is considered to be significant.

DISCUSSION

The small sample sizes from many of the collections have hampered statistical comparisons of the cattle distal metacarpal breadth measurements. Where these have been attempted, the results have shown no statistical differences, but they do all follow a trend. That is, the 1st-4th century bones from the north-west of England show no change in size through time and are similar to those from Iron Age sites elsewhere in the country (such as Thorpe Thewles in the north-east, and Danebury in southern England: Grant 1991). The morphology of the horncores is similarly uniform and can be described as 'Celtic short-horn' type. The cattle metacarpals from sites in or around North Wales (Wroxeter and Segontium) and from south-western England (Exeter and Dorchester: Maltby 1990) show similar patterns. These collections contrast with those from the north-east of England (Corstopitum, Piercebridge and Catterick) and further south. The study also considered material from Chichester (Levitan 1989), Colchester (Luff, unpublished data, see Luff 1993), Baldock (Chaplin & McCormick 1986), Stonea (Stallibrass in press) and Barton Court Farm (Wilson *et al* 1984) all of which demonstrated the presence of larger cattle metacarpals, even in 1st/2nd century material.

Further work is required, particularly with earlier Roman (1st/2nd century) material from north-east England, and later Roman (3rd/4th century) material from north-west England to confirm or test the generality of these conclusions, and many more data are required from Iron Age sites in northern England (especially the western and central part) before any consideration can be made of variations that may have existed within the region prior to its occupation by the Roman army. These variations may involve morphological or congenital traits, besides metrical distinctions. To fully test the impact of the Roman presence on types of livestock and animal husbandry in the region, native (ie Romano-British) sites must also be targeted for investigation but, at the moment, very few data are available.

The author's working hypothesis is that the part of England that lay at the north-west frontier of the Roman Empire was less affected by Roman influences than parts further south and east, and that this can even be seen through a comparison of animal bones from either side of the Pennines. How this relates to any differences that may (or may not) have existed prior to contact with (and invasion by) people under the political control of the Roman Empire cannot be studied through animal bones at the moment due to a lack of Iron Age material. Likewise, the consequences of varying degrees of acculturation during the Roman occupation of part of Britain, and the effects of these on the post-Roman period remain tantalisingly unknown. By the Medieval period, regional types of livestock were becoming well known. Did they have their origins in the Roman period?

ACKNOWLEDGEMENTS AND REQUEST FOR FURTHER DATA

I am extremely grateful to Miranda Armour-Chelu and to Rosie Luff for providing me with their raw data for Wroxeter and Colchester (which have not been included in the published reports). Thanks, too, to various people who have provided me with other unpublished data from smaller collections (which I have not mentioned in this extended abstract, but which will be included in the 'final' (?!) study that I aim to have completed in 1998). I would still be interested to hear from colleagues who know of relevant data sets, whether or not these are published. Although collections of as few as 10 measurable cattle metacarpal distal breadths are not, individually, ideal for statistical comparisons, several such collections certainly do have group value

REFERENCES

Bartosiewicz, L 1987 'Cattle metapodials revisited: a brief review', *ArchaeoZoologia*, I(1), 47–51.

Bökönyi, S 1974 *History of domestic mammals in Central and Eastern Europe*. Budapest: Akademiai Kiado.

Chaplin, R E & McCormick, F 1986 'The animal bones', *in* Stead, I M & Rigby, V *Baldock. The excavation of a Roman and pre-Roman settlement, 1968–72*. London: Society for the Promotion of Roman Studies, Britannia Monograph Series 7, 396–415.

Driesch, A von den 1975 'Die Bewertung pathologisch-anatomischer Veränderungen an vor- und frügeschichtlichen Tierknochen', *in* Clason, A T (ed), *Archaeozoological Studies*. Amsterdam: North Holland Publishing Company, 413–425.

Driesch, A von den & Boessneck, J 1974 'Kritische Anmerkungen zur Widerristhöhenberechnung aus Längenmassen vor- und frühgeschichtlicher Tierknoche', *Säugetierkundliche Mitteilungen*, 22, 325–348.

Grant, A 1989 'Animals in Roman Britain', *in* Todd, M (ed),

Research on Roman Britain 1960–89. London: Society for the Promotion of Roman Studies, Britannia Monograph Series 11, 135–146.

Grant, A with C Rushe & D Serjeantson 1991 'Animal husbandry', *in* Cunliffe B & Poole C, *Danebury. An Iron Age hillfort in Hampshire. Volume 5. The excavations 1979–1988: the finds*. London: Council for British Archaeology. CBA Research Report 73, 447–487.

Higham, C F W 1969 'The metrical attributes of two samples of bovine limb bones', *Journal of Zoology*, 157, 63–74.

Hodgson, G W I 1967 *A comparative analysis of faunal remains from some Roman and native sites in northern England*. Unpublished MSc thesis, University of Durham.

Hodgson, G W I 1969 'Some difficulties of interpreting the metrical data derived from the remains of cattle at the Roman settlement of Corstopitum', *in* Ucko, P J & Dimbleby, G W (eds), *The domestication and exploitation of plants and animals*. London: Duckworth, 347–353.

Howard, M M 1962 'The early domestication of cattle and the determination of their remains', *Zeitschrift für Tierzüchtung und Züchtungsbiologie*, 76, 252–264.

Huntley, J P & Stallibrass, S 1995 *Plant and vertebrate remains from archaeological sites in northern England: data reviews and future directions*. Durham: Architectural and Archaeological Society of Durham and Northumberland Research Report 4.

Levitan, B 1989 'The vertebrate remains from Chichester Cattlemarket', *in* Down, A, *Chichester Excavations VI*. Chichester: Phillimore, 242–276.

Luff, R 1993 *Animal bones from excavations in Colchester, 1971–85*. Colchester: Colchester Archaeological Reports 12.

Mainland, I & Stallibrass, S M 1990 *The animal bone from the 1984 excavations of the Romano-British settlement at Papcastle, Cumbria*. London: English Heritage. Ancient Monuments Laboratory Report Series 4/90.

Mainland, I & Stallibrass, S forthcoming 'The animal bone', *in* Olivier, A C H, Quartermaine, H & Howard-Davies, C L E (eds), *Excavations at Papcastle (Derventio), 1984*. Lancaster: Lancaster University Archaeological Unit, Lancaster Imprints.

Maltby, M 1979 *The animal bones from Exeter 1971–1975*. Sheffield: Department of Prehistory & Archaeology. Exeter Archaeological Reports Vol. 2.

Maltby, M 1990 *The animal bones from the Romano-British deposits at the Greyhound Yard and Methodist Chapel sites in Dorchester, Dorset*. London: English Heritage. Ancient Monuments Laboratory Report Series 9/90.

Meddens, B 1990a *Animal bones from Bainesse Farm, a Roman roadside settlement near Catterick (Yorkshire), excavated in 1980 and 1981*. London: English Heritage. Ancient Monuments Laboratory Report Series 98/90.

Meddens, B 1990b *Animal bones from Catterick Bridge (CEU 240), a Roman town (North Yorkshire), excavated in 1983*. London: English Heritage. Ancient Monuments Laboratory Report Series 31/90.

Noddle, B A 1983 'Size and shape, time and place: skeletal variations in cattle and sheep', *in* Jones M K (ed), *Integrating the subsistence economy*. Oxford: British Archaeological Reports. BAR International Series 181, 211–238.

Noddle, B 1993 'Bones of larger mammals', *in* Casey P J, Davies J L & Evans J, *Excavations at Segontium (Caernarfon) Roman fort 1975–1979*. London: Council for British Archaeology. CBA Research Report 90, 97–118.

O'Connor, T P 1988 *Bones from the General Accident Site, Tanner Row, York*. York Archaeological Trust, and the Council for British Archaeology. The Archaeology of York fascicule 15/2.

Payne, S 1990 *Animal bones from excavations in 1972 at Catterick Site 434, North Yorkshire*. Ancient Monuments Laboratory Reports Series 5/90.

Rackham, D J 1987 'The animal bone', *in* Heslop, D H, *The excavation of an Iron Age settlement at Thorpe Thewles Cleveland, 1980–1982*. London: Council for British Archaeology. CBA Research Report 65, 99–109 and Fiche 5:F1–G14.

Rackham, D J & Gidney, L J unpub. (1984) *Piercebridge Roman fort and environs. An analysis of a sample of animal bones from the collections excavated at Piercebridge Roman fort and vicus. archive report.* Department of Archaeology, University of Durham.

Rackham, D J, Stallibrass, S M & Allison, E P 1991 'The animal and bird bones', in McCarthy, M R, *The structural sequence and environmental remains from Castle Street, Carlisle: excavations 1981–2. Fascicule 1.* Kendal: Cumberland and Westmorland Antiquarian and Archaeological Society Research Series No. 5, 73–88.

Scott, E 1993 *A gazetteer of Roman villas in Britain.* School of Archaeological Studies, University of Leicester: Leicester Archaeology Monographs No. 1.

Stallibrass, S 1991a *Animal bones from excavations at Annetwell Street, Carlisle, 1982–4. Period 3: the earlier timber fort.* London: English Heritage. Ancient Monuments Laboratory Report Series 132/91.

Stallibrass, S 1991b *A comparison of the measurements of Romano-British animal bones from periods 3 and 5, recovered from excavations at Annetwell Street, Carlisle.* London: English Heritage. Ancient Monuments Laboratory Report Series 133/91.

Stallibrass, S 1991c *The animal bones from Church Chare, Chester-le-Street, Co. Durham 1990–91.* London: English Heritage. Ancient Monuments Laboratory Report Series 134/91.

Stallibrass, S 1993a *Animal bones from excavations in the southern area of the Lanes, Carlisle, Cumbria, 1981–1982.* London: English Heritage. Ancient Monuments Laboratory Report Series 96/93.

Stallibrass, S 1993b *Ribchester graveyard excavations, 1989. The reports on the animal bones.* Department of Archaeology, University of Durham: Durham Environmental Archaeology Report 22/95.

Stallibrass, S 1993c 'Animal bone evidence' (pp 74–79), *in* Bishop, M, 'Excavations in the Roman fort at Chester-le-Street (Conganis), Church Chare 1990–91', *Archaeologia Aeliana* (5th series), 21, 29–85.

Stallibrass, S forthcoming a 'The animal bones', *in* McCarthy, M R, *Fascicule 1. The stratigraphic sequence, absolute dating and the environmental remains at the southern end of The Lanes, Carlisle: excavations 1981–2.* Kendal: Cumberland and Westmorland Antiquarian and Archaeological Society Research Series.

Stallibrass, S forthcoming b 'The animal bones', *in* Buxton, K & Howard-Davies, C L E (eds) *Brigantia to Britannia: Excavations at Ribchester 1980 and 1989/90.* Lancaster University Archaeological Unit. Lancaster Imprints.

Stallibrass, S in press 'The animal remains', *in* Jackson, R P J & Potter, T W, *Excavations at Stonea, Cambridgeshire.* London: British Museum monograph.

Teichert, M 1984 'Size variation in cattle from Germania Romana and Germania Liberia', *in* Grigson, C & Clutton-Brock, J (eds), *Animals and archaeology: 4. Husbandry in Europe.* Oxford: British Archaeological Reports. BAR International Series 227, 93–103.

Thomas, R N W 1989 *Cattle and the Romano-British economy: a metrical analysis of size variation.* Unpublished PhD thesis, University of Southampton.

Wijngaarden-Bakker, L H van & Bergström, P L 1988 'Estimation of the shoulder height of cattle', *ArchaeoZoologia*, II, 67–82.

Wilson, B *et al* 1984 'Faunal remains: animal bones and marine shells', *in* Miles, D (ed), *Archaeology at Barton Court Farm, Abingdon, Oxon.* Oxford: Oxford Archaeological Unit Report 3 and London: Council for British Archaeology Research Report 50, Fiche A1–G14.

9. Disturbance and regeneration phases in pollen diagrams and their relevance to concepts of marginality

Kevin J Edwards and Graeme Whittington

Abstract

Palynologically-derived episodes of woodland disturbance and regeneration, and their utility as indicators of possible site marginality, are examined. The concept of marginality is explored in terms of different kinds of environment – physical, human and perceptual. In their turn, 'disturbance' phases are seen as a generic term comprising a hierarchy of disturbance types – *viz.* disturbance, interference and clearance. The nature of 'regeneration phases' is considered, especially with regard to its reflection of marginality and its value as an anthropogenic indicator.

A discussion of the reliability of disturbance and regeneration phases for inferring marginality looks at three aspects: pollen source areas; the aggregating nature of disturbance phases; and forest farming. A woodland utilisation model, contrary to orthodox interpretations, could imply that regeneration phases were not periods of inactivity or marginality.

It is concluded that changes of a floristic nature observable in pollen diagrams are potentially important indicators of marginality. Such marginality is demonstrated in part by woodland disturbance phases, but probably more strongly by intervening regeneration phases.

INTRODUCTION

The aim of this contribution is to examine ways in which episodes of woodland disturbance, and conversely, regeneration, both of which may be established by pollen analysis, can be interpreted as indicators of possible site marginality. Before such a task is attempted, however, it must be prefaced by a definition of marginality. It is seen here as a concept which is derived from environmental factors. These factors may be of a physical nature, for example a hillslope, or human, as in the existence of fields. More important is the coincidence of the two, exemplified by a hillslope cleared of vegetation to accommodate fields. In such a case, marginality may arise from neither the physical nature of the hillslope nor the human development of the fields, but from limitations posed by technology or, even from human perceptions.

Detailed pollen diagrams from appropriate areas have revealed phases of woodland reduction (usually considered to be the result of human activity), and subsequent woodland regeneration, ever since the publication of Iversen's 'landnam' paper in 1941. Such phases are not confined solely to Europe (eg Simmons & Tooley 1981;

Behre 1986; Birks *et al* 1988), but they are also becoming clearly established for other parts of the world (eg Tsukada *et al* 1986; Delcourt *et al* 1986; Edwards & MacDonald 1991).

MARGINALITY

Marginality, in any consideration of human activity, is of major importance, as its reality to a particular group of people may dissuade them from utilizing an environment fully, if at all. This may have socioeconomic consequences which lead to settlement response being placed on a gradient anywhere from 'life being difficult' to 'total abandonment or death', if partial or full out-migration is not embraced (Figure 9.1).

Physical environment

The physical environment may be marginal because of, for example, such factors as the nature of soil, climate and topography. It will also depend upon the activity

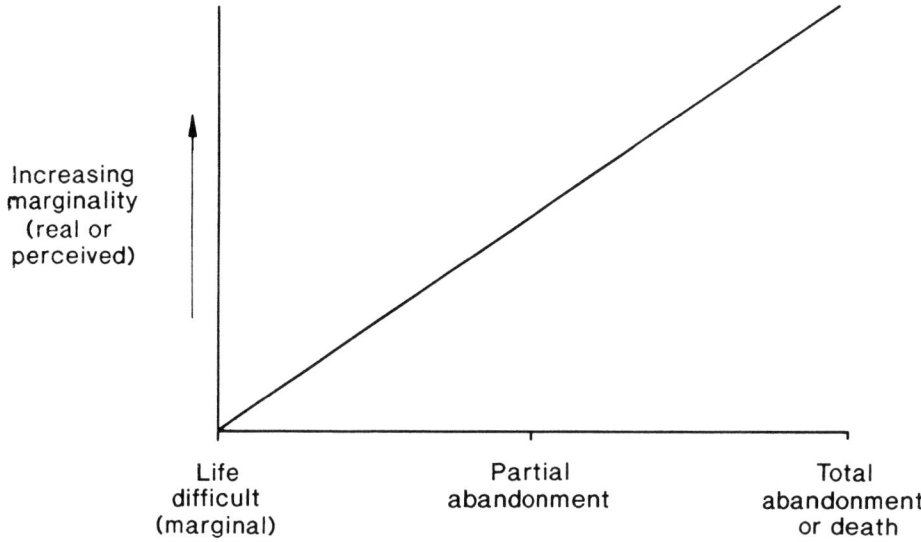

Figure 9.1 Schematic diagram of the response/marginality gradient.

intended for an area; for example, land which is sub-optimal for cereal cultivation may be adequate for pasture. The degree of marginality may further change with development in technology; the acquisition of metal axes may have allowed more extensive and more easily achieved woodland removal, enabling more or different land to be brought into use; developments in ploughing, manuring and crop types may have provided opportunities for arable exploitation of previously unused or pastoral areas.

Human environment

The human environment may be marginal as a result of many different technological and cultural features. A group of people who rely on arable agriculture developed by ploughed strip-cultivation, will suffer from marginality if their population grows and the unoccupied land around them is not amenable to exploitation by their landuse culture. The type of social organization, in terms of settlement form, may also induce marginality; where nucleated settlement is favoured, flatter land may be preferable to more accidented terrain, even though the latter possesses other environmental advantages. In such examples as these, the cultural dimension is seen as being stronger than would be the case where physical environmental determinants and opportunities were paramount; for example, an upland zone may well be desirable if flooding of coastal areas or valley floors was frequent. Defence needs may also render upland zones more desirable; in times of social upheaval, a hillfort could be a powerful stimulant to the development of settlement in its vicinity; locations which might otherwise be sub-optimal for subsistence, as with lake dwellings, could also be favoured.

Perceptual environment

It is impossible, with any degree of certainty, to penetrate the mind of early peoples. At a simple level, it could be maintained that one person's hostile zone is another's refuge or utopia. A rain forest can be perceived as highly desirable by the pygmy but malevolent by the agriculturalist (Turnbull 1961); but can it really be known (*pace* Hodder 1990) whether the Neolithic cultivator viewed the forest as inhospitable, dangerous and marginal, or as a vital resource area for edible plants, timber products and meat supply? Clearance may have turned otherwise marginal areas into useful ones, but soil exhaustion may have rendered them unattractive and marginal again, unless they were also seen as a resource-in-waiting (ie fallow), or land capable of alternative exploitation, such as one of rich browse or, more marginally, poor pasture.

Perception, necessity and complexity can also play a role in determining the apparent marginality of a particular area. The Romans saw certain peripheral areas of their empire in northern Europe as dangerous territory which harboured enemies (Groenman van-Waateringe 1980; Maxwell 1989). Military necessity dictated that these areas, previously agriculturally-rich, should become marginalised in terms both of agriculture and population (Groenman van-Waateringe 1983; Whittington & Edwards 1993). Apparent abandonment of such areas for a number of centuries made them marginal for settlement and agriculture, even for long after the Romans had departed.

DISTURBANCE PHASES

Disturbance phases in pollen diagrams are, by definition, just that – a phase in which some disturbance of the flora is displayed. This paper focuses on disturbance in wooded environments but, in doing this, the impression should not be conveyed that there was no human impact in largely unwooded areas; not only did that occur but it is detectable in pollen diagrams, although sometimes with difficulty (Vorren 1986; Hicks 1988).

Disturbance phases can be seen as the eponymous representative of a hierarchy of disturbance types. 'Disturbance' carries no implications as to cause, but merely indicates that some disruption of an existing vegetational pattern occurred; the disruption may have arisen from such causes as coppicing, pollarding, clearance, browsing, natural fire, disease or windthrow. The term 'interference phase' suggests intent, but does not imply a specific mechanism or cause for disturbance; game-driving, timber-recovery, or the winning of land for building or agriculture may, for instance, have been the reason for interference. 'Clearance' is taken here to indicate intentional removal of vegetation; the cause may be, for example, axe-felling, ring-barking, burning of standing trees or fallen timber, and the reasons for the activity could also be the same as those suggested under 'interference'. 'Landnam' (Old Norse, 'land take') is simply a species of 'clearance' which in its original conception at least (cf Iversen 1941; 1949) denoted a short-term phase of clearance for agricultural purposes (pastoral or arable). Disturbance phases are found at times of both hunter-gatherer and agricultural activity (and almost certainly neither), and their duration can be extremely short (and will often go undetected if sampling resolution is inappropriate) or of extreme longevity (Pilcher *et al* 1971; Edwards 1979).

Assuming that interference or clearance phases are those under consideration, then at the most basic of levels, and for a given time and place, the pollen catchment areas would have included land under some kind of usage – and this may have been marginal. An abundance of cereal or meadowland pollen may characterise a non-marginal environment. A profusion of heath and acid grassland pollen may typify marginality; however, as is suggested by the perceptual factor, some farming communities may have been more satisfied with the yield of animal products (or turf) from such areas – in which case 'marginality' could not be argued without qualification.

REGENERATION

An inherent aspect of this discussion of the disturbance phase is the period during which regeneration took place (Figure 9.2). These periods often intervene between adjacent disturbance episodes (although the latter may consist of successive bouts of more intensive disturbance [Figure 9.3]). The regeneration phase is taken to reflect the regrowth of woodland, whether that be partial or complete. If partial, it may signify that the land has become less intensively utilised, and it may conceivably have become more marginal, or could be perceived as having

become so. A full restoration of arboreal pollen levels may denote a complete regeneration of the woodland cover, thus reflecting the abandonment of a pre-existing agricultural landuse – whether this is, or is not, in response to a change in environmental or social factors, the land has certainly become more marginal compared with its previous level of use.

Thus, quite simply, regeneration phases may show that an area has become marginal for agriculture – they serve potentially, therefore, as a valuable anthropogenic indicator, and may be exploited in the same way as the pollen of *Plantago lanceolata, Ambrosia* or Cerealia.

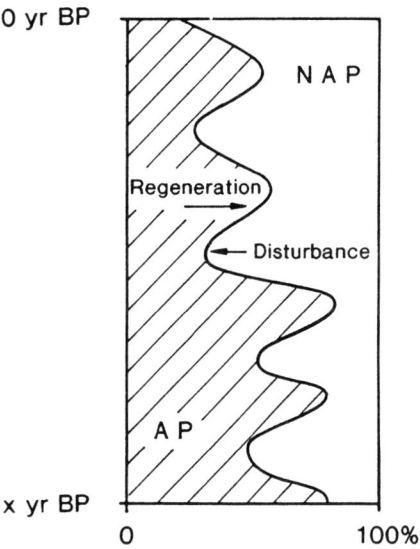

Figure 9.2 Schematic summary pollen diagram of disturbance and regeneration phases (AP = arboreal pollen, NAP = non-arboreal pollen).

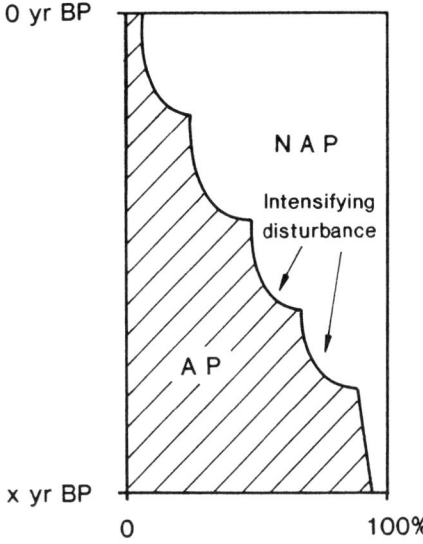

Figure 9.3 Schematic diagram of successive phases of intensifying woodland disturbance (AP = arboreal pollen, NAP = non-arboreal pollen).

HOW MARGINAL?

Assuming the existence of a sufficiently detailed pollen diagram with a satisfactory floristic signal, a combination of disturbance and regeneration phases will provide the data to allow an inference of marginality. Given the cyclical nature of such episodes – recurrent patterns of disturbance and regeneration are common in pollen diagrams (Edwards 1979; Birks *et al* 1988; Whittington *et al* 1991; Whittington & Edwards 1994) – marginality can be posited as a repetitive phenomenon in the lifetime of a site, with there being no necessity for the site to be an upland one.

How reliable is this palynological inference of marginality? Pollen analysis has been established long enough for methodological shortcomings to be appreciated (Faegri & Iversen 1989). In the context of marginality, at least three aspects of this may be commented upon.

i) Pollen source areas

The nature of an area from which pollen is received may change through time (for example as a cleared area is being expanded or reduced, or as shifting cultivation occurs [Turner 1964; Edwards 1979; 1982; 1991]). Although the proximal cause of such activity may be the marginalisation of land, it could also be the case that the growth of woodland fringing a pollen deposition site might be preventing the pollen from cultivated areas reaching the sampling point; in such instances, marginality might be more apparent than real.

ii) The aggregating nature of disturbance phases

A multitude of disturbances, each spatially discrete but in close proximity and also covering many overlapping or consecutive time periods, could give the impression, palynologically, that individual disturbances were long-lived, perhaps embracing centuries or even millennia (Edwards 1979; Smith 1981; Buckland & Edwards 1984). In reality, each disturbance episode may have been of short duration and each may have led to land abandonment and woodland regeneration.

iii) Forest farming

The traditional model of clearance phase interpretation assumes that intervening phases of woodland pollen recovery reflect a regeneration of woodland and land abandonment (cf Berglund 1986). It could be, however, that in some instances the woodland growth is no more than the reflection of pollen accumulation at the fringe of a deposition site, and/or that forest farming (involving say, woodland management, small cereal plots and pasture) was occurring within openings in a regenerated woodland (Göransson 1986; Edwards & McIntosh 1988). This woodland utilisation model would also suggest that depopulation or a decline in agricultural productivity

were not, of necessity, reflected in orthodox interpretations of regeneration phases. These processes are difficult to prove (Edwards 1993; and cf O'Connell 1987), but they do bring to the fore the need to consider alternative interpretations of woodland regrowth. In any case, the possible occurrence of forest farming would still suggest that, whether for physical or human reasons, or both, there had been a change in former landuse practices.

CONCLUSIONS

Changes of a floristic nature which can be observed in pollen diagrams are potentially important indicators of marginality. Such marginality, demonstrated in part by woodland disturbance phases, and probably more strongly by intervening regeneration phases, may, in turn, also indicate the disruption of an existing cultural system. This does not necessarily mean that methods of exploitation are inevitably becoming less productive; that may be the case where, for example, soil deterioration leads to falling yields and land abandonment. In other instances, however, marginality may be induced, and thus become apparent in the disturbance and regeneration phases, by a forced change in, or an adaptation of an existing culture, or the adoption of a technological innovation. Disturbance and regeneration phases in pollen diagrams can stem from a variety of factors, and while marginality is involved, it can be a manifestation of positive as well as negative influences. The ideas presented here should not be seen as prescriptive; with appropriate circumspection, they are intended to provide a plausible basis for the interpretation of marginality from pollen diagrams.

REFERENCES

Behre, K-E (ed) 1986 *Anthropogenic indicators in pollen diagrams.* Rotterdam: A.A. Balkema.

Berglund, B E 1986 'The cultural landscape in a long-term perspective. Methods and theories behind the research on land-use and landscape dynamics', *Striae,* 24, 79–87.

Birks, H H, Birks, H J B, Kaland, P E & Moe, D (eds) 1988 *The cultural landscape – past, present and future.* Cambridge: Cambridge University Press.

Buckland, P C & Edwards, K J 1984 'The longevity of pastoral episodes of clearance activity in pollen diagrams – the rôle of post-occupation grazing', *Journal of Biogeography,* 11, 243–249.

Delcourt, P A, Delcourt, H R, Cridlebaugh, P A & Chapman, J 1986 'Holocene ethnobotanical and paleoecological record of human impact on vegetation in the Little Tennessee River Valley, Tennessee', *Quaternary Research,* 25, 330–349.

Edwards, K J 1979 'Palynological and temporal inference in the context of prehistory, with special reference to the evidence from lake and peat deposits', *Journal of Archaeological Science,* 6, 255–270.

Edwards, K J 1982 'Man, space and the woodland edge: speculations on the detection and interpretation of human impact in pollen profiles', *in* Bell, M & Limbrey, S (eds), *Archaeological aspects of woodland ecology*, Oxford: BAR International Series, 5–22. (=Brit Archaeol Rep Int Ser, 146).

Edwards, K J 1991 'Spatial scale and palynology: a commentary on Bradshaw', *in* Harris, D R & Thomas, K D (eds), *Modelling ecological change*, London: Institute of Archaeology, University College London, 53–59.

Edwards, K J 1993 'Models of mid-Holocene forest farming for north-west Europe', *in* Chambers, F M (ed), *Climate change and human impact on the landscape*. London: Chapman and Hall, 133–145.

Edwards, K J & MacDonald, G M 1991 'Holocene palynology: II. Human influence and vegetation change', *Progress in Physical Geography*, 15, 364–391.

Edwards, K J & McIntosh, C J 1988 'Improving the detection rate of cereal-type pollen grains from *Ulmus* decline and earlier deposits from Scotland', *Pollen et Spores,* 30, 179–188.

Faegri, K & Iversen, J 1989 *Textbook of pollen analysis*, 4th edition by Faegri, K, Kaland, P E and Krzywinski, K. Chichester: John Wiley & Sons.

Göransson, H 1986 'Man and the forests of nemoral broad-leaved trees during the Stone Age', *Striae*, 24, 145–152.

Groenman-van Waateringe, W 1980 'Urbanization and the North-West frontier of the Roman Empire' *in* Hanson, W S & Keppie, L J F (eds), *Roman Frontier Studies XII*. Oxford: BAR International Series, 1037–1044. (=Brit Archaeol rep Int Ser, 71).

Groenman-van Waateringe, W 1983 'The disastrous effect of the Roman occupation', *in* Brandt, R & Slofstra, J (eds), *Roman and Native in the Low Countries*. Oxford: BAR International Series, 147–157. (=Brit Archaeol Rep Int Ser, 184).

Hicks, S 1988 'The representation of different farming practices in pollen diagrams from northern Finland', *in* Birks, H H, Birks, H J B, Kaland, P E & Moe, D (eds), *The cultural landscape – past, present and future*. Cambridge: Cambridge University Press, 189–207.

Hodder, I 1990 *The domestication of Europe: structure and contingency in Neolithic societies*. Oxford: Basil Blackwell.

Iversen, J 1941 'Landnam i Denmarks stenalder (Land occupation in Denmark's Stone Age)', *Danmarks Geologiske Undersøgelse* II, 66, 1–68.

Iversen, J 1949 'The influence of prehistoric man on vegetation', *Danmarks Geologiske Undersøgelse* IV, 3, 1–23.

Maxwell, G S 1989 *The Romans in Scotland*. Edinburgh: The Mercat Press.

O'Connell, M 1987 'Early cereal-type pollen records from Connemara, western Ireland and their possible significance', *Pollen et Spores*, 29, 207–224.

Pilcher, J R, Smith, A G, Pearson, G W & Crowder, A 1971 'Land clearance in the Irish Neolithic: new evidence and interpretation', *Science*, 172, 560–562.

Simmons, I G & Tooley, M J (eds) 1981 *The environment in British prehistory*. London: Duckworth.

Smith, A G (with Grigson, C, Hillman, G & Tooley, M J) 1981 'The Neolithic', *in* Simmons, I G and Tooley, M J (eds), *The environment in British prehistory*. London: Duckworth, 125–209.

Tsukada, M, Sugita, S & Tsukada, Y 1986 'Oldest primitive agriculture and vegetational environments in Japan', *Nature*, 322, 632–634.

Turnbull, C 1961 *The forest people*. London: Chatto and Windus.

Turner, J 1964 'The anthropogenic factor in vegetational history. I. Tregaron and Whixall Mosses', *New Phytologist*, 63, 73–90.

Vorren, K-D 1986 'The impact of early agriculture on the vegetation of Northern Norway – a discussion of anthropogenic indicators in biostratigraphical data', *in* Behre, K-E (ed), *Anthropogenic indicators in pollen diagrams*. Rotterdam: A.A. Balkema, 1–18.

Whittington, G & Edwards, K J 1993 '*Ubi solitudinem faciunt pacem appellant*: the Romans in Scotland, a palaeoenvironmental contribution', *Britannia*, 24, 13–25.

Whittington, G & Edwards, K J 1994 'Palynology as a predictive tool in archaeology', *Proceedings of the Society of Antiquaries of Scotland*, 124, 55–65.

Whittington, G, Edwards, K J & Cundill, P R 1991 'Late- and postglacial vegetational change at Black Loch, Fife, eastern Scotland – a multiple core approach', *New Phytologist*, 118, 147–166.

10. Shredding and the production of winter fodder in northern Greece. An interim statement on the archaeological detectability of shredding

John Tierney

Abstract

The survival of traditional woodland management practices in marginal locations provides an opportunity for experimental and ethnoarchaeological investigations. The practice of shredding trees for the production of leafy hay by the villagers of Plikati, northwest Greece, has been recorded. Shredding involves the repeated removal of the tree's side-branches. A dendrochronological analysis of samples taken from shredded and non-shredded trees reveals that distinctive patterns of stress and recovery are evident in the growth curves of managed trees. Some of the results from this analysis are presented here, indicating that trees which have undergone similar management regimes can be identified using standard dendrochronological techniques of analysis. Furthermore, in exceptional circumstances, it may even be possible to match samples from managed trees with unmanaged trees in a dated assemblage.

INTRODUCTION

During research into montane pastoralism in northwest Greece, Halstead (1990) noted the continued practice of shredding trees for the production of winter fodder by the villagers of Plikati. A multi-disciplinary ethnoarchaeological project was devised by Dr Halstead, which aimed, by the application of entomological, palynological and dendrochronological techniques of analysis, to assess the archaeological detectability of the practice. This paper is concerned with the detection of shredding by an analysis of the tree-rings of shredded and unshredded trees. As with other studies of modern woodland management (eg Crone 1987; Bridge *et al* 1986; Rasmussen 1990), it is intended that by an examination of tree-ring patterns, wood managed in a similar way in the past can be identified and that a greater understanding of the economic importance of woodland management be gained. The latter factor will be fully addressed in Halstead (forthcoming) as will the complete tree-ring analysis.

The practice of woodland management for leafy hay production is known to have survived until recent times in Scandinavia, Germany (Austad 1988; Ellenberg 1989; Rasmussen 1990), Italy and France (Halstead pers comm).

Its practice in antiquity has been hypothesized by Troels-Smith (1960) and suggested to explain curious morphology and patterns in tree-ring (Rackham 1988; Robinson & Rasmussen 1989) and palynological studies (Andersen 1988).

WOODLAND MANAGEMENT AS PRACTISED IN PLIKATI

The village of Plikati lies at *circa* 1200 m above sea level, half way up Mount Grammos, on the Greek-Albanian border. Within living memory, the village has been inhabited by sedentary mixed farmers who have cultivated on a small scale and kept a few domestic animals (cf Halstead 1990). The village is under snow every winter for several months and livestock must be stall fed on grassy and leafy hay, the latter cut principally from the deciduous Turkey oak (*Quercus cerris* L.). Leafy hay is harvested in late summer/early autumn (ie, towards the end of the growing season) from trees growing around the village fields and also from an area of communal woodland set aside by the village for this purpose some 60 years ago.

Trees are ideally said to be cut on a cycle of 3 to 4 years, but there is considerable variability.

METHODOLOGY

Samples for tree-ring analysis were chosen haphazardly from three areas of woodland known to have different management histories (Table 10.1). These areas of woodland were identified by an examination of woodland structure, tree morphology, and most importantly, following information given by the villagers. Areas A and B were shredded continuously until recently (ie, the late 1970s/ early 1980s) although the shredding of trees in Area B is supposed to have been abandoned at an earlier date (early to late 1970s). Area C trees are from the upper slopes of the valley and were apparently never managed.

It was intended to sample living trees with a Swedish Pressler corer, but this plan had to be abandoned when the corer broke during the sampling of the first tree. Fortuitously, the areas of woodland marked for sampling had been thinned by the villagers in 1988 and 1989 and discs of wood were taken from the stumps of the felled trees. The author was present during the felling of the Group C trees and observed their form, while the morphology of the Group A and B trees was observed from the wood left beside the stumps. None of the sampled trees appeared to differ from those growing in the vicinity.

After rough and fine sanding, each sample was measured using standard dendrochronological techniques (Hillam 1985) with an attempt being made to measure two radii from each sample. In some cases it was possible to measure only one radius, while in others it was necessary to measure up to five different paths through the lateral transverse section of the sample. If more than one radius was successfully measured a mean of the radii was formed. Shredding produces groups of very narrow growth rings which are difficult to measure. Therefore, complete cross-sections of the trees are preferable to samples taken with the Pressler corer because the probability of obtaining measurable radii is increased. Once measured, the tree-ring patterns were graphed on a logarithmic scale and compared visually and statistically using the student 't' test.

RESULTS AND DISCUSSION

Shredding has a number of effects on the growth of oak trees in Plikati. Defoliation of a tree necessitates that the trees' food stores are used to grow twigs and leaves rather than in lateral growth of the trunk. The accumulation of food stores can also be disrupted by defoliation depending on the timing of the shredding episode.

Late shredding affects tree growth in the following year, while shredding during the growing season will also affect growth in the same year. It is hypothesized that should

Sample area	A	B	C
No. of samples taken	30	31	13
No. of samples successfully measured	21	25	11

Table 10.1 Number of samples taken from the three sample areas.

Group	Average ring width (cm)	Mean diam. (cm)	Mean age (yrs)	Mean no. of sapwood years
A	1.3	24	75	32
B	1.4	25	76	33
C	2.0	43	83	32

Table 10.2 Attributes of the sampled trees.

Sample	B5	B7	B10	B11	B31	CM
A3					3.6	
A6					3.8	
A9					4.4	
A13				4.3		
A14						3.7
A18			3.7		3.8	
A21	4.0					
A22		3.5				4.8
A25						3.7

Table 10.3 't' values of significant cross-matches between the three sample groups A, B & C.

shredding occur before the development of Lammas shoots the effects on tree-ring growth both in the year of shredding and the following year will be less than if it follows the Lammas flushing. Lammas shoots *sensu strictu* occur in early August, although replacement of leaf cover can occur from mid-July onwards (Gradwell 1974).

The effects of shredding on tree growth from the three sample groups is obvious when mean diameter and average ring width are compared (Table 10.2). The unshredded trees of Group C have approximately 35% greater wood production than the shredded trees despite the groups having similar age structures.

The assumption that regular shredding increases variability in ring width to the extent that trees whose ring patterns are affected by human interference should not cross-match with trees whose growth ring patterns are predominantly controlled by climate, is supported by the analysis of Groups A, B and C (see Table 10.3). All but one of the C group samples cross-matched at significant levels (ie, a 't' value greater than or equal to 3.5) at the appropriate years, allowing one master mean curve to be produced. The A group samples cross-matched to produce four subgroups and the B group produced five subgroups.

Study	Species	% of defoliation	% increment loss	Average % increment loss
Plikati				
A3	Oak	100	53	59
A6	"	100	60	
A9	"	100	64	
B13	"	100	35	37
B31	"	100	38	
Lejre*	Ash	100		58
Hareskonen*	"	50 every year		65
Rackham 1976	Elm	?		89
Semevskii	?	90		59

*Table 10.4 Percentage of increment loss from studies of natural and artificial defoliation. (*Rasmussen 1990)*

The formation of subgroups, based in this case on our knowledge of the provenance of the samples, supports the idea that trees which have undergone predominantly similar management regimes should be detectable using dendrochronological techniques. None of the B group samples cross-matched with the C master while three of the A group samples had viable cross-matches. Twelve individuals from the A and B groups matched on eight separate occasions.

As an illustration of the potential of standard dendro-chronological techniques in the analysis of shredded trees, one subgroup from each of groups A and B will be discussed. The recognition of shredding episodes in these subgroups is assisted by our knowledge of the provenance of the samples and by the villagers' testimony. The occurrence of a shredding episode is inferred from a tree-ring curve when a large increment loss not corresponding to a similar loss or trough in the control C master curve is noted. Once troughs interpreted as being the result of shredding were identified, the length of the cycle of cutting was also calculated. The amount of increment loss for a narrow ring interpreted as being caused by shredding is expressed as a percentage of the previous years growth. This percentage increment loss has been used as a measure of the effects of defoliation on annual increment in studies of both natural and artificial defoliation (Semevskii *in* Gradwell 1974; Rasmussen 1990). The percentage of increment loss for troughs suggested to have been caused by shredding has been calculated for the five samples described below. These are presented in Table 10.4 along with measurements from other studies of defoliation. These figures indicate that percentage increment loss cannot be used to differentiate human defoliation from that caused by natural agencies.

Subgroup Alpha 1

This subgroup consists of three individuals which cross-

Sample	A3	A6
A6	8.1	
A9	8.2	7.6

Table 10.5 't' values of significant cross-matches between individuals in Subgroup Alpha 1.

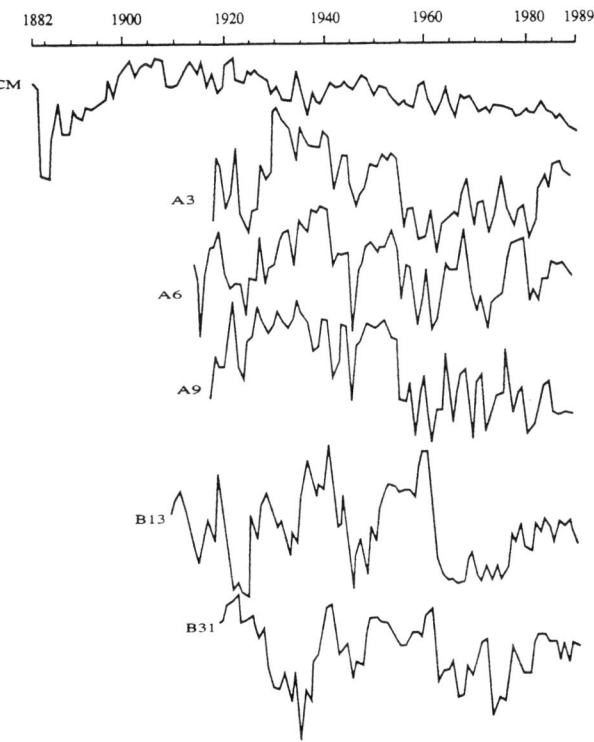

Figure 10.1 Tree-ring curves of samples in subgroups Alpha 1 and Beta 1.

matched with high 't' values (Table 10.5), as well as matching visually (Figure 10.1). The three samples are from the most northerly section of the common wood, ie, closest to the village, and they are situated within 40 m of the main roadway which runs through this section of the woodland. Tree A6 is approximately 20 m south of Tree A3 while Sample A9 is 30 m further upslope to the east. All three samples cross-matched significantly with other samples but only these three satisfied the criteria of matching both visually and statistically to produce a subgroup.

All of the trees had significant troughs in the first three years of growth which could be the product of natural defoliation or some other environmental stress factor. All three samples were shredded when six years old and then at an average four year interval thereafter (Table 10.6). Sample A3 may not have been shredded from 1927 to 1940, although the troughs evident in 1933 and 1935 could be the products of shredding before the flushing of the

Lammas shoots. There was a gap in the 1950s and an eight year gap from 1961 to 1968. Sample A9 was intensively shredded throughout its life except for a ten year period from 1945 to 1954. All three samples were shredded twice during World War II, in 1940 and 1944 while the apparent absence of a shredding episode from 1945 to 1954 could have resulted either from the harvesting of leafy hay from another part of the wood or the disruption of the village economy by the Greek Civil War.

Subgroup Beta 1

This subgroup consists of two individuals B13 and B31 which cross-matched with a 't' value of 3.9 with both sequences ending in 1989. Tree B13 was approximately 200 m southwest of B31 and is situated on a more gradual slope.

Both samples have shorter average shredding intervals than subgroup Alpha 1 (Table 10.7) which may indicate that different farmers harvested this area of woodland. Sample B13 may have been defoliated from the age of three years onwards, except for an eleven year gap from 1949 to 1959 when the average ring width was almost twice that of the previous decade. Sample B31 may have been shredded from the age of five years onwards except for a fourteen year gap from 1946 to 1959 when the average ring width was almost twice that of the following 29 years. Paradoxically, both of these samples may have been shredded during the 1980s despite the fact that the B group trees appear to have been neglected for longer than the A group.

CONCLUSIONS

The physical effects of shredding upon a tree are the same regardless of the nature of the economic system of which the shredding is part. For this reason, experimentation (Rasmussen 1990) and the ethnoarchaeological investigation of traditional woodland management practices are valid methodologies for the study of past woodland management practices.

The taphonomic factor that managed trees are often not suited for use in construction affects the archaeological detection of shredding. Logs and twigs from shredded trees are used as firewood in Plikati. The problems of differentiating natural from human defoliation (both of which are often of a cyclical nature) and the archaeological detection of shredding can probably be addressed with the application of an extensive sampling strategy. Assuming that natural defoliation affects fewer trees in a wood than the implementation of a woodland management strategy, the detection of contemporaneous stress-related cycles in trees from a large area of woodland or from a large number of archaeological samples may be indicative of human defoliation. The ability of shredded A group samples to cross-match with the non-shredded C master curve (see

Sample	Average cycle length	Range
A3	4	1–10
A6	4	1–8
A9	4	1–14

Table 10.6 Number of years between shredding episodes for individuals in Subgroup Alpha 1.

Sample	Average cycle length	Range
B13	2	1–11
B31	3	1–14

Table 10.7 Number of years between shredding episodes for individuals in Subgroup Beta 1.

Table 10.3) reveals the potential of shredded wood to be not only identified but also anchored in time.

While the practice of shredding survives mostly in marginal and peripheral areas, its detection in archaeological contexts should not be taken to indicate the presence of a marginal economy, or an economy attempting to adapt to a marginal environment. The survival of the practice as part of diversified montane economies throughout Europe may not reflect the distribution of the practice in antiquity.

ACKNOWLEDGEMENTS

This research has been partially funded by the British Academy and I would like to thank both Paul Halstead and Martha Hannon for editing an early draft of this paper.

REFERENCES

Andersen, S T 1988 'Changes in agricultural practices in the Holocene interglacial in a pollen diagram from a small hollow in Denmark', *in* Birks, H H, Birks, H J B, Kaland, P E & Moe, D (eds), *The cultural landscape, past, present and future.* Cambridge: Cambridge University Press, 395–409.
Austad, I 1988 'Tree pollarding in western Norway', *in* Birks, H H, Birks, H J B, Kaland, P E & Moe, D (eds), *The cultural landscape, past, present and future.* Cambridge: Cambridge University Press, 11–31.
Bridge, M C, Hibbert, F A & Rackham, O 1986 'Effects of coppicing on the growth of oak timber trees in the Bradfield Woods, Suffolk', *Journal of Ecology*, 74, 1095–1102.
Crone, B A 1987 'Tree-ring studies and the reconstruction of woodland management practices in antiquity', *in* Jacoby, G C & Hornbeck, J W (eds), *Proceedings of the international symposium on ecological aspects of tree-ring analysis*, Palisades, New York: Tree-Ring Laboratory, Lamont-Doherty Geological Observatory, 327–336.
Ellenberg, H 1988 *Vegetation ecology of central Europe.* Cambridge: Cambridge University Press.
Gradwell, G R 1974 'The effects of defoliators on tree growth', *in* Morris, M G & Perring, F H (eds), *The British Oak: Its history*

and natural history. Faringdon: E W Classey Ltd, 182–193.

Halstead, P 1990 'Present to past in the Pindhos: diversification and specialisation in mountain economies', *Rivisti di Studi Liguri*, A, 46, 1–4, 61–80.

Hillam, J 1985 'Dendrochronology: How to make a date with a tree', *in* Phillips, P (ed), *The archaeologist and the laboratory*, CBA Research Report No. 58, 17–23. London: CBA, 17–23.

Rackham, O 1976 *Trees and woodland in the British landscape*. London: Dent.

Rackham, O 1988 'Trees and woodland in a crowded landscape – the cultural landscape of the British Isles', *in* Birks, H H, Birks, H J B, Kaland, P E & Moe, D (eds), *The cultural landscape, past, present and future*. Cambridge: Cambridge University Press, 53–79.

Rasmussen, P 1990 'Pollarding of trees in the neolithic: often presumed – difficult to prove', *in* Robinson, D (ed), *Experimentation and reconstruction in environmental archaeology*, Symposia of the Association for Environmental Archaeology No. 9. Oxford: Oxbow Books, 77–99.

Robinson, D E & Rasmussen, P 1989 'Botanical investigations at the neolithic lake village at Weier, north east Switzerland: leaf hay and cereals as annual fodder', *in* Milles, A, Williams, D & Gardiner, N (eds), *The beginnings of agriculture*, Oxford: BAR International series, 149–163 (=Brit Archaeol Rep Int Ser, 496).

Troels-Smith, J 1960 'Ivy, mistletoe and elm. Climate indicators – fodder plants. A contribution to the interpretation of the pollen zone border VII-VIII', *Danmarks Geologiske Undersogelse* II. Raekke 4 (4).

11. A study of anthropogenic activity and pedogenesis from the 2nd millennium BC to the 2nd millennium AD at Lairg, northern Scotland

Timothy G Acott

Abstract

A soil micromorphological analysis was carried out on a multi-period archaeological site at Lairg in northern Scotland. The field survey combined with the micromorphological data indicates a considerable change in the pedological environment through time. Evidence from soils buried by prehistoric monuments suggests that erosion of the soil profile, probably caused by cultivation, was occurring. By the 2nd millennium AD there had been a reduction in the amount of erosion and a deepening of A horizons had occurred. The causes of these changes will not be known until the full post-excavation analysis is corroborated with the soil micromorphological data. However, it does seem likely that the changes in the pedological environment were a result of changes in landscape management practices.

INTRODUCTION AND OBJECTIVES

The objective of this paper is to examine evidence for human impact on pedogenesis using preliminary results from a multi-period archaeological site at Lairg in Sutherland, northern Scotland. An initial survey of the Lairg site, carried out in 1989 by AOC (Archaeological Operations & Conservation, Historic Scotland), identified 653 archaeological features. Evidence for human activity dates from the Neolithic to the Post-Medieval period representing 4000–5000 years of land-use. Agricultural activity on the site is restricted by a combination of climatic and pedological factors. The viability of cereal cultivation is limited by the length of growing season, annual temperatures and amount of precipitation. In addition, the soils in the area are not particularly good agricultural soils. Large areas of the site are waterlogged and where the soils are more freely draining podzolisation occurs. It is likely that these factors will have always made the area marginal in terms of agricultural activity.

This study was implemented to investigate how changing anthropogenic activity has affected the pedological environment at the Lairg site. Lairg is the ideal study area owing to the high concentration of archaeological monuments and the long period of site occupation. The preliminary results and hypothesis presented in this paper are derived from an intensive soil micromorphological analysis of buried soils.

SITE DESCRIPTION

The site is situated approximately 2.5 miles south of Lairg in Sutherland, northern Scotland, within a series of broad terraces rising from the River Shin. The River Shin forms an important access route for a number of roads, a railway and a line of pylons. There are only a few minor roads on the west side of the river which is dominated by coniferous woodland extending to an altitude of approximately 200 m. The woodland includes Gruids Wood, Raemore Wood and Braemore Wood. The eastern side of the valley has few trees apart from a band of coniferous and deciduous woods along the lower slopes. The vegetation on the east side of the valley is mainly dominated by heather, bracken and areas of improved grassland. The survey area is bisected by the Allt na Fearna Mor which flows from the moorland into the River Shin.

The local soils are included in the Arkaig association (Futty & Towers 1982). The structures of soils in this association are generally weak. Along the valley sides there are areas of enclosed and semi-enclosed mires and encroaching hill peat. Within the survey area the soils are dominated by peaty podzols, gleys and freely drained brown podzols. In places along the Shin Valley the soils have very dark humose topsoils which have probably been reclaimed from peat. Gleys occurring over a coarse textured stony till are common. Towards the higher slopes of the valley, and on the moorland, soils are dominated by peat, peaty gleys and some peaty podzols. Iron pan development can be seen in places often associated with gleying. Where the land has not been reclaimed heather moors dominate. There are limited areas of fluvioglacial gravels and river alluvium in the valley bottom.

The solid geology is dominated by siliceous granulites of the Precambrian Moine series. Owing to glacial processes the dominant soil parent material is glacial till. The till is composed of material derived from the Precambrian Moine series and some granite and hornblende schists. The glacial till is generally very stony with textures varying between loamy sand or sandy loam.

ANTHROPOGENIC ACTIVITY AND PEDOGENESIS

To assess the development of soil through time it is necessary to consider pedogenesis as one part of a larger ecosystem in which all the components interact. These interactions can be described in terms of five factors; climate, vegetation, organisms, parent material and time. Of these time is probably the only truly independent variable (Fitzpatrick 1983). Human activity is usually included with organisms although this does not necessarily reflect the large impact that humans have had on the soil system (Davidson 1982).

The temporal sequence of buried soils at the Lairg site provides a framework for studying soil development in an area with a long history of use by humans. An important question is the extent to which anthropogenic activity has influenced pedogenesis. It is possible to make assumptions about soil development not affected by anthropogenic activity by using conditions during Pleistocene interglacials as an analogue (Roberts 1991). If this method is adopted then a cycle with five phases can be identified; protocratic, mesocratic, oligocratic, telocratic and cryocratic (Roberts 1991; Inversen 1958). In summary this represents a transition from open woodland to mixed deciduous woodland, with brown forest soils, to heath and conifer vegetation, with leaching of soils and the formation of podzols. As glacial conditions return retrogression occurs and arctic mineral soils develop. This sequence suggests that soil development can be considered in terms of cyclical changes. It is suggested (Bridges 1978; Askew *et al* 1985) that during the Holocene, in many upland regions, soil development can follow developmental pathways towards podzolisation and gleying.

Human activity can interact with pedogenesis in many ways. In the early part of the Holocene anthropogenic activity would probably have modified the soils indirectly through the manipulation of vegetation. Later, as cultivation increased, soils and geochemical environments would have been directly affected. It is suggested (Ball 1975) that human activity has tended to direct pedogenesis towards leaching, acidity, podzolisation, gleying and peat formation through post-glacial times. This trend, similar to the suggested cycle of events during a Pleistocene interglacial, implies the effect of human impact has been mainly one of accelerating a natural process.

Using anthropocentric value judgements soil can be considered as a resource and human activity can be judged to be either detrimental or beneficial (Bidwell & Hole 1965). Beneficial effects include addition of materials that improve the soil environment, resulting in increased nutrient status to balance leaching. Detrimental effects can include activities that promote gleying, podzolisation and soil erosion.

Smith (1975) considered early agriculture and soil degradation and examined the assumption that agriculture in general had necessarily been detrimental to soil processes. In considering this topic he did not confine himself specifically to talking about prehistoric agriculture. His general conclusion was that as a result of agricultural practises humans caused manipulation and erosion of soils and contributed to the enhancement of natural soil processes. The general effect was to create high energy environments where low energy, more stable systems once existed.

Romans & Robertson (1983) presented evidence for the development of soils in Scotland. Brown forest soils had developed on fluvioglacial sands around the main mass of the Grampian mountains by 5000 years ago. The presence of deciduous woodland dominated by oak is indicated at some sites. During the same period evidence from a site on Arran showed a podzol profile had developed (Mackie 1966). In many upland regions of the Scottish highlands deforestation had occurred before the 2nd millennium BC (Askew *et al* 1985). However, at lower altitudes areas of continuous forest cover remained into the Bronze Age allowing time for the development of leached forest soils (Askew *et al* 1985). Results of pollen analysis, being carried out by Melanie Smith for the Lairg area, will provide important information about the history of vegetation in the area.

The Lairg site offers a long period where human activity and soil development have been interacting. An examination of buried soils dated between the second millennium BC and the 2nd millennium AD allow these interactions to be studied.

METHODS OF ANALYSIS

The main analytical technique in this study is soil micromorphology which is being increasingly used as a diagnostic tool on many archaeological sites. A number of general papers concerned with evaluating this discipline

Field results of prehistoric buried soils

Monument 62 (M62) is a cross contour dyke. The dyke belongs to a land-use phase contemporary with Monument 64. The dominant vegetation surrounding the monument is grass and bracken. Excavation revealed a buried podzol profile (Figure 11.1). This is the only monument excavated with an organic layer preserved above the A horizon. The profile consists of a buried organic layer overlying a non-humose A horizon. Below this is an E horizon which itself overlies a Bs horizon.

Monument 64 (M64) is a large embanked and platformed hut circle set within a parallel cross contour field system. Apart from some *Juncus* growing within the centre of the hut circle, suggesting localised waterlogged conditions, the dominant vegetation is grass and bracken. A trench excavated through the wall on the eastern side of the hut circle showed a buried A horizon overlying a Bg horizon.

Monument 505 (M505) is a prehistoric field boundary. A trench excavated in an east west direction across the structure showed a buried podzol profile (Figure 11.2) consisting of an A horizon, an E horizon and a mottled Bs horizon.

Monument 659 (M659) is an early prehistoric hut circle. A brown podzol is buried by the structure. Ard marks are present in the top of the Bs horizon.

Monument 504 (M504) is a hut circle dated to the Iron Age. The vegetation immediately around the monument is dominantly heather with some grass. The excavated trench revealed a brown podzol underlying floor deposits in the internal part of the structure. There are ard marks in the top of the Bs horizon.

Field results of 2nd millennium AD buried soils

Monument 75 (M75) is a dyke that formed the boundary between an area of improved grassland and land dominated by heather. To the north of the excavated trench there is standing water. The area around the trench is dominated by mosses, grass and *Juncus*. The buried profile consisted of a humose A horizon overlying a non-humose A horizon. The B horizon shows indications of gleying.

Monument 127 (M127) is a rectangular structure. Excavation revealed a buried brown podzol underlying a thin stratified sediment interpreted as an anthropogenic deposit predating construction (Figure 11.3).

Monument 164 (M164) is a dyke which formed part of an addition to M75. Directly below the structure is an organic layer overlying an A horizon (Figure 11.4).

Monument 975 (M975) is an enclosure dyke. The vegetation surrounding the dyke consists predominantly of bracken with some grass. The profile is a buried brown podzol consisting of an A horizon overlying a Bs horizon. The top of the buried A horizon is a very dark greyish layer (10YR 3/2), bounded by a thin iron pan above and below. This is interpreted as being the result of the localised reduction of a buried organic-rich horizon. As the organic matter decomposed, reducing conditions were created mobilising iron which was illuviated when it came into contact with the oxygen-rich soil above and below (Limbrey 1975).

Monument 1069 (M1069) is a rectangular structure surrounded by grass and bracken. A trench was dug through an inner wall of the structure showing that a brown podzol is buried by the monument.

Table 11.1 Summary descriptions of buried soils.

have been published (Davidson *et al* 1992; Goldberg 1983). The publication 'Soils and micromorphology in archaeology' (Courty *et al* 1989) provides a concise discourse on the different ways in which this technique can be used to improve archaeological interpretations.

The method used for making the thin sections was based on that described by Murphy (1986). The stony nature of the material meant that Kubiena tins could not always be used for sampling. Instead large blocks of soil were cut from the profile and wrapped in a heavy gauge aluminium sheet. These were then wrapped with tape and placed in plastic bags to prevent moisture loss. In the laboratory blocks up to 10 x 10 x 10 cm could be made. Water was removed from the sample using acetone exchange and the samples impregnated using an epoxy resin.

To implement the soil micromorphological analysis a temporal framework was created using radiocarbon dating of contexts. If this was not available the age of monuments was estimated by comparison to other structures of known age. Sampled monuments were divided into those dated to the 1st and 2nd millennium BC and those dated to the 2nd millennium AD. One soil profile which was not buried by any monument was also studied.

RESULTS AND DISCUSSION

The results and discussion are divided into two sections. Firstly Table 11.1 summarises the field description results of buried profiles. A discussion is then presented based upon the results of the thin section analysis and standard field descriptions.

Analysis of thin section samples taken from glacial till

Thin section samples were taken from glacial till below M504, M505 and M975. The characteristic feature of the glacial till is the abundance of cappings present on the top of rock fragments or link cappings within the groundmass. Cappings are a common feature of soil that has been effected by periglacial action and are among the most reliable climatic indicators found in the soil (Catt 1991). Cappings have been identified elsewhere in Scotland from soils of the Ettrick association (Romans & Robertson 1975) and in the Alpine soils of North East Scotland (Romans *et al* 1966; Romans & Robertson 1974).

The cappings observed in the samples from Lairg are divided into two categories; firstly those from the glacial till which are undisturbed and secondly those from A horizons and B horizons which have been fragmented and mixed

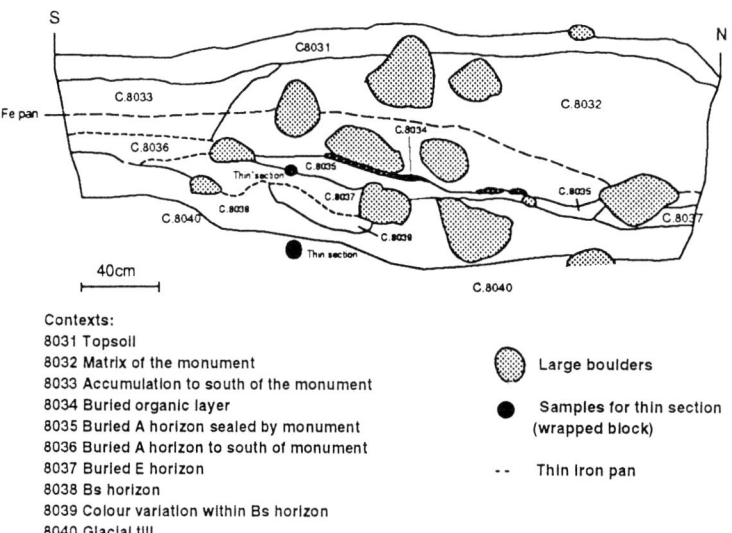

Contexts:
8031 Topsoil
8032 Matrix of the monument
8033 Accumulation to south of the monument
8034 Buried organic layer
8035 Buried A horizon sealed by monument
8036 Buried A horizon to south of monument
8037 Buried E horizon
8038 Bs horizon
8039 Colour variation within Bs horizon
8040 Glacial till

⬡ Large boulders

⬤ Samples for thin section
(wrapped block)

-- Thin iron pan

Figure 11.1 Monument 62, prehistoric cross contour dyke.

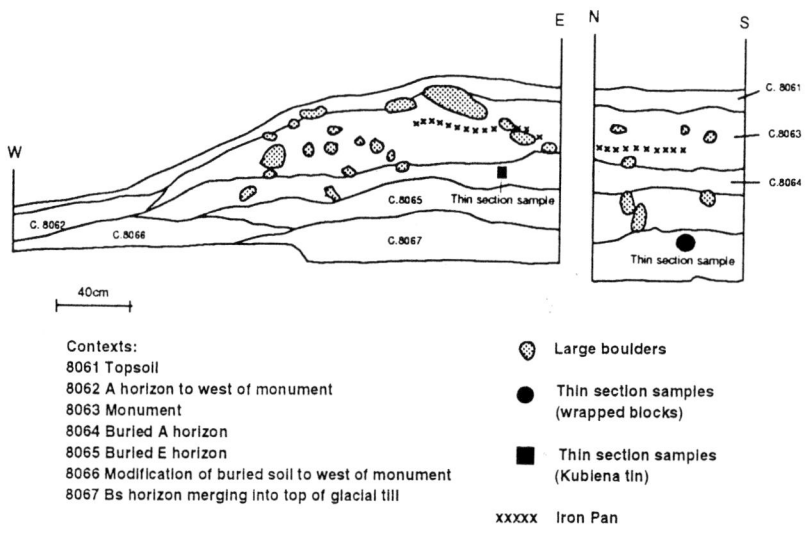

Contexts:
8061 Topsoil
8062 A horizon to west of monument
8063 Monument
8064 Buried A horizon
8065 Buried E horizon
8066 Modification of buried soil to west of monument
8067 Bs horizon merging into top of glacial till

◉ Large boulders

⬤ Thin section samples
(wrapped blocks)

■ Thin section samples
(Kubiena tin)

xxxxx Iron Pan

Figure 11.2 Monument 505, prehistoric field boundary.

throughout the profile. The undisturbed cappings are useful because they provide a lower limit for profile disturbance (Table 11.2). The depths of the profiles were all shallow. It is not known what the maximum depth of these profiles would have been during prehistory but it is possible they would have been deeper under a forest vegetation. Subsequent erosion of the soil profile is suggested.

Analysis and discussion of prehistoric buried soils

In order to consider the impact that humans have had on the soil system it is necessary to evaluate the condition of the soils prior to anthropogenic activity. There was no buried profile that could be definitely placed into this category although two sites had buried podzol profiles preserved. One was located below a cross contour dyke (M62) and the other below a prehistoric field boundary

(M505). Samples were taken to determine whether the podzolisation was a post-depositional effect or alternatively whether genuine podzols had been preserved under early monuments.

A thin section analysis of the buried A horizon below M505 showed an abundance of coarse and fine carbonized material often associated with fragmented ferruginous nodules. There were no carbonized residues in the E horizon. The Bs horizon consisted of monomorphic translocated material of a similar type to that described by De Connick (1980) and De Connick & Righi (1983). This was unusual because elsewhere Bs horizons consisted of polymorphic material. The only place where a similar morphology was observed was at the base of Bs horizons just above and within undisturbed glacial till. This suggests that significant erosion had occurred under M505 which had truncated the old profile removing the Bs horizon and

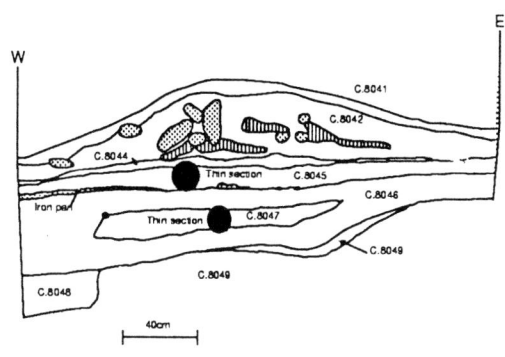

Contexts
8041 Topsoil
8042 Turf wall with stone revetment on west face
8043 Orange material within wall
8044 Buried organic layer
8045 Stony anthropogenic layer
8046 Buried A horizon
8047 Colour variation within buried A horizon
8048 Bs horizon
8049 Glacial till

● Thin section samples
(wrapped blocks)

◁▭▷ C.8043

◉ Large boulders

*Figure 11.3 Monument 127, 2nd millennium AD
rectangular structure.*

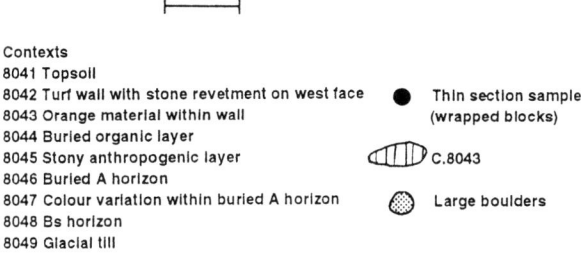

■ Thin section samples
(Kubiena tin)

◉ Large boulders

ı|ı|ı Areas of organic material
within the dyke

Contexts
3031 Topsoil
3032 Matrix of monument
3033 Buried A horizon
3034 Buried B horizon
3035 Soil to east of monument

*Figure 11.4 Monument 164, 2nd millennium AD turf and
stone dyke.*

exposing the surface of the till. A field boundary, M505, was then constructed on the truncated profile. An E horizon subsequently developed at the top of the Bs horizon.

Due to the likelihood that the E horizon under M505 was post-depositional it is possible that the podzol under M62 was also a post-depositional effect. However, the morphologies of the A horizon and the Bs horizon are quite different. The A horizon under M62 has minimal carbonized residues and the Bs horizon consists of many small pellets. It is interesting to note that the E horizon is present under M62 where a layer of organic material is also preserved. Does this therefore represent an example of an early soil with minimal direct human impact which already had complete podzol development? It is possible that a complete podzol profile had developed prior to human activity, perhaps even beneath the woodland that once covered the area. Considering that the evidence is

Monument	Depth (cm) to undisturbed cappings from old ground surface
504	45
505	40
975	45

Table 11.2 Depth of undisturbed cappings from old ground surfaces.

		Depth of buried A horizon (cm).
Monuments dated to 2nd millennium BC	62	10–15
	64	15
	505	5–15
	659	5–10
Monument dated to 1st millennium BC	504	10–15
Monuments dated to 2nd millennium AD	75/4	25
	127	36–44
	1069/2	26
	975/3	20
	975/2	25–30

Table 11.3 Thickness of buried A horizons in prehistoric and 2nd millennium AD soils.

derived from only one buried profile these results can only be speculative.

Fragmented ferruginous nodules are found in some buried prehistoric cultivated horizons. These nodules are composed of a strongly impregnated groundmass, sometimes combined with large fragments of charcoal. The precise elemental composition of these features is not known but they resemble fragmented iron pan. They were found under M64 and M505. Both of these monuments were dated to the Bronze Age. The context of M505 is known to belong to an early period of land-use on the site. The fragmented nodules are also found in two undated contexts buried by an accumulation of soil. Although these contexts were undated it is possible that they are also associated with prehistoric activity on the site. The fragmented nodules were not observed in any contexts dating to the 2nd millennium AD. It is possible that they are part of an early period of pedogenesis on the site.

Some general comments can be made about the morphology of buried prehistoric A horizons. The A horizons were poorly sorted with an average coarse:fine ratio of 60:40 (50 μm threshold). There was little organic or fungal material present in any of the contexts studied. The depth of the prehistoric A horizons is illustrated in Table 11.3; the average depth is 11.5 cm. Field evidence for agricultural activity during this period included ard marks in the top of Bs horizons and the formation of lynchets. Erosion around some of the prehistoric hut circles had truncated the soil profile and left the monuments raised slightly above the surrounding land-surface.

Some conclusions about the effect of anthropogenic activity on the soil system during the prehistoric period can be summarised:

1. Erosion was caused by cultivation;
2. Soil erosion has caused truncation of profiles resulting in the complete removal of Bs horizons in places;
3. The buried soils were mainly brown podzols although it is likely that these were truncated profiles.

Analysis and discussion of 2nd millennium AD buried soils.

There were no sites identified from the 1st millennium AD that were considered suitable for soil micromorphological analysis. This reflects the lack of visible remains dating to this period. The reason for this is thought to be either a genuine rarity of monuments or because evidence is being masked by later agricultural activity. All of the 2nd millennium AD soils studied pre-date the clearances and large-scale sheep farming in the area.

The average depth of buried A horizons of the 2nd millennium AD is 27.7 cm, some 16.6 cm deeper than the prehistoric A horizons. A micromorphological analysis shows that the A horizons are better sorted than the prehistoric horizons and there are lower coarse:fine ratios, on average approximately 40:60. There is only one example of a capping incorporated into an A horizon dated to the 2nd millennium AD. This suggests that there has been no mixing of material derived from underlying horizons for a long period of time. All of the excavated buried profiles are brown podzols and there is no recorded example of E horizon development. There is a greater abundance of fungal material associated with these later contexts, particularly in the form of bright and dark rings.

Thin section samples were taken from turves that formed part of a wall of a dyke constructed in the 2nd millennium AD. In the organic part of the turf charcoal is distributed in layers. In the underlying mineral soil there are only one or two fragments of fine carbonized material. This suggests there was little mixing of the surface organic layer and the mineral layer below it. A similar distribution of carbonized material is present in the organic layer preserved below the monument and the buried A horizon.

Thin section analysis of an undated deep accumulation of soil, located in an area of improved land, indicates that biological activity had homogenised the entire deposit. Close to the surface, fragments of burnt peat and peat which had been partially modified by biological activity were observed. The A horizon is well sorted and has a coarse:fine ratio that varies between 35:65 and 50:50.

A summary of results from the analysis of soils dated to the second millennium AD is as follows:

1. Erosion was no longer operating on the scale recorded in the Bronze Age;
2. Buried A horizons were deeper than in the Bronze Age;
3. The concentration of fungal material had increased since the Bronze Age;
4. Charcoal was concentrated in surface organic layers.
5. Less material had been incorporated into the A horizon from underlying contexts.

HUMAN IMPACT ON PEDOGENESIS SINCE THE BRONZE AGE

The results presented above can be used to discuss the impact of humans on pedogenesis. Soil is a medium that constantly tries to achieve a dynamic equilibrium with its environment. If any of the soil forming factors alter then changes in the pedological environment will also occur. Since the Bronze Age there was an overall deepening of A horizons. A dominant environmental process in the Bronze Age was the erosion of soil caused by cultivation. By the 2nd millennium AD soil erosion was no longer occurring to such an extent and there is evidence of an increase in the mass of the soil system. It is possible that manuring contributed to the deepening of the soils but there was no micromorphological evidence to support this. Evidence of manuring practises might have been lost owing to the acidic nature of the soils.

Spore cases of vesicular arbuscular mycorrhizae (VA) were present in many contexts dating to the 2nd millennium AD. It is suggested that these features, also referred to as bright and dark rings (Romans & Robertson 1983), might be associated with certain types of grazing. The VA group, part of the endomycorrhiza group, is one of the most widespread types of soil fungi and is thought to be an important component in helping plants absorb nutrients from relatively infertile soils (Brady 1990). Fungi rely on organic matter in the soil for their energy and carbon. The abundance of bright and dark rings might have increased during the second millennium AD as a result of an increase in organic inputs to the soil.

The concentration of carbonized material in the buried surface organic layers could be explained if burning of surface vegetation was occurring. Burning would encourage more nutritious grasses to grow and improve the land for pasture.

It seems likely that the changes in the soil environment were linked to a change in the way in which the land was managed. A reduction in the intensity of cultivation combined with areas being left as pasture for extended periods is a possible explanation.

CONCLUSION

The impact of humans on upland soil is sometimes regarded as being detrimental. A transition of brown earths to podzols with increased leaching and loss of nutrients caused by human activity is often suggested. At Lairg the legacy of the prehistoric farmer was one of erosion and profile truncation caused by agricultural practices. This was no longer occurring by the 2nd millennium AD. Brown podzols were still the main type of profile but erosion had been reduced and no E horizon development was observed.

Despite the evidence for soil erosion during the prehistoric period it should not be assumed that this made the area more marginal for agriculture. The condition of the soils prior to the Bronze Age is not known. If the soils were already in a degraded state then erosion of the soil profile might have released more bases from depth. Additionally, where agricultural activity has stopped stagnopodzol profiles have tended to form. This suggests the possibility that human activity is actually making the area more viable for agriculture rather than less.

It is evident that human activity can considerably modify the pedological environment. However, it is too simplistic to regard this modification solely in terms of degradational effects. Humans interact with the environment through social systems and cultural filters in an evolving dynamic relationship. It is this potential of dynamic response that allows humans to adapt to changing environmental circumstances. On the evidence available it cannot be suggested that a degradation in the pedological environment caused the changes in landscape management, but it is reasonable to suggest that adaptability to changing environments contributed to the long period of occupation observed at Lairg.

ACKNOWLEDGEMENTS

This work was completed as part of a SERC/CASE studentship based in the Department of Environmental Science at the University of Stirling and sponsored by Historic Scotland through AOC. My thanks to those at AOC, in particular Dr S Carter. Thanks to Prof D Davidson, for continued support, and Mrs M MacLeod for generous help in making the thin sections.

REFERENCES

Askew, G P, Payton, R W & Shiel, R S 1985 'Upland soils and land clearance in Britain during the second millennium BC', *in* Spratt, D & Burgess, C (eds), *Upland settlement in Britain, the second millennium BC and after*, Oxford: BAR British Series, 5–33. (=Brit Archaeol Rep Brit Ser, 143).

Ball, D F 1975 'Processes of soil degradation: a pedological point of view', *in* Evans, J G, Limbrey, S & Cleere, H (eds), *The effect of man on the landscape: The highland zone*, CBA Research Report, 11, 20–27.

Bidwell, D W & Hole, F D 1965 'Man as a factor of soil formation', *Soil Science*, 99, 65–72.

Brady, N C 1990 *The nature and properties of soils.* New York: Macmillan.

Bridges, E M 1978 'Interaction of soil and mankind in Britain', *Journal of Soil Science*, 29, 125–139.

Catt, J A 1991 'Soils as indicators of Quaternary climatic change in mid-latitude regions', *Geoderma*, 51, 167–187.

Courty, M A, Goldberg, P & Macphail, R I 1989 *Soils and micromorphology in archaeology.* Cambridge: Cambridge University Press.

Davidson, D A 1982 'Soils and man in the past', *in* Bridges, E M & Davidson, D A (eds), *Principles and applications of soil geography.* London: Longman, 1–27.

Davidson, D A, Carter, S P & Quine, T A 1992 'An evaluation of micromorphology as an aid to archaeological interpretation', *Geoarchaeology*, 7 (1), 55–65.

De Connick, F 1980 'Major mechanisms in the formation of spodic horizon', *Geoderma*, 24, 101–128.

De Connick, F & Righi, D 1983 'Podzolisation and the spodic horizon', *in* Bullock, P & Murphy, C P, (eds), *Soil micromorphology*, Vol. 2, Berkhamstead: AB Academic Publisher, 389–416.

Fitzpatrick, E A 1981 *Soils: Their formation, classification and distribution'.* Harlow: Longman Scientific and technical.

Futty, D W & Towers, W 1982 *Soil and landuse capability for agriculture, northern Scotland.* Aberdeen: The Macaulay Institute for Soil Research.

Goldberg, P 1983 'Applications of micromorphology in archaeology', *in* Bullock, P & Murphy, C P (eds), *Soil micromorphology.* Berkhamstead: AB Academic Publishers, 139–150.

Inversen, J 1958 'The bearing of glacial and interglacial epochs on the formation and extinction of plant taxa', *Uppsala Universiteit Arssk*, 6, 210–15.

Limbrey, S 1975 *Soil science and archaeology.* London: Academic Press.

Mackie, E W 1966 'New excavations on the Monomere neolithic chambered cairn, Lamlash, Isle of Arran', *Proceedings of the Society of Antiquaries of Scotland*, 97, 1–34.

Murphy, C P 1986 *Thin section preparation of soils and sediments.* Berkhamstead: AB Academic Publishers.

Roberts, N 1991 *The Holocene, an environmental history.* Oxford: Blackwell.

Romans, J C C & Robertson, L 1974 'Some aspects of the genesis of alpine and upland soils in the British Isle', *in* Rutherford, G K, (ed), *Soil Microscopy.* Kingston, Canada: Limestone Press, 498–510.

Romans, J C C & Robertson, L 1975 'Some genetic characteristics of the freely drained soils of the Ettrick in East Scotland', *Geoderma*, 14, 297–317.

Romans, J C C & Robertson, L 1983 'The environment of north Britain: Soils', *in* Chapman, J C & Mytum, H C (eds), *Settlement in north Britain 1000 BC – AD 1000*, Oxford: BAR British Series, 55–80. (=Brit Archaeol Rep Int Ser, 118).

Romans, J C C, Stevens, J H & Robertson, L 1966 'Alpine soils of north east Scotland', *Journal of Soil Science*, 17, 184–199.

Smith, R T 1975 'Early agriculture and soil degradation', in Evans, J G, Limbrey, S & Cleere, H (eds), *The effect of man on the landscape: The highland zone*, Research Report, 11, 27–36.

12. Beyond the fringe? Recognising change and adaptation in Pictish and Norse Orkney

Julie M Bond

Abstract

The site of Pool, Sanday, Orkney has a long settlement sequence, from the Neolithic period to at least the 11th century AD. Study of the economic evidence from this island site is focused on the detection of change in the subsistence base over the life of the settlement. Is any such change the result of innovation by a native or Norse population or is it, as is often claimed, the result of long-term adaption to a marginal environment? The paper concentrates on the evidence for change in arable cultivation in the area, rather than the faunal remains which are the basis for study in many of the more western settlements.

INTRODUCTION

Marginality, like so much in archaeology, is a matter of perception and perspective. The further away we are, in time or space, the more distorted that perspective is likely to be. An area or culture may be perceived as geographically, environmentally or economically at the periphery, and this may be as much a political or value judgement, as one based on solid fact. For us as archaeologists, this causes many problems since we cannot observe our chosen subjects at first hand, but instead must rely on the distorted perspectives offered by archaeological survival and recovery, and sometimes by literary sources. This paper concerns an attempt to deal with such problems of perception in the study of possible change and innovation in the subsistence base of Pictish and Norse Orkney.

THE GEOGRAPHICAL AND ENVIRONMENTAL POSITION OF ORKNEY

From London, or Edinburgh, Orkney is geographically marginal, unsuited to many of the crops grown further South. Kirkwall is nearer to Oslo and the Arctic Circle than to London, on much the same latitude as Leningrad or the southern tip of Greenland (Figure 12.1). At midsummer the sky is never really dark, whilst in midwinter the sun is above the horizon for less than 6 hours a day. Snow rarely lies long in the winter, but the summers are cool; the equivalent, it is said, of Lapland or Alaska (Berry 1985, 20). Rain falls on an average of 241 days a year, though when driven by an Orkney gale it is more likely to move horizontally than vertically. Gales occur throughout the year; in January 1952 a gale removed 7000 poultry houses and 86,000 hens from their homes in Orkney, since when Orcadian farmers seem to have concentrated on cattle.

Whereas lowland England has a growing season of 7 to 8 months, Orkney has a season of 5 to 6 months. This has consequences not only for the cereal crops, but for the pasturage and fodder vital to the survival of stock through the winter. The present day start of the barley harvest is in September (Coppock 1976, 19), but 17th century records from Orkney show that the crops sometimes were not harvested until well into October, and that in a bad year they would still not be fully ripe when harvested (Shaw 1980, 93, quoting Shirreff 1814). This must also have been the case in 18th century Orkney; Patrick Fea, farmer of Stove on Sanday noted in 1768 that he had started the barley harvest on September 1st and finished on the 12th of October '... the best and earliest I ever remember.' In the years covered by his farm diary, Fea was not always so lucky; in both 1771 and 1772, the harvests began on September 21st and 17th respectively, and ended on November 2nd (Marwick 1930, 67).

Wheat seems never to have been a commercially successful crop in Orkney, although it has been tried. James Wallace, minister of Kirkwall in the late 17th century, described Orkney as '... destitute of wheat, rye and pease ...' (Wallace 1693, 13) and Barry (1805, 357) commented that '... few attempts have yet been made to raise either rye or wheat.' The growing of crop plants in a complex society is influenced as much by the economic situation as by environmental factors. The author of a modern agricultural atlas of Scotland states that wheat reaches its northern economic limit in Scotland and is not grown in the north of the country, because '... the late onset of spring and inadequate summer warmth are major handicaps, for the *quality* of grain suffers from a late harvest' (Coppock 1976, 55, my italics). Presumably when prime grain for the market is not required, the viable area for wheat growing is more extensive than that seen in the modern period, though still restricted by environmental factors. Conversely, when the market forces of a more complex society are taken into account, even a small change in the environment might lead to quite drastic shifts in the types of crops produced.

Famines were not unknown in the Northern Isles in the 17th and 18th centuries, and although, like the Irish Great Famine, some were undoubtedly caused or exacerbated by human agency, others were genuinely due to climatic conditions; the years 1782 to 1785 saw cool and rainy summers and early snowfall damaging the late harvests. 'Farms that usually produced 100 bolls oatmeal, did not produce 20 bolls, and the lesser farms proportionally deficient.' (Thomson & Graham 1978, [Statistical Account 1791–1799], xii, 71). Yet this was as nothing compared to the years 1635/6, still remembered 150 years later in the islands as when '... three or four thousand people perished of famine in the Orkney islands.' (Statistical Account 1791–1799, 71). A contemporary account makes it clear that this was due to the weather; 'For this last harvest, before the cornes were fully rypped and cutt doun, suche tempestuous and bitter weather blew from the ocean upon these parts that the cornes were so blasted as they never filled ... quhilk hes caused so great dearth and famine in these parts that multitudes dee in the opin feilds and there is none to burie thame ...' (Register of the Privy Council of Scotland, V [1633–35] quoted in Thomson & Graham 1978, xii).

To some extent, records like these have led to a form of environmental determinism; the school of thought that argues that Orcadian subsistence strategies were fixed in their course many centuries ago, and stayed that way until relatively recently; that nothing much has changed since then because nothing much can. As Ritchie says; 'Whatever barbarities existed in social behaviour and tribal ritual of which no trace remains, practical daily life can have been little different from the basic Orcadian pattern that survived until recent times' (Ritchie 1985, 52–3). Ironically it is a view that has probably been fostered by the excellent records of Orcadian life in the 18th, 19th and early 20th centuries, which, together with the drystone buildings so reminiscent of techniques seen at Neolithic Skara Brae, fix an idea of both remote and recent past as primitive and virtually unchanging. The writers who described what they saw as ancient forms of agriculture still being practised in Orkney in the eighteenth and nineteenth centuries had mixed motives. Some travellers undoubtedly meant to make their own travels seem more daring, by emphasising the primitive nature of the country through which they passed (eg Scott 1814). Others, such as some of the contributors to the Statistical Account (1791–1799), hoped to encourage reform of what they saw as outdated farming practices, which had '... hardly changed over the centuries' (Thomson & Graham 1978, xiv). That agriculture in the Northern Isles was still so basic, was probably not due solely to the exigencies of soil or climate, but also to the heavy hand of the landlord, who expected rents to be paid in kind with up to a third of the grain crop and 'on call' labour at harvest or kelping, leaving little room for experimentation or improvement (Thomson & Graham 1978, xiv–xv; Napier Commission 1884, eg 1437–1438, 1445–1446, 1449–1451).

When compared to other Viking settlements, in the Faroes, Iceland, or Greenland, rather than lowland Britain, Orkney appears a far from marginal environment. Compared even with its northern neighbour, Shetland, Orkney has always had a better record of success in farming; in 1931 Orkney had 37.3% of its land under arable cultivation, compared to 3.4% in Shetland. This is the root of the traditional description of an Orcadian as a 'farmer with a boat' compared to the Shetlander's 'fisherman with a plough' (Fenton 1978, 2). From this viewpoint, Orkney would have been seen by Viking settlers as a fertile land, better suited to agriculture than much of the Scandinavian homelands.

THE HISTORICAL POSITION OF ORKNEY

Politically, there are different viewpoints of marginality. Prior to the Viking settlements, Orkney seems to have been a minor power in the developing Pictish kingdom. In Adamnan's account of St Columba's visit to the court of the Pictish king Bridei, he meets the regulus, or subject-king, of Orkney and the point is clearly made that Orkney, though part of the Pictish world, is on the periphery (Ritchie 1985, 185). Yet in the Viking world the Orkney Earldom was centred on the important route between Scandinavia, the Western Isles and Ireland, and the rule of the Earls reached through Shetland, Orkney and the Hebrides to the Isle of Man (Crawford 1987, 75–76). In the present day, our vision of the geographical marginality of Orkney is coloured by the political fact

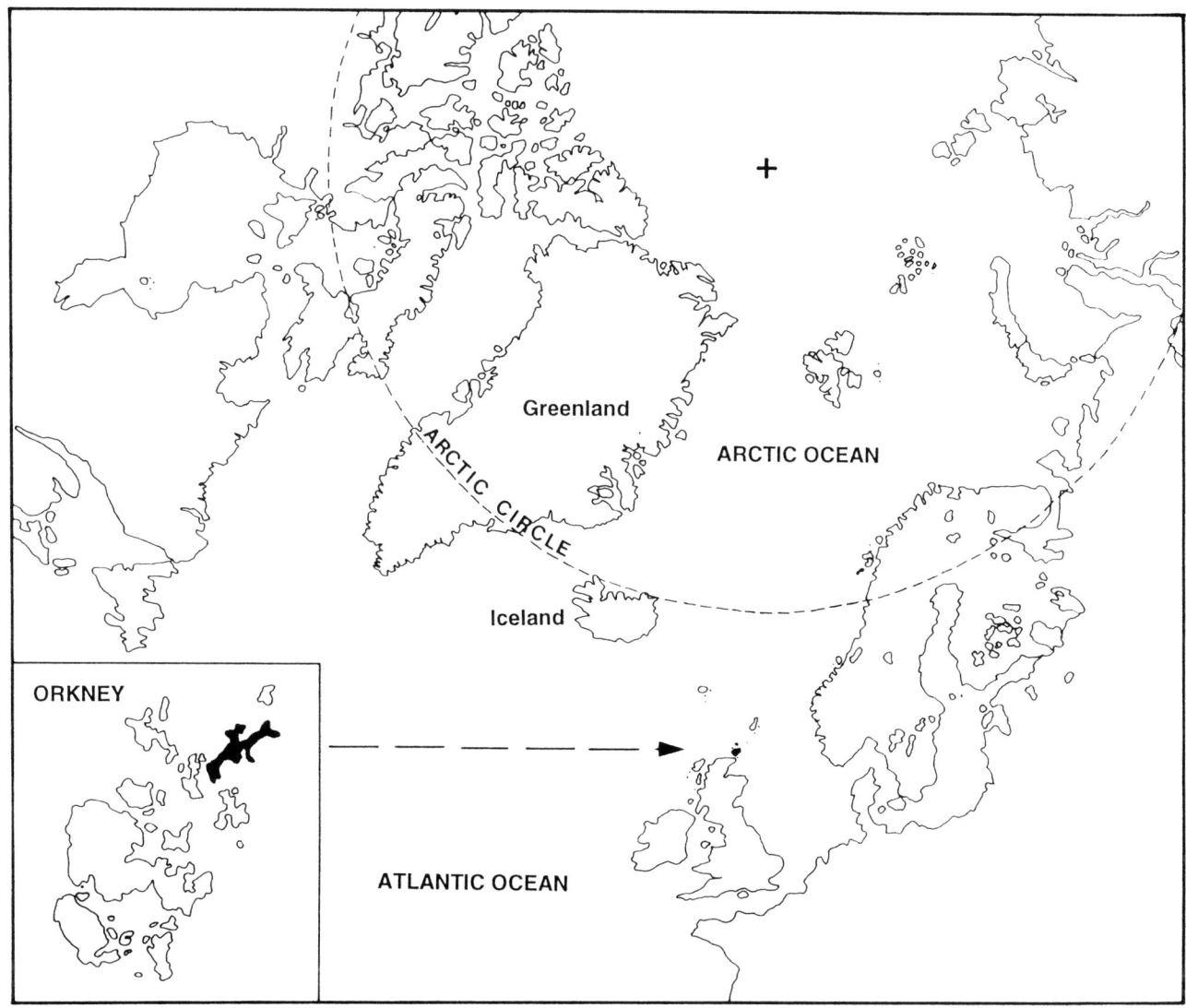

Figure 12.1 The geographical position of Orkney; Orkney group shown enlarged, bottom left (Sanday in black).

of its rule from Edinburgh and, ultimately, London, which again places it firmly on the periphery of maps of the British Isles.

THE POSSIBILITIES FOR CHANGE IN THE ORCADIAN ECONOMY

Though our view of Orkney as a marginal area of Britain should be subject to qualification, nevertheless, the restraints of environment and geography were and are real. This paper concerns work centred on the island of Sanday, rich in archaeological sites and monuments but only 13 miles long and, in most places, about a mile wide (Figure 12.1). The restricted range of habitats and area available for settlement on such an island must itself add another layer of constraints, and make change and development somewhat harder (Bond 1994).

If the constraints of environment and resources are

real, change over time might well be slow and subtle, and more subject to these forces than to the will of farmers. What hope have we, then, of seeing change or adaptation in the archaeological environmental record? If anywhere, it should be visible in the upheaval in Orcadian society with the arrival of Viking settlers in the islands in the 9th century AD. It might be expected that the influx of not only political but also cultural changes would see an echo in changes in even the most well-adapted agricultural system.

THE ISLAND OF SANDAY AND THE SETTLEMENT SITE AT POOL

Sanday is a small island; one of the northernmost of the Orkney group and one of the least typical, being for the most part very low-lying (below the 5 m contour) and covered with sand. The southern end, where the settlement site of Pool is situated (Figure 12.2), conforms rather more closely to the general pattern of the Orkney landscape; there are gently sloping hills and the influence of sand movement is less obvious though still present. There are no large stretches of machair grassland in this area of Sanday, but the backslope of the hill above Pool is, at 50 m above sea level, the highest moving dune system in Britain (Mather *et al* 1974, 96–97).

The site itself lies on a partially-enclosed bay next to a former loch (Figure 12.2). The 'soft' parts of the Sanday coastline show evidence of many changes; there are narrow spits either of shingle with a sand covering, or of pure sand, which often form across a shallow bay or between islands. Where these spits, known locally as 'ayres', develop to completely enclose an area from the sea, the resultant fresh or brackish-water lagoon is known as an 'oyce'. Over time the oyce may fill with windblown sand or may silt up, to create new land (Mykura 1976, 4). The head of Pool bay seems to have been one of these lagoons, now an eroding peat-filled basin covered with wet pasture. The settlement of Pool is on slightly higher ground to the north of this basin. The bay is divided, and partially enclosed, by a natural rock formation which shows some evidence of human efforts to turn it into a pier (Hunter pers comm).

The site came to be noticed because it too is eroding into the bay; first recorded by the Regional Archaeologist, Raymond Lamb, it was subsequently excavated by John Hunter and a Bradford University team over 6 seasons between 1983 and 1988, funded by the then Scottish Development Department (Hunter *et al* forthcoming).

The origins of the settlement date to some time before 3400 BC, with Unstan and Grooved Ware pottery present (Hunter & MacSween 1991), and after a hiatus when a turf layer developed over the site, settlement evidence resumes in the 3rd to 5th centuries AD, continuing until the middle of the 11th century (Hunter *et al* 1993). It is the later Pictish phases, the Viking phases and the period of change, adaptation and rebuilding between these two distinctive cultural traditions (termed the 'interface phases' by the excavator) which concern us here.

Structural evidence from Pool shows an expansion of the site in the 6th century AD (Hunter 1990, 184) with the addition of chambers and passageways to the existing buildings, and extremely tightly-laid flagged floors. A rectangular 'courtyard', some 20 m in length, was subsequently added. It contained not only an orthostat with an ogham inscription, but also a symbol stone, laid face down in the paving (Hunter 1990, 185). Paving was also laid down around the buildings, effectively sealing the entire Neolithic mound and providing a visually uniform surface throughout the settlement. A triangular area in the centre of this paving was revetted and used as a midden, in contrast to the previously more casual arrangements (Hunter 1990, 187 & Illus 10.9, 188).

The settlement appears to have begun to contract in the

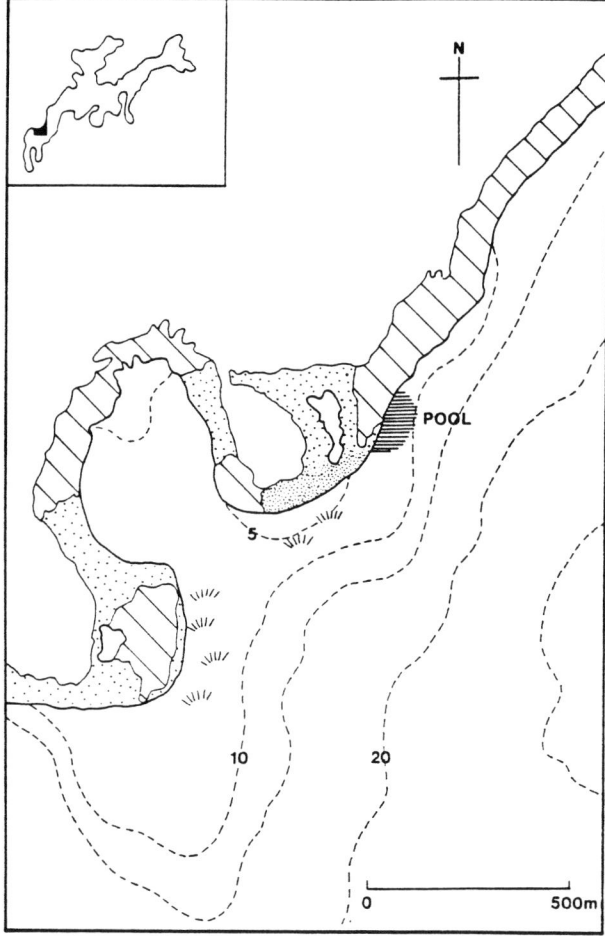

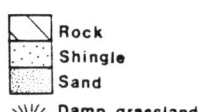

	Rock
	Shingle
	Sand

\\\\\// Damp grassland

Figure 12.2 The settlement site at Pool, Sanday (contours in metres).

seventh century AD; buildings fell out of use and doorways were blocked, midden spilled from the central walled area and spread across the paving. The first evidence of Scandinavian activity is the levelling of much of this area, with the infilling of disused buildings. A sub-rectangular structure with a central hearth was created, partly by utilising existing wall-lines (Hunter 1990, 189). Oddly, the circular building which had been at the heart of the Iron Age settlement was still in use, surviving as part of the building complex until the 11th century AD. The artefacts from these levels were a mixture of both native and Scandinavian types, in what Hunter describes as a phase of cultural interface (Hunter 1990, 189). Later buildings and artefacts on the site are more obviously Scandinavian in influence.

It is clear, then, that the changes seen on this site are neither simple nor unambiguous. Scandinavian influence on the site is preceded by a period of contraction and decline, and continuity of both buildings and artefactual types can be demonstrated. The question must be asked, do changes in building style and cultural artefacts go hand-in-hand with changes in subsistence strategies? Or do any recognisable changes in the subsistence base have their origins in the same factors which led to the contraction of the settlement, in the 7th century AD?

VIKING SETTLEMENT IN ORKNEY

Orkney differs from the Viking settlements in the North Atlantic, in Greenland, Iceland or Faroe, not only because it is slightly more temperate but also in having a native population with a viable farming system already in place when the Norse arrived in the 9th century AD. This is vividly illustrated by a number of sites such as Pool on Sanday, Saevar Howe (Hedges 1983b), Buckquoy (Ritchie 1976) and the Birsay area sites (Hunter 1986; Morris 1989), all on Mainland Orkney, where Pictish houses and artefacts are physically overlain by their Viking equivalents. Anna Ritchie, at Buckquoy, argued for a relatively peaceful takeover, citing Pictish artefacts found inside Norse buildings, though the overall position in the Northern and Western Isles is probably more patchy and complex, as Morris has argued (Morris 1985, 216). At Pool, the 'interface' phase has a native building, already apparently standing for several hundred years, re-used as a part of a Viking longhouse, and the re-use of other standing walls in new buildings.

The environmental evidence from these sites should add an entirely new facet to the arguments and poses the further question; in a marginal area where environmental constraints are active, how much change is possible within the system, given that cultural and technological changes to the agricultural base might be expected from a different group of people?

ARCHAEOZOOLOGICAL EVIDENCE FROM POOL

The faunal remains from Pool are similar to those from other comparable Orcadian sites of the same period such as Buckquoy (Ritchie 1976) and the Birsay Bay sites (Morris 1989). There is a high proportion of cattle and a relatively high proportion of pig, and like the other sites, there is no real indication of change in the major domesticates between the Pictish and Norse periods (Table 12.1).

Ageing data (toothwear and epiphysial) from the Pool cattle show a high proportion of very young or neonate cattle in both Pictish and Norse phases, paralleled at other similar sites, with intensification of this trend from the interface into the Viking period. An increase in dairying is an obvious interpretation. There is a tiny but apparent increase in the numbers of horse, and the presence of young and neonate bones in the Viking phases suggests they were bred at the settlement (Bond & Serjeantson, in Hunter *et al* forthcoming).

Like other Orcadian sites, but unlike the North Atlantic settlements, the numbers of wild animals utilised at Pool are very low; red deer never forms more than 2% of the mammal bone collection, and seal is virtually insignificant

Phase	Horse	Cattle	Sh/g	Pig	Cat	Otter	R/deer	Seal
6.1 Pictish	0.0	49.8	36.4	11.3	0.0	0.4	2.1	0.0
6.7 Pictish	1.4	39.2	46.9	10.4	0.5	0.4	1.7	0.0
7.1 Interface	0.9	34.1	51.5	10.7	0.2	0.1	1.9	0.6
7.2 Interface	4.0	36.8	48.1	9.2	0.5	0.4	1.0	<0.1
8.1 Norse	4.6	35.7	49.8	9.1	0.1	0.2	0.7	<0.1
8.2 Norse	1.8	35.3	51.2	11.3	0.4	<0.1	<0.1	<0.1

KEY: Sh/g = sheep/goat, R/deer = Red deer

Table 12.1a Pool animal bone; number of identified specimens as percentages.

Phase	Horse	Cattle	Sh/g	Pig	Cat	Otter	R/deer	Seal
6.1 Pictish	0.0	40.0	37.1	17.1	0.0	2.9	2.9	0.0
6.7 Pictish	5.6	32.4	50.6	16.7	4.6	2.8	7.4	0.0
7.1 Interface	5.3	35.5	33.6	18.3	1.2	1.2	3.1	1.9
7.2 Interface	10.2	31.4	31.7	18.0	2.4	1.8	4.0	0.6
8.1 Norse	10.9	33.9	32.8	19.5	0.8	1.6	0.8	0.0
8.2 Norse	6.1	34.9	36.4	21.2	1.5	<0.1	<0.1	<0.1

KEY: Sh/g = sheep/goat, R/deer = Red deer

Table 12.1b Pool animal bone; minimum number of individuals as percentages.

(Table 12.1). In comparably-dated settlements in Iceland, birds and walrus played a much more important part, whilst the Norse in Greenland were heavily dependant on seal and caribou, with even major farms producing up to 25–30% seal bone (Amorosi 1991, 277, Table 2; McGovern 1990, 342).

It is probable that red deer were not living on such a small and densely populated island as Sanday, but Eday, with a much higher proportion of hill-land, is only a short boat journey away (though across an admittedly-dangerous tidal race). There is little data on when red deer may have died out in the Northern Isles, but the numbers do seem to decline towards the end of this period.

Seal makes only an erratic appearance in the record, and then mostly in the 'interface' phases.

Fish and wild birds were utilised; this material has been examined by Rebecca Nicholson and Dale Serjeantson respectively (in Hunter *et al* forthcoming). Nicholson suggests that there is a change in fishing strategy between the late Iron Age and Viking phases, with increased specialisation and an emphasis on medium to large gadids (cod family)(Nicholson pers comm). There is also a change in seabird exploitation, with gannet becoming the commonest bird in the Viking period proper, taking over from cormorant or shag in the 'interface' period. This agrees with the evidence for seabird exploitation from other sites in the Northern and Western Isles (Serjeantson 1988).

There is also some exploitation of shellfish, but the numbers are not large compared to the animal bones. There is no archaeological evidence to indicate whether shellfish were used as fish bait (Fenton 1978, 528), as a seasonal food supplement, or as a 'bad year' resource, utilised when other foods were scarce. The minor but persistent presence of the Iceland mussel, *Arctica islandica* (L.), as well as limpets and periwinkles (*Patella vulgata* L. and *Littorina littorea* (L.)), indicates something other than collection for fish bait. Whilst limpets can be found in the near vicinity, the nearest source of *Arctica* today is Stove Bay, on the south side of the island, which has the requisite firm sandy base in the lower reaches of the intertidal zone (Tebble 1976, 92).

Both limpets and periwinkles were traditionally regarded as famine-food in the Northern Isles; a Shetland saying, to 'gang i'da wylk (winkle) ebb' means to be reduced to poverty (Fenton 1978, 542). In 1762 a complaint was made that kelp-burning had left the limpet-rocks unprotected from the sun, so that the limpets fell off in the heat and the poor were deprived of food (Fenton 1978, 542).

Nevertheless, all of these wild animals, whilst showing the broad resource base available, form a smaller proportion of the diet than one might expect, given their abundance. The cliffs of the Northern Isles are famous for their birdlife today, and both grey and common seals are ubiquitous; the bar and bay at Pool are favourite haul-outs for them. The under-utilisation of these wild resources suggests that there was for the most part a comfortable margin of subsistence, rather than the large

numbers of wild animals seen as 'stress indicators' in some of the North Atlantic settlements (eg, see Amorosi 1991, 276–7).

The faunal evidence suggests that a similar strategy employing a broad economic resource base is present in both pre-Viking and Viking phases at Pool, but with some intensification in the areas of cattle dairying and fishing from the 'interface' onwards. The continuing but weak presence of seal and red deer throughout this period suggests that, like the shellfish, they were seen as a useful supplement or fall-back rather than as a major resource.

ARCHAEOBOTANICAL EVIDENCE FROM POOL

Arable crops in Orkney

Sites with good macrobotanical assemblages for both the Pictish and Viking periods in Orkney are few and far between, but nevertheless, the basic pattern is known; six-row barley, mostly a naked form in the early prehistoric period and later hulled, has been the major crop throughout the archaeological and historical record, with cultivated oats appearing at some point between the Iron Age and the Viking period.

Orkney is too far north and too wet for successful wheat production, and there is no evidence that it was ever widely grown although there have been occasional, mostly Neolithic, finds. There are some grains from Neolithic Isbister (Lynch 1983), from the bottom of the midden at Skara Brae (Maclean & Rowley-Conwy 1984) and from the middle Neolithic phases at Pool.

Cultivated oats are first definitely identified in the Birsay Bay area in the Pictish-Viking period (Donaldson & Nye 1989, 262–66; Morris 1989, 229), and flax first makes its appearance in the area at the same time, although a seed identified from Warebeth Broch on Mainland Orkney, and another from Crosskirk Broch, Caithness, suggests the crop had a greater antiquity in the area (Dickson & Dickson 1984; Dickson 1989).

The macrobotanical evidence from Pool

The evidence from Pool gives a more detailed view of the sequence for one site. The macrobotanical samples produced a sizeable set of data relating to charred material from the Pictish, Interface and Viking phases at Pool. Since the objective is the study of change over time, on a site which had a restricted environmental catchment and a similar mix of context types throughout the phases, the data are presented here as changes in the percentage presence, the ubiquity, of major species over time (Pearsall 1989, 212–217; Popper 1988, 60–64) (Figure 12.3). Sample selection and methodology, and a more detailed account of the botanical and faunal material from Pool, can be found in the author's thesis and the forthcoming site report (Hunter *et al* forthcoming; Bond 1994).

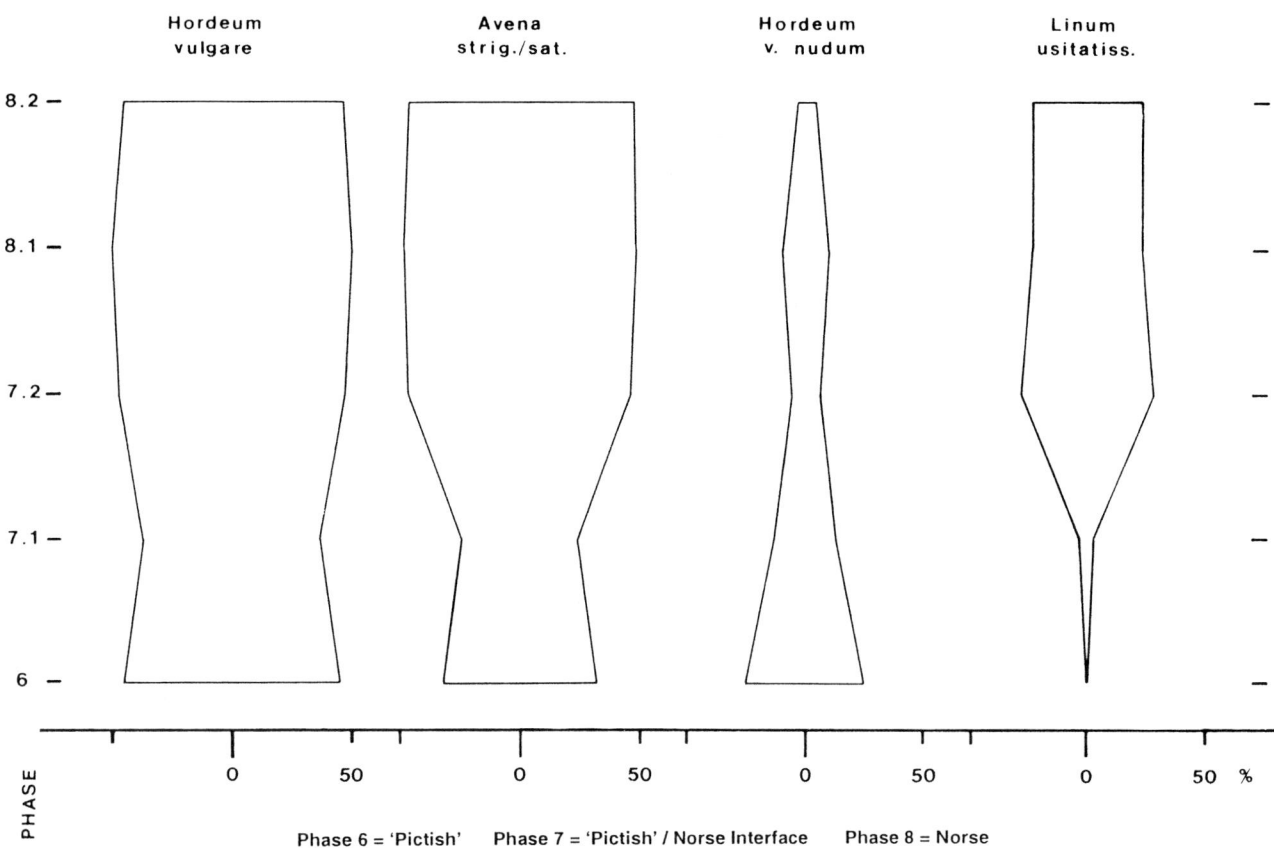

Figure 12.3   Changes in the percentage presence scores of the main crop species over time (Phase 6 = 'Pictish', Phase 7 = 'Pictish'/Norse interface, Phase 8 = Norse).

Cultivated oat, *Avena strigosa* or *A. sativa*, is first identified at Pool in Phase 5, which is probably 4th to 5th century AD in date, and is therefore not a Viking introduction. Cultivated oats vastly outnumber identified wild oats, *A. fatua*, throughout the Pool record, the wild oat first being positively identified in Phase 6, the Late Iron Age or Pictish period. By contrast flax, *Linum usitatissimum*, seems to be a genuine Viking introduction to the site, appearing in the 'Interface' phase, with the first indications of Viking presence. The fact that this crop is around in Orkney before this date, yet only appears in the Pool record at this time, is intriguing and suggests a real change, perhaps an intensification, in the agricultural system there. Flax is traditionally a crop which exhausts the soil, and its processing is labour-intensive, although the soil and climate of Sanday would be ideally suited to it (Bond & Hunter 1987). The general increase in ubiquity of all crop plants over the Viking period seems genuine, and not a consequence of sample type (Figure 12.3). The overall impression is of an increase in the intensity of agricultural production in the Viking period.

Non-crop species
The non-crop species too, show interesting changes over time, with *Urtica urens* (small nettle) and *Raphanus raphanistrum* (wild radish) first appearing in Phase 7.2,

the 'Interface', along with a diversity of other weeds too infrequent for the calculation of ubiquity scores. Other species such as *Montia fontana* (blinks), present since the Neolithic, decline greatly in importance, whilst the ubiquity scores of *Rumex* species (dockens) and *Stellaria media* (chickweed) rise. *Spergula arvensis* (corn spurrey), often regarded as a common weed of flax crops, appears here before the introduction of flax, and seems instead to follow the cereal curves. Two groups of weeds increase in importance at Pool during Phases 7 and 8; one, including *Urtica urens* and *Spergula arvensis*, prefers light, sandy soils. The other, represented chiefly by *Stellaria media* and *Raphanus raphanistrum*, prefers a rich, well-cultivated soil (Bond 1994; Hunter *et al* forthcoming). These two groups may indicate the taking-in of poorer sandy land to accommodate increasing cereal production. *Avena strigosa* is known for its ability to produce a crop on poor sandy soils. Barley, however, produces greatly increased yields on well-manured and tilled land (Shaw 1980, 98–109) and therefore may have continued to be grown on the infield.

Taphonomic factors in the interpretation

The 'dip' in all ubiquity scores in Phase 7.1, the first Viking presence, caused some concern (Figure 12.3).

Because it was possible that contexts from the margins of the settlement area were responsible for this, the figures were re-calculated without these samples but the relative values remained much the same. The range and proportion of context types are also much the same as in other phases. The number of seeds per litre is also low for most of the samples in this phase. Phase 7.1 is defined by the erection of new, Viking structures and the incorporation or demolition of Pictish ones – many of the midden contexts have the appearance of dumps associated with rubble – so it would seem that there are two main possibilities. These are, firstly, that there is genuinely less agricultural activity, which is a possibility given the events in progress (although the first presence of flax in this phase might be seen to argue against this), or, secondly, that the low numbers and ubiquity scores are due to heavy destruction through weathering and redeposition, in turn due to the major and widespread destruction and rebuilding in this phase. It is a salutary reminder that the environmental data can never be separated from the archaeology.

CONCLUSIONS

Whether Orkney is regarded as a marginal environment or not, it seems that the system was neither incapable of nor immune to change, either in the Pictish or the Norse periods, as the botanical evidence shows. Cultivated oats were introduced to the settlement in the Pictish period, and the first indications of flax-growing coincide with the appearance of Scandinavian features in the archaeological record. The weed flora also shows evidence of change, and suggests the farming of lighter, sandier soils in the Norse period. Intriguingly, work on the Pool material since the first delivery of this paper suggests that, although the percentage ubiquity scores of both barley and cultivated oats rise in the Norse period, the percentage dominance of oats (ie, the number of samples in which oat grains are numerically dominant) also increases relative to barley. It could well be that more land, of a lighter nature, was being taken into cultivation to increase oat production (Bond 1994).

Although there is no substantial change in the basic faunal data between the Pictish and Norse phases at Pool, there is evidence for an increase in dairying, and for a concentration on gadids in fishing. Both these developments suggest, like the botanical evidence, real changes towards increased production between the Pictish and Norse periods.

It seems as if there are real, traceable changes between the Pictish and the Viking settlements, although perhaps not as dramatic nor as obvious as one might wish.

There is sometimes a feeling in Viking period archaeology that cattle are the centre of the Viking economy, that, as one writer puts it, they are 'more status-giving' than other forms of livestock (Christensen 1991, 161). If that is so, it would appear to be an attitude the settlers shared with their Pictish predecessors, since none of these studies have found a discernable change in the proportions of cattle and sheep, between the two periods in Orkney. It has been justly said that 'The farmers of Sogn, Hordaland and Rogaland in western Norway who made up the bulk of the Norwegian emigrants to Scotland were pastoral farmers whose life-style was based on the raising of cattle and sheep with a little growing of oats or barley where possible' (Crawford 1987, 37). Yet these were the farmers who in Orkney seem to have appreciated the potential of the Sanday soils for flax-growing and other arable improvements.

The economic evidence from Pool has shown that Orkney is not so marginal an environment that the options for subsistence are as limited as has sometimes been assumed. Change was possible, within the constraints of environment and technology. The 'traditional' crofting agriculture of the region may well have had much in common with past practices, but to see it as the direct equivalent of earlier systems, with its different social organisation and cultural perspectives, is misleading. So too is the assumption that similar agricultural practices pertained across the entire area settled by the Vikings. Comparison of data from sites across this area shows that whilst faunal subsistence may well have been of prime importance in the western Atlantic, and fishing an important part of other Northern Isles economies (eg Bigelow 1984), on Sanday at least the main thrust of agricultural change and intensification took place in that area where most change was possible; arable agriculture. Only by quantitative evaluation of all detectable aspects of the economy can such change be recognised, allowing a more intriguing question – the purpose of this increased production – to be posed.

The importance of arable production in Orkney is perhaps unconsciously reflected in the Orkneyinga Saga, now considered to have been largely written before AD 1190 (Gudmundsson 1993, 206), with its account of the life of Svein Asleifarson, one of the last of the Orkney Vikings;

> 'This is how Svein used to live. Winter he would spend at home on Gairsay, where he entertained some 80 men at his own expense ... In the spring he had more than enough to occupy him, with a great deal of seed to sow, which he attended to carefully himself. Then when that job was done, he would go off plundering in the Hebrides and in Ireland, on what he called his 'spring trip'; then back home just after midsummer, where he stayed till the cornfields had been reaped and the grain was safely in.' (Palsson & Edwards 1978, 190–191).

ACKNOWLEDGEMENTS

The excavation at Pool and the post-excavation processing was undertaken as part of a programme financed by the Scottish Development Department (now Historic Scotland). Analysis of the environmental data was undertaken as part of an SERC research programme. The author wishes to thank Becky Nicholson, Dale Serjeantson and Camilla Dickson for generously allowing access to information in advance of publication. The staff of the Pool and Tofts Ness excavation and post-excavation programmes have been the providers of much invaluable discussion, and in particular, John Hunter has given advice and comment, although the views expressed here are entirely the responsibility of the author. Since the first presentation of this paper, work on the site of Pool has been completed; interested readers are referred to the author's thesis (Bond 1994) for a fuller account. The author is grateful to Terry O'Connor and Gerry Bigelow for much helpful advice and comment on that work.

REFERENCES

Amorosi, T 1991 'Icelandic archaeofauna: a preliminary review', *in* Bigelow, G F (ed), 1991, 273–283.

Barry, G 1805 *The history of the Orkney Islands.* Edinburgh: Constable & Co. (Facsimile edition, Edinburgh 1975: James Thin).

Berry, R J 1985 *The natural history of Orkney.* London: Collins.

Bigelow, G F 1984 *Subsistence in late Norse Shetland; an investigation into a northern island economy of the Middle Ages.* PhD thesis, University of Cambridge.

Bigelow, G F (ed) 1991 *The Norse of the North Atlantic,* Copenhagen: Acta Archaeologica. (=Acta Archaeologica, 61 (1990)).

Binford, L 1981 *Bones. Ancient men and modern myths.* New York: Academic Press.

Bond, J M & Hunter, J R 1987 'Flax-growing in Orkney from the Norse period to the 18th century', *Proceedings Society Antiquaries Scotland,* 117, 175–181.

Bond, J M 1994 *Change and continuity in an island system; the palaeoeconomy of Sanday, Orkney.* Phd thesis, University of Bradford.

Burl, A 1984 'Report on the excavation of a Neolithic Mound at Boghead, Speymouth Forest, Fochabers, Moray, 1972 and 1974', *Proceedings Society Antiquaries Scotland,* 114, 35–73.

Christensen, K M B 1991 'Aspects of the Norse economy in the Western Settlement in Greenland', *in* Bigelow, G F (ed), 1991, 158–165.

Coppock, J T 1976 *An agricultural atlas of Scotland.* Edinburgh: John Donald

Crawford, B E 1987 *Scandinavian Scotland.* Leicester: Leicester University Press.

Dickson, C 1989 'Human coprolites', *in* Bell, B & Dickson, C, 'Excavations at Warebeth (Stromness Cemetery) Broch, Orkney', *Proceedings Society Antiquaries Scotland,* 119, 115–122.

Dickson, C A & Dickson, J H 1984 'The botany of the Crosskirk site', *in* Fairhurst, H, 1984, 147–155.

Donaldson, A M & Nye, S 1989 'The botanical remains', *in* Morris, C D, *The Birsay Bay Project Vol 1. 1976–82,* Durham: University of Durham Department of Archaeology Monograph, 262–67. (=Univ Durham Dept Arch Monograph 1).

Fairhurst, H 1984 *Excavations at Crosskirk Broch, Caithness,* Edinburgh: Society of Antiquaries of Scotland Monograph. (= Soc Antiq Scot Monograph 3).

Fenton, A 1978 *The Northern Isles: Orkney and Shetland.* Edinburgh: John Donald.

Gudmundsson, F 1993 'On the writing of the Orkneyinga saga', *in* Batey, C E, Jesch, J & Morris, C D (eds), *The Viking age in Caithness, Orkney and the North Atlantic.* (Proceedings of the Eleventh Viking Congress). Edinburgh: Edinburgh University Press.

Hedges, J W 1983a *Isbister, a chambered tomb in Orkney,* Oxford: BAR British Series (=Brit Arch Rep Brit Ser, 115).

Hedges, J W 1983b 'Trial excavations on Pictish and Viking settlements at Saevar Howe, Birsay, Orkney', *Glasgow Archaeological Journal,* 10, 73–124 & microfiche.

Hunter, J R 1986 *Rescue excavations at the Brough of Birsay 1974–82,* Edinburgh: Society of Antiquaries of Scotland Monograph. (=Soc Antiq Scot Monograph 4).

Hunter, J R 1990 'Pool, Sanday: A case study for the Late Iron Age and Viking periods', *in* Armit (ed), 1990, 175–193.

Hunter, J R, Bond, J M & Smith, A N 1993 'Some aspects of early Viking settlement in Orkney', *in* Batey, C E, Jesch, J & Morris, C D (eds), *The Viking age in Caithness, Orkney and the North Atlantic.* (Proceedings of the Eleventh Viking Congress). Edinburgh: Edinburgh University Press.

Hunter J R, Dockrill S J, Bond J M & Smith A N forthcoming *Archaeological investigations on Sanday, Orkney,* Edinburgh: Society of Antiquaries of Scotland Monograph.

Hunter, J R & MacSween, A 1991 'A sequence for the Orcadian Neolithic?', *Antiquity,* 65, 911–914.

Lynch, A 1983 'The seed remains', *in* Hedges, 1983a, 171–183.

Maclean, A C & Rowley-Conwy, P A 1984 'The carbonised material from Boghead, Fochabers', *in* Burl, 1984, 69–71.

Marwick, H 1930 'An Orkney Jacobite Farmer', *Journal of the Orkney Agricultural Discussion Society,* 5, 64–75.

Mather, A S, Smith, J S & Ritchie, W 1974 *Beaches of Orkney.* Aberdeen: Aberdeen University Dept of Geography.

McGovern T H 1990 'The archaeology of the Norse North Atlantic', *Annual Review Anthropology,* 19, 331–351.

Morris, C D 1985 'Viking Orkney: a survey', *in* Renfrew, C, 1985, 210–242.

Morris, C D 1989 *The Birsay Bay Project Vol 1. 1976–82,* Durham: University of Durham Department of Archaeology Monograph. (=Univ Durham Dept Arch Monograph 1).

Mykura, W 1976 *British regional geology; Orkney and Shetland.* Edinburgh: HMSO.

Napier Commission 1884 *Evidence taken by Her Majesty's Commissioners of Inquiry into the condition of the crofters and cottars. Vol 2.* HMSO: British Parliamentary Papers. (= Parliamentary Papers, Agriculture Vol 23).

Palsson, H & Edwards, P (Trans.) 1978 *Orkneyinga Saga.* London: The Hogarth Press.

Pearsall, D M 1989 *Palaeoethnobotany: a handbook of procedures.* London: Academic Press.

Popper, V S 1988 'Selecting quantitative measurements in palaeoethnobotany', *in* Hastorf, C A & Popper, V S, *Current Palaeoethnobotany – analytical methods and cultural interpretations of archaeological plant remains.* Chicago: Chicago University Press, 53–71.

Renfrew, C (ed) 1985 *The Prehistory of Orkney.* Edinburgh: Edinburgh University Press.

Ritchie, A 1976 'Excavation of Pictish and Viking-Age farmsteads at Buckquoy, Orkney', *Proceedings Society Antiquaries Scotland,* 108, 174–227.

Ritchie, A 1985 'Orkney in the Pictish kingdom', *in* Renfrew, C 1985, 183–204.

Scott, Sir W 1814 *Northern Lights, or a voyage in the lighthouse yacht to Nova Zembla and the Lord knows where in the summer of 1814.* New edition pub. 1982, Hawick: Byway Books.

Serjeantson D 1988 'Archaeological and ethnographic evidence for seabird exploitation in Scotland', *Archaeozoologia,* II/1.2, 209–224.

Shaw, F J 1980 *The Northern and Western Islands of Scotland; their economy and society in the seventeenth century.* Edinburgh: John Donald.

Shirreff, J 1814 *A general view of the agriculture of the Orkney Islands*. Edinburgh: Board of Agriculture.

Statistical Account 1791–1799. Sir John Sinclair (ed) *The Statistical Account of Scotland 1791–1799*. Facsimile edition 1978, Vol 19, Orkney and Shetland. With a new introduction (pages ix–lx) by W P L Thomson & J J Graham. Wakefield: E P Publishing.

Tebble, N 1976 *British bivalve seashells*. Edinburgh: HMSO.

Thomson, W P L & Graham, J J 1978 'Introduction', *in* Statistical Account 1791–1799 (Sinclair), ix–lx.

Wallace, J 1693 *A description of the Isles of Orkney*. Edinburgh: John Reid.

13. Early land management at Tofts Ness, Sanday, Orkney: the evidence of thin section micromorphology

Ian A Simpson

Abstract

Soils buried by calcareous wind blown sands at the Tofts Ness peninsula, Sanday, Orkney, provide the opportunity to examine land management activity associated with Neolithic, Bronze and Iron Age settlement. Using the technique of thin section micromorphology, soils of the Neolithic and early Bronze Age landscape show evidence of vegetation clearance, tillage and manuring with midden material. Soils of the late Bronze and Iron Age landscape are developed on wind blown sand deposits and show evidence of improvement with midden material and the introduction of plaggen manuring.

INTRODUCTION

The island of Sanday to the north-east of the Orkney archipelago is, at first sight, an unlikely place from which to be able to identify and discuss the responses of early society to a soil environment with significant limitations for arable activity. The medieval rentals of the island demonstrate that Sanday was undoubtedly the wealthiest of the north isles of Orkney and that this wealth was based on the island's suitability for growing barley. While the skat (tax) value in the rentals for Stronsay and Westray, two islands comparable in size to Sanday, amounted to no more than 13 urislands each, the skat value of Sanday was some 36 urislands (Davidson *et al* 1983). Evidence of major arable activity is not confined to the historical period. Analysis of the faunal and botanical remains from the recent excavations at Pool, in the south west of Sanday, have demonstrated a healthy pastoral-arable economy during the Neolithic, Iron Age and Norse periods (Hunter *et al* in prep).

Despite these observations, there are areas and time periods in Sanday where prehistoric arable activity was marginal because of soil environment stresses. One such area is the Tofts Ness peninsula located at the north eastern end of Sanday and bound by an inland loch to the south (HY 754 472). The land is low lying with much of the area less than 5 m above sea level. Extensive deposits of calcareous wind blown sand have buried much of the area, contributing to the peninsula having one of the richest concentrations of archaeological deposits in the Northern Isles (Lamb 1980). Good site survival and fossilisation of associated landscapes under the sand has resulted in an outstanding opportunity to examine early society-land relationships on Sanday.

Exploratory excavations at Tofts Ness by Dockrill *et al* between 1985 and 1990 (Dockrill 1986; 1987; Hunter *et al* in prep) have yielded a wide range of archaeological and environmental information from the Neolithic, Bronze and Iron Ages. On the basis of the settlement sites so far examined, occupation patterns are interpreted as being from *circa* 3000 BC – 1500 BC and then from 1000 BC to the final abandonment of the area *circa* 400 BC.

It is entirely feasible, however, that other settlement sites within the area will have different occupation patterns. Current understanding of soil formation is that soils associated with the earlier phases of occupation were developed on glacial till derived from Old Red Sandstone. Environmental deterioration occurred during the later occupation phases, with sand blow occurring from the late Bronze Age and covering much of the soil developed on glacial till. Further environmental deterioration occurred from *circa* 1000 BC onwards with rising groundwater levels resulting in periodically waterlogged soils.

The analyses of buried soils and sediments by thin section micromorphology forms part of the overall excavation strategy for Tofts Ness. The specific objectives of this work are to identify soil conditions and soil management activity associated with the Neolithic, Bronze and Iron Age landscapes, thus improving our understanding of early land management at Tofts Ness.

MATERIALS AND METHODS

Soils and sediments of the buried landscape were surveyed within the vicinity of four settlement mounds at Tofts Ness, Mounds 11, V, W and X. Three basic soil types were identified: soils developed on glacial till of Old Red Sandstone origins; soils developed on calcareous wind blown sand deposits; and anthropogenic soils. Undisturbed samples of representative soils around settlement mounds, dated by radiocarbon, thermoluminescence and association with settlement features of known cultural age, were collected in Kubiena tins for the preparation of thin sections. Three samples were collected from Mound 11 (Samples C2033, C2037 and C1694), three from Mound V (Samples V1921, V19 and V1913) and one each from Mounds W (Sample W1) and X (Sample X1). This gave five samples from the Neolithic/early Bronze Age, two from the late Bronze Age and one from the Iron Age occupation phases.

Thin sections were prepared at the micromorphology laboratory, University of Stirling following the procedures of Murphy (1986). Key points in this process are the replacement of soil water by acetone to prevent shrinkage of the sample, impregnation with resin and cutting to 30 μm thickness, the optimum thickness for mineralogical investigation. Thin sections were described using an Olympus BH-2 petrological microscope and by following the procedures of the International Handbook for Thin Section Description (Bullock *et al* 1985). This allows systematic description of soil microstructure, basic mineral components, basic organic components, groundmass and pedofeatures. Interpretation of the descriptions rests on the accumulated evidence of other workers, notably Courty *et al* (1989), MacPhail *et al* (1990) and Romans & Robertson (1975) as well as Simpson (1985).

RESULTS AND DISCUSSION

Summaries of the thin section descriptions are provided in Tables 13.1 to 13.3 and full descriptions are available in Hunter *et al* (in prep) or from the author. These sections contain a range of features indicative of past environments which allow interpretation of soil development and management practices.

Soils and sediments of the Neolithic/early Bronze Age landscape

Soils developed on glacial drift deposits of Old Red Sandstone origin associated with the Neolithic/early Bronze Age occupation period demonstrate multiple phases of development and varying degrees of human modification. Sample C2033 is a silty clay loam formed on medium to fine glacial drift deposits. This soil is overlain by calcareous wind blown sands and there is a sharp boundary between these deposits. Key processes in the pedogenic develop-

Section	Basic mineral components (C/F Limit 10 μm)									
	Coarse material									Fine material
	Quartz	Feldspar	Biotite	Rock frag	CaCO₃	Calc. oxal.	Phytoliths	Diatoms	Bone	
C2033 Upper	*	*		*	****					Light brown/ White limpid Greyish brown; dotted limpidity.
Lower	***	*		*	*		*			
C1694	**	*		*	****		*			Light brown, reddish brown O.I.L; speckled limpidity
C2037	***	*	*	*	***		*	*		Brown/grey; dotted limpidity.
V1921	***	*		*	***		*			Brown, greyish brown, reddish brown; dotted limpidity.
V1913	*	*		*	****					Brown; dotted limpidity.
V19	**	*	*	*	***	**	*			Reddish brown, greyish brown, grey, red O.I.L.; speckled, cloudy and dotted limpidities.
W1	***	*		*	*		*			Light brown, brown; cloudy limpidity.
X1	***	*		*	**		*		*	Brown, grey brown; dotted limpidity.

Frequency classes refer to the proportionate area of thin section (Bullock *et al* 1985)
* very few   ** few   *** frequent/common   **** dominant/very dominant

Table 13.1 Tofts Ness: Summary of micromorphological characteristics; basic mineral components.

ment of the underlying silty clay loam soil are leaching of iron from sandstones, biological activity, evidence for land clearance and further redistribution of iron through periodic waterlogging. The occurrence of a few sandstones with depletion and iron stained rims, together with a medium to course sub-angular blocky soil structure suggests an acid brown earth moving towards podzolic conditions. The presence of phytoliths suggests an acid grassland vegetation cover. Biological activity within this early soil was moder-

Section	Basic organic components (C/F Limit 5 – 10 cells)				
	Coarse material			*Fine material*	
	Organ tissue	Lignified tissue	Parenchymatic tissue	Amorphous	Cell residue
C2033 Upper Lower		*		*	
C1694		*		**	*
C2037		*		***	*
V1921	*		*	*	
V1913		*		*	*
V19	*		*	***	
W1				*	
X1		*		**	*

Frequency classes refer to the proportionate area of thin section (Bullock *et al* 1985)
* very few   ** few
*** frequent/common   **** dominant/very dominant

Table 13.2 Tofts Ness: Summary of micromorphological characteristics; basic organic components.

ate with a number of lignified plant residues, excremental pedofeatures and medium brown organic pigment intermixed with fine inorganic material. These features of biological origin are similar to those of a contemporary imperfectly drained podzol under uncultivated rough grazing in the West Mainland of Orkney (Simpson 1985). Phytoliths and fine charcoal material are incorporated into both the groundmass of the soil and into the few incomplete silty clay infills that are evident.

Also present are a very few amorphous organomineral fragments that are distinctly red in oblique incident light. This indicates that vegetation clearance by burning took place at some point in the development of this soil prior to burial by sand. Whether this burning was deliberate for management purposes or accidental cannot be ascertained. The infill material suggests that the soil surface was bare for a period of time after burning with a small amount of slaking taking place. Evidence of slaking is limited and not as extensive as that found in the thin section associated with Area W (Sample W1) for which arable activity can be inferred. It would appear therefore that there has been little if any human disturbance of the soil profile. Periodic waterlogging of the soil has occurred as evidenced by the redistribution of iron into mottles. These occur across the groundmass and infill material and there are also iron hydroxide/oxide hypocoatings associated with root channels. This evidence suggests a later date of origin than for the features described above, indicating an increase of soil wetness with time and subsequent development towards a gley soil.

The upper part of this thin section is dominated by wind blown sand with an apedal single grain structure. On the

Section	Microstructure	Groundmass			Pedofeatures			
		Coarse material arrangement	Fine material arrangement (b fabric)	Related distribution	Excremental	Textural	Amorphous and cryptocrystalline	Depletion
C2033 Upper	Apedal single grain	Random	Stipple	Monic to enaulic				
C2033 Lower	Subangular blocky	Random	Stipple, mozaic crystallitic	Porphyric	*	*	**	*
C1694	Bridged grain	Random	Stipple, mozaic	Enaulic	**			
C2037	Apedal crack	Random	Stipple	Porphyric	**	*	*	
V1921	Crack	Random; locally clustered	Stipple	Porphyric; locally enaulic	*	**	**	*
V1913	Intergrain microaggregate	Random	Crystallitic	Enaulic	*			
V19	Vughy	Random to locally clustered	Stipple, mozaic, crystallitic	Porphyric	**			
W1	Subangular blocky	Random	Stipple, mozaic	Porphyric	*	***	*	*
X1	Intergrain channel and chamber	Random	Stipple, porostriated	Porphyric	**	*	*	

Frequency classes refer to the proportionate area of the thin section Bullock *et al* 1985)
*very few   ** few   *** frequent/common   ****dominant/very dominant.

Table 13.3 Tofts Ness: Summary of micromorphological characteristics; microstructure, groundmass, pedofeatures.

evidence of this single thin section, deposition of sand was rapid and in a single episode. Some of the sand material was incorporated into the lower part of the section and has fallen into voids. Chemical and mechanical weathering of the calcareous sand has, in the lower part of the section, resulted in a predominantly stipple speckled b fabric with mosaic speckled b fabrics in places.

Sample V1921 is similar to Sample C2033. It is sandy silt loam in texture formed on medium to fine textured glacial drift deposits derived from Old Red Sandstone. Calcareous wind blown sands were deposited on this soil at a later date and some of this later material has been incorporated into the underlying soil matrix. Post-depositional processes evident in this thin section are redistribution of iron, biological activity and the translocation of silty clay material. Sandstone depletion rims indicate the beginnings of a podzolic soil, although not as marked as in Sample C2033. A limited range of plant residues, excremental pedofeatures and fine organic material is evident in thin section. Biological activity was therefore moderate but does not demonstrate the enhanced levels associated with the incorporation of midden material. Fragmented phytoliths, fine charcoal material incorporated within silty clay infill material and a very few red amorphous organomineral fragments indicate limited burning of vegetation and an exposed soil surface for a period of time early in the development of the soil. This resulted in a small amount of slaking taking place, predominantly through root channels. As with C2033, the level of slaking does not support an interpretation of arable activity. Amorphous and cryptocrystalline pedofeatures indicate further redistribution of iron as a result of periodic waterlogging. These features occur across the groundmass and some of the textural pedofeatures thus formed at a later stage in the evolution of the soil.

One soil thin section (Sample W1) was examined from around Settlement Mound W and three distinct but overlapping phases of soil development can be identified. These phases can conveniently be described as (i) pre-cultivation phase, (ii) cultivation phase and (iii) post-cultivation phase. Medium textured glacial drift deposits of Old Red Sandstone origin form the parent material for this soil. Indication of soil conditions prior to cultivation is provided by iron stained rims on sandstones, evidence of a medium subangular blocky soil structure prior to void infilling, and the presence of a very few phytolith fragments. Iron stained rims of sandstone occur only rarely, but nevertheless are consistent with acid brown earth conditions. Such conditions would have continued into the late cultivation phase when calcareous wind blown sand deposits would have contributed to rising soil pH levels. The presence of phytoliths within the section indicate a predominantly acid grassland vegetation cover and medium sub-angular blocky structures are not inconsistent with this view. Phytoliths can be introduced to the soil by deposition of ash and manure but are not present in sufficient concentrations to have been introduced in this way. A more likely explanation

for the presence of these smooth phytoliths is that clearance of the land for cultivation involved burning the pre-cultivation grassland vegetation cover with subsequent phytolith incorporation into the soil. Further evidence of surface vegetation burning is the occurrence of a very few red (in oblique incident light) organomineral fragments. Indirect evidence of cultivation is the silty clay, non-laminated void infill material within which are fragmented phytoliths and fine charcoal material. These pedofeatures are common in this thin section and such features are indicative of exposed soil conditions and declining structural stability to which a reduced organic matter content would contribute. The physical process forming these features involves the structural break-up of surface soil aggregates, slaking and mobilisation of water saturated soil followed by deposition against a barrier or when the soil dried out. Such processes are common under arable land management. Associated with this process of soil slaking are a few vesicles that may have been formed by trapped air. In addition, the very few intra-aggregate straight planes partially infilled with silty clay material may be attributed to ploughing activity. However, given the limited evidence for ard cultivation, the lack of depletion features in peds and the accumulation of impure silty clay in voids it is clear that depth of cultivation was limited and confined to a surface zone immediately above the thin section sample point.

It appears unlikely that there was much, if any, additional material applied to this cultivated soil in an effort to raise nutrient levels. The fine charcoal and phytoliths identified in the section are consistent with the vegetation clearance phase of soil development rather than anthropogenic deposition. Furthermore the lack of structural stability that contributed to the formation of the silty clay infill suggests declining organic matter status rather than the enhancement expected when manuring takes place. This interpretation is further supported by the limited occurrence of excremental pedofeatures in the section. Deposition of the predominantly calcareous wind blown sand material commenced during the cultivation phase of soil development and this material was incorporated into the existing soil matrix. Size and number of wind blown sand grains increase to a maximum of 700 μm towards the top of the profile, pointing to an increasing intensity of sand deposition towards the end of the cultivation period which eventually covered this cultivated land.

Two pedofeatures have formed across some of the silty clay infills thus post-dating cultivation activity. Root channels are partially infilled with dark brown amorphous organic material, considered to be root remains, while iron oxide/hydroxide pedofeatures indicate the onset of wetter conditions. These latter features are of particular significance in that their major period of formation was probably after the formation of at least some of the textural infill material associated with agricultural activity.

A number of factors would have limited the agricultural productivity of this soil. The medium textured soil would have been moderately workable to early cultivation implements with the related factor of soil wetness possibly becoming more prevalent as early as the late cultivation period. Soil nutrient status would have been moderate at the commencement of agricultural activity given the inherent fertility of the Old Red Sandstone drift parent material and the initial input of ash from burnt vegetation cover. As agricultural activity continued, organic levels in the soil declined in the absence of manuring. This contributed to declining nutrient status and declining structural stability. With increasing soil wetness and declining organic levels this soil was significantly reduced in its agricultural quality over time prior to it being covered by wind blown sands.

Deposition of anthropogenic material and of wind blown sands are the primary processes associated with the formation of the soil represented by Sample C2037, with the former dominant. Pedogenic processes identified in this section are biological activity, the formation of textural pedofeatures, weathering of calcium carbonate sand deposits and periodic wetting and drying of the soil. Anthropogenic deposits are both mineral and organic in origin. Basic mineral components included single grain quartz, feldspar and biotite together with sandstone rock fragments of up to 2 cm in diameter. These components exhibit a similar frequency pattern to the underlying C2033 material.

A second type of mineral material deposited was ash residue, evidenced by the fine grey brown material of dotted limpidity, with associated phytoliths, and the presence of amorphous organomineral material which is red in oblique incident light. Preservation of organic material is limited in this thin section with charcoal, amorphous black coarse material and fungal spore remains being the only basic organic component, together with brown organomineral fine material. Fungal spores, excremental pedofeatures and intra-aggregate channels indicate that biological activity was high in this soil, responding to substantial organic inputs. Without morphological evidence, identifying the source of the material is speculative although the frequency pattern is similar to the midden sample (V19). Irrespective of the origins of this midden material, its application would have significantly enhanced the nutrient status of this cultivated soil. A negative aspect of this form of land management, in view of the spore material evident in thin section, may have been crop disease.

The other type of material deposited to form this soil was calcareous wind blown sands. Deposition of calcareous wind blown sand would appear to have been contemporaneous with anthropogenic deposition and thus represents a period of sand erosion and deposition considerably earlier than the main late Bronze Age period of deposition. Weathering of this material is evident from the stipple speckled b fabric, with rainwater and organic acids contributing to this process. Given the similar basic mineral and organic components of this section to the underlying mineral soil and the midden sample, it is apparent that this deposit is homogenised soil and midden material. The most

likely explanation for this homogenisation is cultivation. Evidence supporting cultivation is provided by the thin juxtaposed silt and organomineral layer pedofeatures. These are formed by surface micro-erosion and downward translocation (slaking) and are often characteristic of arable soils with periodically bare cultivation surfaces. As with other soils of the Neolithic/early Bronze Age landscape soil wetness conditions would have become increasingly prevalent, despite the raised ground surface, as evidenced by diffuse mottles of up to 2 mm diameter and the very few diatoms. Alternatively, the diatoms could have been introduced with the anthropogenic sediments. Related to wetness is soil workability which would have become moderately difficult with the tools available to the early inhabitants of Tofts Ness.

The vughy micro-structure, together with enhanced levels of both basic organic components and excremental pedofeatures, in Sample V19 confirms the midden origin of this material. Although the basic organic components are poorly preserved, inorganic materials of biological origin remain to provide some indication of the midden's organic composition. Calcium oxalate spherulites and a very few fragmented phytoliths, common throughout this thin section, indicate the faecal material of herbivorous domestic animals. The other coarse mineral material associated with inorganic materials of biological origin was most likely ingested from the soil surface by livestock. If this were the case then domesticated animals were grazing vegetation growing on soils developed on glacial drift and calcareous sands. In view, however, of the volume of calcareous sands in thin section it is also likely that wind blown deposits have contributed to the calcareous sand content. Ash material is the second major component of this midden, evidenced by the red fine material obvious in oblique incident light. In plane polarised light this material is greyish brown and brown with crystallitic stipple and mosaic speckled b fabrics. It is also associated with a very few fragmented phytoliths and fine charcoal material. Quartz grains are frequently embedded in the red fine material indicating that the most likely source of the fuel material was turf which had formed on soils developed from glacial drift and that burning temperatures were less than 500°C. There is no evidence of laminated micro-depositional patterns in this thin section, which are often seen in midden deposits. The high level of biological activity and/or slow midden deposition rates with prolonged surface disturbance has removed any trace there might have been of micro-deposition.

Soils of the late Bronze Age/Iron Age landscape

Sample V1913 represents two phases of moderately sorted calcareous wind blown sand deposition, unmodified by human activity. The sands in both parts of the thin section have been partly weathered resulting in an intergrain micro-aggregate structure. The lower part of the section is less weathered than the upper part; these are characterised by few granular intergrain peds and frequent sub-angular blocky to granular intergrain peds respectively. A more concentrated zone of biological activity between the upper and lower parts of the section is evidence that some degree of sand stabilisation did occur.

Two distinctive processes are associated with Sample C1694. One is the deposition of moderately sorted wind blown calcareous sands which form the inorganic parent material. The second is cultivation activity, with evidence of both tillage and manuring. Secondary processes are weathering of the wind blown sand material and biological decomposition of organic material. The sand material is dominated by sub-rounded and sub-angular calcium carbonate with few single quartz grains, compound grains and rock fragments. The size range of these basic mineral components is between 60 µm and 750 µm. The origin of this material, and how far it may have travelled, is unknown but in view of the section's uniformity it is likely that deposition was over a short period of similar wind conditions. Cultivation may, however, have served to remove any evidence of periodicity in the deposition of windblown sand. Chemical alteration of the calcium carbonate has resulted in a stipple speckled b fabric of the groundmass together with minute dispersed residues. Rainfall and dissolved organic acids from organic sources would have contributed to calcium carbonate weathering. The presence of fine organomineral material with fine charcoal gives an enualic related distribution to the groundmass. This, together with enhanced amounts of quartz, indicates that the wind blown sand has had additional material applied to it.

Additions appear to have included ash and manure; stipple speckled and mosaic speckled b fabrics associated with phytoliths are indicative of ash deposition. These same features without phytolith inclusions indicate weathered calcium carbonate material. Reddish brown fine material observed in oblique incident light also points to the addition of a fine ash material to this sandy soil. Evidence for manure application comes from the amounts of quartz, which are greater than unmodified sands (V 1913) and similar to the amounts of quartz found in the midden material (V 19). This material would have contained a significant organic content as evidenced by the enhanced biological activity seen in the thin section. Any structural remains of organic material that might have been included in additions have been decomposed by meso-fauna and micro-organisms. There are enhanced numbers of excremental pedofeatures many of which have been degraded by micro-organism activity. The uniform distribution of thin section features suggest a sand well-mixed by tillage implements. Organic additions and the weathering of calcium carbonate sand resulted in a loamy sand texture and, almost certainly, enhanced nutrient status. A loamy sand texture improves the available water capacity by around 6% compared to sand while still retaining a soil of easy workability for cultivation implements. Even allowing for this improvement in available water holding capacity it is likely that there would have

been a soil moisture deficit in some summers. Erosion would have remained the most significant risk to arable activity. Sand blow may have been reduced by the use of organic additions and by enclosures, otherwise it is likely that there would have been loss of seeds and seedlings together with abrasion of remaining plants. The limitations of drought and erosion notwithstanding, management of this soil by the early inhabitants of Tofts Ness turned it from a soil that was unviable for cultivation to one that was marginally viable.

One thin section, Sample X1, was examined from a dark brown soil sandwiched in bands of sand around Mound X. The context of this soil, between bands of sand, clearly indicates that it is anthropogenic in origin, purposely created by the movement of inorganic and organic soil forming materials from one part of the landscape to another. The frequency of single mineral grain quartz is consistent with the quartz frequencies of soils developed on glacial drift deposits elsewhere in Tofts Ness (see C2033, V1921, W1). This suggests that turves have been stripped from mineral soils, transported and used to form the basis of the anthropogenic soil around Mound X. Apart from the vegetation associated with the turves, midden material provided the basic organic components of this soil. Organic material is poorly preserved but the range and frequency of the remaining material is consistent with a midden origin. Enhanced frequencies of excremental pedofeatures and the presence of a burnt bone fragment also supports a predominantly midden origin for the organic components. The predominantly homogenised nature of the soil in this thin section suggests a substantial level of disturbance. Soil structure is not, however, highly perforated, suggesting that mechanical disturbance by cultivation rather than high biological activity was responsible for soil homogenisation. The limited heterogeneity that does exist further indicates cultivation activity. These features are the small areas of complex packing voids, within which are ultra fine granules, and a very few non-laminated incomplete silt infills. These indicate fragmentation of artificial clods by cultivation implements and slaking of a periodically exposed soil surface. Moderately sorted wind blown sand material is common throughout this sample, contributing to the inorganic materials forming this soil. Some of these sands may have been introduced with turf and midden material but it is more likely that wind erosion and deposition of sands was ongoing during the formation and cultivation of the soil. Indeed, given the context of this soil, it is evident that without the anthropogenic soil forming process it is unlikely that any form of arable activity would have been possible in this area. The anthropogenic soil would prove less susceptible to soil erosion and have greater soil water retention during the growing season than the surrounding sands. Eventually however, these efforts, as with other land management efforts by the early inhabitants of Tofts Ness, were covered by wind blown sands.

CONCLUSIONS

Analysis of soil thin sections from Tofts Ness has demonstrated a range of land management activity associated with early occupation (Figure 13.1). Land management activity within the Neolithic/early Bronze Age landscape, as evidenced in Samples C2033 and V1921, included vegetation clearance and a short period when the soil

surface was exposed. Similar micromorphological features are evident in Sample W1 but here there has been greater movement of silty material suggesting a greater level of soil disturbance which can be related to tillage. In none of these samples is there any evidence of additions to the soil. This is in contrast to Sample C2037 where midden material was applied and incorporated into the mineral soil. Based on the evidence of Sample V19, this midden material was

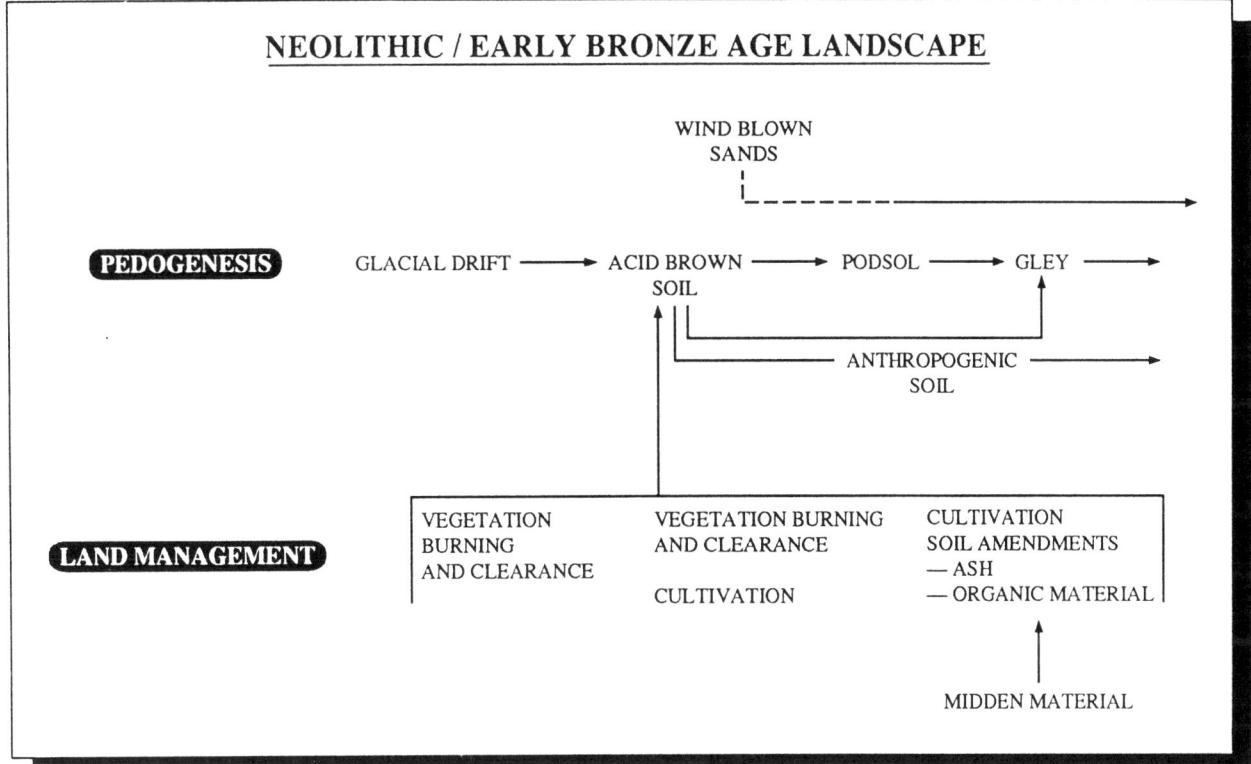

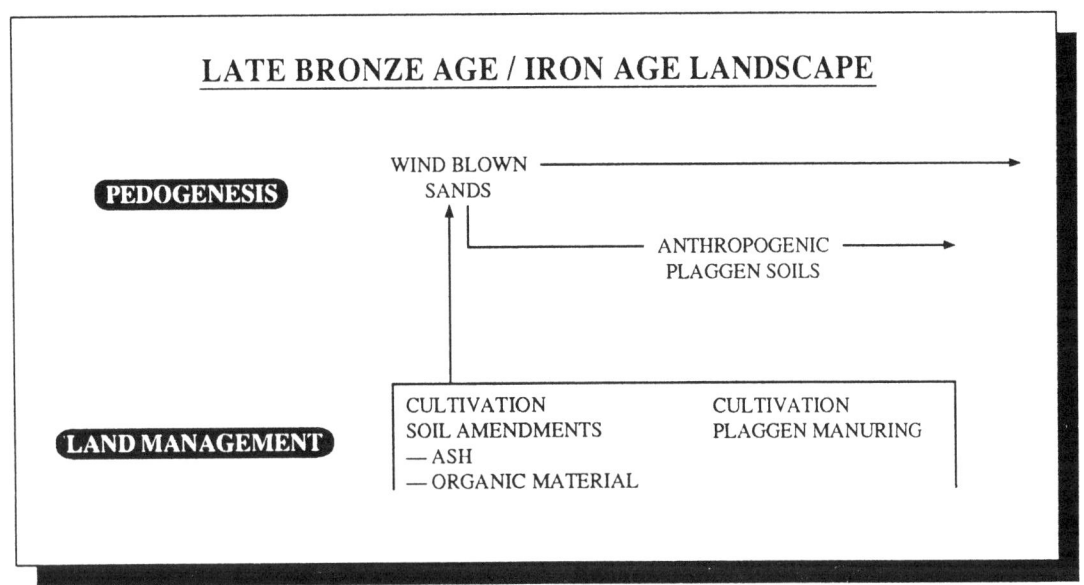

Figure 13.1 Tofts Ness thin section analyses: summary interpretation.

dominantly herbivorous faecal material and ash. In Samples W1, C2037 and V19 there is incorporation of wind blown sands into the matrix of the soil, suggesting that wind erosion of calcareous sands was significant well before the major period of sand deposition commenced during the late Bronze Age. Redistribution of iron in thin section suggests that soils became increasingly wet over this time period.

Although the chronology of occupation in the Tofts Ness area is not yet sufficiently refined, it is tempting to postulate evolution of land management during the Neo-lithic/early Bronze Age period from simple vegetation clearance, through a greater intensity of tillage, to the use of midden material and intensive tillage on arable soils. This would appear to coincide with deteriorating environ-mental conditions in the form of increasing sand blow and soil wetness and may also be related to decline in soil nutrients and organic status resulting from tillage. Thus, while soils of the Neolithic and early Bronze Age were not inherently marginal for arable activity, there is evidence of development towards marginality as a result of changes in the natural environment as well as tillage activity. Midden material was then used on arable soils to overcome this drift towards soils of poorer arable quality. Whether this greater land management activity also coincided with increasing population pressure may become apparent from current archaeological analyses. The alternative hypothesis to that advanced above is that these three land management activities were contemporaneous and represent different parts of an organised land economy.

Attempts were also made to manage the highly marginal wind blown sand deposits of the late Bronze Age and Iron Age landscape for agricultural purposes. Within Sample C1694 midden material was applied to the sand deposits which were cultivated. This form of soil enhancement would have helped to stabilise the sand, improve soil water holding capacity during the growing season and enhance soil nutrient status. In Sample X1, between sand deposits, there is evidence of plaggen manuring. Turves were removed from one part of the landscape, transported and deposited in this arable area along with midden material. The result of this practice was to create an artificially improved soil within an area of sand deposits thus allowing agricultural activity in this area.

Late Bronze and Iron Age land management clearly required the careful management of the wind blown sands to maintain arable activity. Application of midden material to arable soils was probably a familiar technique given its occurrence in the earlier Tofts Ness landscape. Plaggen manuring, however, represents a new innovation in the Iron Age and thus places this anthropogenic soil as one of the earliest plaggen soils in north west Europe. Clearly these land management practices during the late Bronze Age and Iron Age turned an environment that was unviable for crops into one that was just capable of supporting arable activity until the abandonment of the area.

ACKNOWLEDGEMENTS

Thanks to Steve Dockrill for collecting the samples; Mrs M Macleod for thin section preparation; Mrs I Mack and Mrs M McDonald for secretarial support and Bill Jamieson for cartography.

REFERENCES

Bullock, P, Federoff, N, Jongerius, A, Stoops, G, Tursina, T & Babel, U 1985 *Handbook for soil thin section description.* Wolverhampton: Waine Research Publications.

Courty, M A, Goldberg, P & Macphail, R I 1989 *Soils and micromorphology in archaeology.* Cambridge: Cambridge University Press.

Davidson, D A, Lamb, R & Simpson, I 1983 'Farm mounds in north Orkney: A preliminary report', *Norwegian Archaeological Review*, 16, 39–44.

Dockrill, S J 1986 *Excavations at Tofts Ness, Sanday. Interim.* Bradford: School of Archaeological Sciences, University of Bradford.

Dockrill, S J 1987 *Excavations at Tofts Ness, Sanday. Interim Report.* Bradford: School of Archaeological Sciences, University of Bradford.

Hunter, J R, Dockrill, S J, Bond, J M & Smith, A N (in preparation) *Archaeological investigations in Sanday.* Edinburgh: Monograph of the Society of Antiquaries of Scotland.

Lamb, R G 1980 *An archaeological survey of two of the north isles of Orkney; Sanday and North Ronaldsay.* Edinburgh: Royal Commission on the Ancient and Historical Monuments of Scotland; HMSO.

Macphail, R I, Courty, M A & Gebhardt, A 1990 'Soil micro-morphological evidence of early agriculture in north west Europe', *World Archaeology*, 22, 53–69.

Murphy, C P 1986 *Thin section preparation of soils and sediments.* Berkhamsted: AB Academic Publishers.

Romans, J C C & Robertson, L 1975 'Some genetic characteristics of the freely drained soils of the Ettrick Association in East Scotland', *Geoderma*, 14, 297–317.

Simpson, I A 1985 *Anthropogenic sedimentation in Orkney: The formation of deep top soils and farm mounds.* Unpublished PhD thesis, University of Strathclyde, Glasgow.

14. The use of peat and other organic sediments as fuel in northern Scotland: identifications derived from soil thin sections

Stephen Carter

Abstract

This paper describes the use of soil thin sections as an aid to the identification of fuels used on two prehistoric sites in northern Scotland. A variety of organic sediments occur in this area but they have rarely been identified from archaeological excavation. These results show that thin sections can be used to identify carbonized fuel fragments and mineral ash from different organic sediments.

INTRODUCTION

The availability of fuel is an important factor in the limiting of human settlement and activity throughout the world. A shortage of fuel can occur whenever human demand exceeds supply but some areas are more susceptible than others. In northern Scotland environmental factors have placed severe limits on tree growth throughout prehistory. As a result, man has been forced to exploit a variety of alternative fuels including peat, turf and other organic matter rich sediments. These alternative resources have not been uniformly available, either spatially or temporally, and therefore fuel supply has been a variable constraint on human settlement.

Until recently, the identification of these fuels from archaeological contexts has generally been restricted to the recognition of 'peat ash', a catch-all term that disguises the potential variation. This paper presents examples of how soil thin sections can provide information about the types of fuel used in the past by analysis of the carbonized and ashed residues of burning. More precise identifications of fuel will lead to a better understanding of fuel availability and exploitation.

ORGANIC SEDIMENTS IN NORTHERN SCOTLAND

Organic sediments that are potential fuels may be found in any of four general situations, limited by climatic or topographic variables:

Freshwater organic sediments

These include organic lake muds and fen peats. They are restricted to well defined basins and may be difficult to exploit until the basin has largely infilled. They will always have had a limited distribution but became more accessible later in prehistory.

Climatic peats

These are acid peats that accumulate in a high water-table maintained by high precipitation and humidity rather than groundwater. They include blanket and raised bog peat and are restricted to areas with a suitable climate. They have become increasingly abundant through the Flandrian as a result of soil and climate change.

High groundwater soils

These include peaty and humic gley soils which have a surface horizon rich in organic matter. Their distribution is determined by topography and climate so they grade into freshwater sediments and climatic peats. Gley soils are most common in areas with poorly draining subsoils and a low moisture deficit.

Mor humus soils

These include podzolic soils under heath vegetation which have a surface organic horizon. Unlike the humic and peaty gley soils, their distribution is not controlled by high groundwater and they are widely distributed on coarse textured acidic soil parent materials. They have probably become more abundant through prehistory as heath vegetation has become more widespread, a process linked to soil development, human intervention in the landscape and climate change.

THE IDENTIFICATION OF FUEL RESIDUES IN THIN SECTION

Soil thin sections are 30 μm thick samples of a sediment. They preserve the organic and mineral components in their true relative positions, and this arrangement creates the fabric and microstructure of the sediment. The identification of fuel residues depends largely on analysis of the components rather than the fabric or microstructure although these will be relevant to an understanding of the formation of the sediment that contains the fuel residues. Existing information about burning and ash in thin section is summarised by Courty *et al* (1989, 105–113) but they make little reference to the types of fuel being considered here.

Fuel is unlikely to survive in an unburnt state in archaeological sediments unless they have remained water-logged. The two types of material likely to be encountered are mineral ash produced by the oxidation of the fuel during combustion and carbonized fuel fragments produced by combustion under reducing conditions. Most of the waste produced under normal conditions will be oxidised so fuel residues should be dominated by ash rather than carbonized fuel. The quantity of mineral ash remaining after complete combustion will vary with the fuel used. Purely organic blanket or raised bog peats may yield less than 5% ash by weight but turves and lake muds could yield in excess of 50%. Mineral components in ash may include both mineral grains and biogenic silica (phytoliths and diatoms). Carbonized organic components may include identifiable plant tissues (stems, roots, fungal spores) and more or less unidentifiable amorphous material. Mineral components may be embedded in carbonized fragments.

EXAMPLES

Recent excavations at two sites in the Northern Isles of Scotland have presented the opportunity to examine fuel residues in thin section: these are at Upper Scalloway in Shetland and St. Boniface Church in Orkney.

Upper Scalloway (HU 406 399)

An Iron Age settlement was discovered during the building of houses and rescue excavation followed immediately in the winter of 1989/90 (Sharples 1990). Up to 2 m of sediments were recorded in a service trench on the south side of the site. These contained frequent fragments of carbonized material and the sediments were interpreted in the field as containing 'peat ash'. Three soil thin sections were prepared from them.

The thin sections contained frequent fragments of carbonized peat up to 23 mm across. A range of types could be recognised in the larger fragments based on the presence or absence of mineral grains, diatoms, fungal spores and higher plant tissue residues (these are illustrated in Figures 14.1 to 14.3). Quantification of the mineral content of these fragments (Table 14.1) showed that most derived from purely organic peats (less than 2% mineral grains by area after carbonization) rather than organo-mineral sediments.

The majority of the carbonized material consisted of

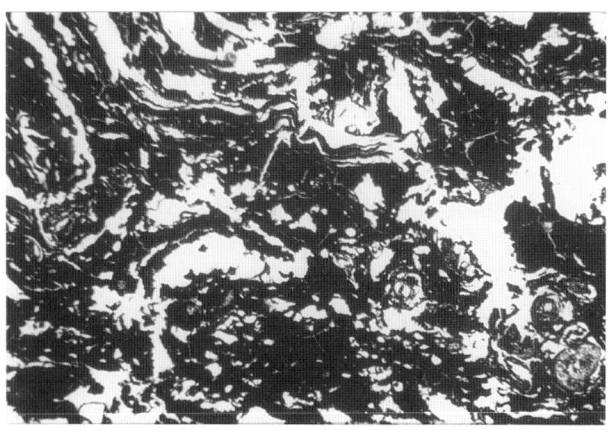

Figure 14.1 Upper Scalloway: Carbonized fibrous peat (width 8 mm).

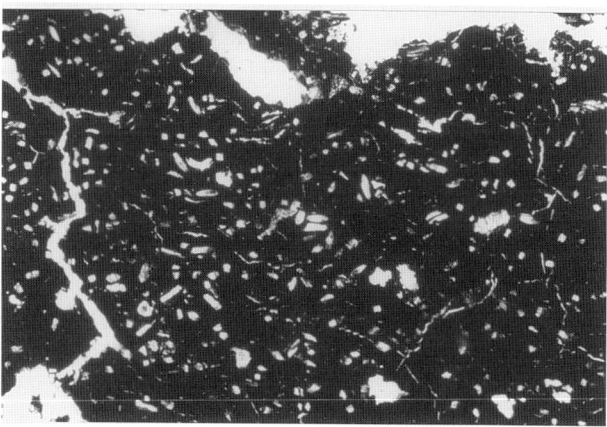

Figure 14.2 Upper Scalloway: Carbonized amorphous peat with pennate diatoms (width 2 mm)

small, amorphous, black fragments (less than 500 μm) with no identifiable components. The microstructure of the sediments was indicative of a high level of invertebrate burrowing (primarily by earthworms) and this had clearly fragmented the carbonized peat.

Carbonized peat made up only a minority of the sediments as a whole which were dominated by sand sized mineral grains (quartz and mica) and larger rock fragments. On their own, these components were not evidence of ash but in incident light the sediment was seen to be reddish brown and highly reflective. This is typical of a sediment that has been burnt and indicates that a significant proportion of the mineral components are derived from fuel ash.

St. Boniface Church (HU 487 527)

The site at St. Boniface Church, Papa Westray, Orkney was partially excavated in 1990 because of coastal erosion

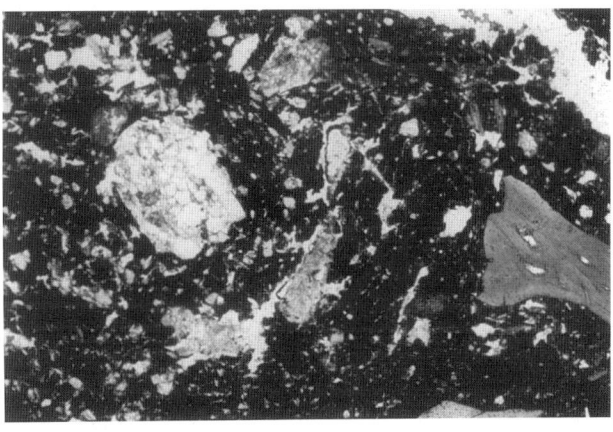

Figure 14.3 Upper Scalloway: Carbonized amorphous peat with abundant mineral grains (width 4 mm)

(Lowe 1990). It is a complex Iron Age settlement mound partially overlain by a deep accumulation of sediments of Late Iron Age and Norse date; these sediments, at least 3 m deep, were found to consist almost completely of fuel ash. A series of ten thin sections were prepared spanning all of the major stratigraphic units in this accumulation.

A range of types of carbonized fuel was present, similar to that found at Upper Scalloway, but the proportions of the different types showed great variation. This is illustrated in Tables 14.1 and 14.2 with data for percentage mineral grains and maximum mineral grain size in carbonized fragments. The thin sections are arranged on the tables in

	Thin Section	Number of Fragments	Mineral Content		
			<2%	2–10%	>10%
Upper Scalloway					
	130	112	82	10	8
St. Boniface					
	111	44	25	39	36
	102	47	12	47	41
	103	58	25	53	22
	104	27	41	44	15
	105	55	51	36	13
	106	28	75	25	0
	107	11			
	108	42	64	36	0
	109	27	82	7	11
	110	7			

Table 14.1 Percentage by area of mineral content in carbonized fuel fragments larger than 1 mm (percentage in each abundance class). Gaps in the table reflect insufficient data in a thin section.

Thin Section	Maximum Grain Size				
	None	50 μm	250 μm	500 μm	2 mm
111	2	16	68	14	0
102	6	9	68	13	4
103	3	11	79	7	0
104	0	26	59	15	0
105	15	27	42	16	0
106	4	46	50	0	0
107					
108	32	22	39	7	0
109	26	33	33	8	0
110					

Table 14.2 St. Boniface: maximum grain size of mineral components in carbonized fuel fragments larger that 1 mm (percentages in each size class). Gaps in the table reflect insufficient data in a thin section.

stratigraphic order and the data reveal consistent trends with the fragments becoming increasingly mineral rich and the mineral grains increasing in size up through the stratigraphy (Figures 14.4 and 14.5). This pattern may indicate a change from dominant pure organic peat in Thin Section 109 to dominant turf or organic sediments by Thin Sections 102 and 111.

As at Upper Scalloway, the sediments were dominated by sand sized mineral grains which cannot be conclusively linked to fuel ash. However, the St. Boniface sediments also contained notable concentrations of biogenic mineral components: phytoliths, diatom frustules and chrysophyte cysts (cysts of certain unicellular freshwater algae). It is thought that these derive from organic sediments and they have been concentrated by combustion. The relative abundance of these components is shown in Table 14.3 and a clear distinction can be made between upper part of the stratigraphy (Sections 111 – 105) and the lower (Sections 106 – 110). This could result from the use, later in the sequence, of organic freshwater muds rich in phytoliths, diatoms and chrysophytes.

DISCUSSION

The two examples presented above illustrate the information that soil thin sections may provide about fuel types. The potential for geographical and temporal variation in fuel availability is realised in the examples and the differences between the two sites are consistent with our understanding of contemporary fuel resources. Shetland had extensive areas of blanket peat by the Iron Age, providing a source of pure organic fuel. Papa Westray in Orkney has no blanket peat at present and the results from St. Boniface appear to show the shift from peat (in the strict sense) in the late Iron Age to other fuels (perhaps including turf and freshwater organic sediments) in the Norse period.

The analysis of carbonized fuels in thin section offers a number of potential advantages over the identification of carbonized plant remains from hand specimens. Firstly, in thin section it is possible to categorise fused or partially humified organic matter that does not separate readily into identifiable components. Secondly, small fragments (less than 500 µm) that are unlikely to be recovered by hand, but may constitute the bulk of the carbonized material, can be examined. Thirdly, different types of carbonized material can be quantified on the thin section. Despite these potential advantages of thin section analysis, the specific identifications of plants gained from macrofossil analysis provide valuable information about the source of the fuel (Camilla Dickson, this volume). Clearly, the two techniques are offering complementary information about fuel type.

In the identification of fuel ash a more fundamental problem is encountered as it seems that the bulk of the mineral residues are ubiquitous materials like quartz grains. It is therefore difficult to demonstrate that a sediment is

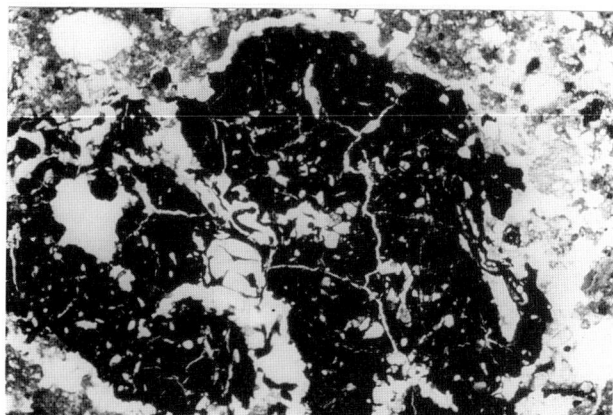

Figure 14.4 St. Boniface: Carbonized amorphous peat with few mineral grains (width 4 mm)

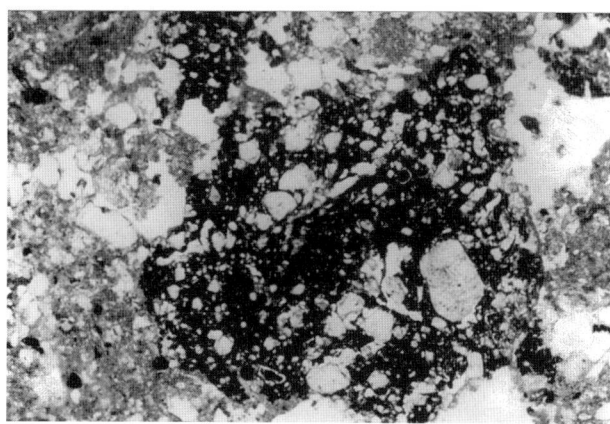

Figure 14.5 St. Boniface: Carbonized amorphous peat with abundant mineral grains (width 4 mm)

Thin Section	Phytoliths	Diatoms	Chrysophytes
111	**	**	*
102	***	**	*
103	***	*	**
104	***	**	**
105	**	*	*
106	*		
107	*		
108	*		
109	**	*	
110	**	*	

*Table 14.3 St. Boniface: relative abundance of three types of biogenic silica (* = few, ** = frequent, *** = abundant).*

ash at all. At very high temperatures silica will melt, producing a distinct vitrified silicate ash, but in the absence of this, the colour of fine mineral material in reflected light and the abundance of biogenic silica may indicate the presence of ash. Measurement of the magnetic properties of a sediment should confirm that it has been heated. Similarly, detailed identification of the biogenic silica may help to locate the source of the fuel from which it derives.

CONCLUSION

Soil thin sections are a useful technique for the study of fuel residues derived from organic sediments. It is possible to identify different types of fuel and propose sources for them. Future improvements of the technique will come from parallel analysis of carbonized plant remains, diatoms and phytoliths, and the preparation of reference thin sections from known sediment types.

ACKNOWLEDGEMENTS

The excavations and thin section analysis discussed in this paper were funded by Historic Scotland. I would like to thank the project directors, Niall Sharples (Upper Scalloway) and Chris Lowe (St. Boniface), for assistance with interpretation of these sites.

REFERENCES

Courty M A, Goldberg P & Macphail R 1989 *Soils and micromorphology in archaeology*. Cambridge: Cambridge University Press.

Lowe C E 1990 'St. Boniface Church', *Discovery and Excavation in Scotland* (1990), 45–47.

Sharples N 1990 'Upper Scalloway', *Discovery and Excavation in Scotland* (1990), 48–49.

15. Past uses of turf in the Northern Isles

Camilla Dickson†

Abstract

Well preserved turf fragments from a post medieval turf roof in Shetland proved rich in macroscopic remains. The habitats of the plants indicated that the turves had been cut from grass and/or heathland. Similar plant remains of Late Iron Age date suggest that turf was also used to supplement peat for fuel in Orkney. Present day heathy turf from Orkney contained similar plant assemblages to those found in the hearths. Recent uses for turf in the Northern Isles suggest possible ways in which turves were used in the past.

INTRODUCTION

This paper sets out to show that macroscopic plant remains may demonstrate the past uses of turf. At present this evidence comes from Orkney and Shetland, the Northern Isles of Scotland. Turf has been widely used in historic times especially in areas where wood, peat or compost were in short supply. Only recently has turf been recognised in Scottish archaeology. Stephen Carter has given the evidence from soil thin sections, also in this volume. The identification of seeds and other plant remains, both burnt and unburnt, may indicate not only that turf was used but also show the type of vegetation that the turf was cut from and possible uses for the turf.

Turf is a surface soil held together with roots of grasses and other plants. It can be largely mineral or mainly organic in composition. Turf was cut from heathy ground and also dry clay ground. This scalping of the fertile ground surface led to soil impoverishment and sometimes erosion and is generally no longer practised. The cutting techniques, sizes and shapes of turves for roofing in the Northern Isles are described by Fenton (1978).

The use of turf for building has continued up to the present time in Orkney and Shetland. As a building material turf is versatile, useful for infilling odd shapes such as those at gable ends as well as for roofing. Turf was formerly used over roofing timbers and usually covered with a thatch of oat straw. In Scandinavia and the Faroe Islands an impermeable layer of birch bark was laid under the turves. In Orkney thin flagstones were sometimes similarly used. The turf then became a living thatch and grasses, more tolerant of extremes of temperature and drought than other plants, would become the dominant vegetation. In the Faroe Islands this grassy roofing is even cut and used as fodder.

When a house roofed in this way fell into disuse the roof would fall in and the turves disintegrate. In time soil biota would attack the organic matter and the binding roots and delicate grass seeds would decay away. The only recognisable remains could be the mineral components of the soil with the turves no longer recognisable as such.

ROOFING TURF

In Shetland and also frequently in Orkney, turf which was covered with a thatch of oat straw could keep its original seed bank and the more durable seeds and other plant remains survive. The following example from Shetland illustrates this. The remains of a building, probably a teind barn, in use from the sixteenth to eighteenth centuries, was excavated at Kebister on the Shetland mainland. From the house debris a turf block measuring 24 x 20 cm and 5 cm thick was recovered with other turf fragments and associated soil. Pieces of wood, assumed to be roofing purlins (the main cross pieces), were also recovered. The whole was interpreted as part of a collapsed roof.

HEATH AND GRASSLAND			
Bryum sp. (1)	o–r	cf *Hypnum cupressiforme* (1st)	o–r
Calluna vulgaris (1)	o–r	*Juncus bufonius* (s)	r–fq
Carex binervis (n)	o–r	*J. effusus* (s)	o–oc
C. demissa/lepidocarpa (n)	o–r	*J.* subg. *Septati* (s)	o–ab
C. dioica (n)	o–r	*J. squarrosus* (s)	oc–fq
C. cf *echinata* (n)	o–r	*Molinia caerulea* (e)	o–fq
C. flacca (n)	oc–ab	*Montia fontana* (s)	o–r
C. nigra (n)	o–r	*Polytrichum* cf *commune* (1st)	o–oc
C. panicea (n)	o–r	*Potentilla erecta* (a)	o–oc
C. pilulifera (n)	r–fq	*Rhytidiadelphus squarrosus* (1)	o–r
cf *Danthonia decumbens* (c)	o–r	*Rumex acetosella* agg. (n)	o–r
Eriophorum sp. (n)	o–r	*Selaginella selaginoides* (m)	oc–ab
Eurhynchium praelongum (1)	o–r	*Sphagnum* sp. or spp. (1)	o–oc
Hylocomium splendens (1st)	o–r	*Thuidium tamariscinum* (1st)	o–fq
ARABLE AND WASTE PLACES			
Spergula arvensis (s)	o–r	*Urtica dioica* (n)	o–r
Stellaria media (s)	o–r		
MARITIME TURF			
cf *Puccinellia maritima* (c)	o–r	*Sagina* cf *maritima* (s)	o–r
WOOD OR CHARCOAL			
Betula (w)	o–r	cf *Quercus* (ch)	o–r
cf *Calluna* (w, ch)	o–r		

KEY: o, absent; r, rare; oc, occasional; fr, frequent; ab, abundant; a, achene; c, caryopsis; ch, charcoal;
e, epidermis; fst, fruitstone; l, leaf or leafy; m, megaspore; n, nutlet; s, seed; st, stem

Table 15.1 Turf fragments and associated soil from 16th–18th century roofing at Kebister, Shetland.

The turf consisted of amorphous organic matter and mineral material with earthworm egg capsules present in some abundance in three out of four samples. This suggested that earthworms may have been largely responsible for breaking down the plant material. The mineral fraction was composed of clay, silt, sand and gravel.

The list of plant taxa from a small portion of the turf and other turf fragments together with associated soil is shown in Table 15.1 (tabulated in full in Dickson *in* forthcoming a). The nomenclature follows Clapham *et al* (1987) for the flowering plants and Smith (1978) for the mosses. The term seed is used for both fruits and seeds in this paper. None of the seeds were burnt. The species richness of the turf is similar to that of the present vegetation. Nearby limestone produces lime-rich drainage water which maintains a herb-rich grassland at the present time. The present day flora has been investigated by Dr Sandra Nye (in Owen & Lowe forthcoming) and includes dry to wet heath and dry to wet grassland.

Grass species are not well represented in the turf fragments; only the tough bases of *Molinia caerulea* (purple moor-grass) stems and rare seeds of cf *Danthonia decumbens* (cf heath grass) were recognised. The delicate

leaves of mosses were only found in one exceptionally well-preserved turf fragment; moss leaves would probably usually decay away as would the delicate seeds of grasses.

Juncus bufonius (toad rush) and *J. effusus* (soft rush), which represent damp places, are present as well as *Juncus squarrosus* (heath rush), a plant of heathy habitats. There were abundant megaspores of *Selaginella selaginoides* (lesser clubmoss) which commonly grows in damp pasture on Shetland. *Selaginella*, with megaspores which seem very resistant to decay, is frequently found in archaeological contexts on Shetland. The presence of eight species of *Carex* (sedge) nutlets is particularly interesting. Five of these species are recorded from heavily grazed fen on Shetland (Spence 1979, Table 10) but none of these sedges are aquatic species. All are species of grass and/or heathland of varying degrees of dampness. The mosses could well have grown in similar habitats. Seeds of arable land and from maritime turf are not surprising inclusions. Wind is a constant feature of the Shetland climate and the site is both near to former habitations and to the sea. Similarly the tiny wood and charcoal fragments probably blew onto the turf from other parts of the settlement. Conditions seem to have been particularly favourable for the preservation of these plant remains but more usually moss leaves would decay away along with less durable seeds such as the seeds of most species of grasses. The more resistant seeds such as those of *Potentilla erecta* (common tormentil) and *Carex* species and megaspores of *Selaginella* would be among the more durable remains.

TURF IN HEARTH AND FLOOR CONTEXTS

Evidence for more ancient turves comes from Howe in Orkney, a habitation site in use for about a thousand years, from the Early to Late Iron Age, the latter being the equivalent of the Pictish period (Dickson 1994). At Howe, hearth contexts of Late Iron Age date produced a number of taxa of grassland and heathland plants in addition to occasional barley and arable weed seeds; all these remains were burnt.

The results from five contexts, each from a different hearth, are shown in Table 15.2. Only contexts with at least five taxa of grassland or heathland plants or similar evidence have been selected. Heather, present as charcoal, was found in all contexts. *Carex* species are well represented and these sedges are again heath and grassland species. *Danthonia*, *Juncus squarrosus* and *Potentilla erecta* are all heathland plants. Rhizome fragments, probably of grasses or sedges, are present in two contexts as is burnt plant material, unfortunately not further identifiable.

This mixture of heathy plants in hearth contexts could be explained as heathy turf, perhaps used as fuel in addition to peat. This suggestion is strengthened by Fenton's comments (1978, 207). Fenton notes that turf was used in those Orkney islands which were short of peat for fuel. The turves were known as back peats; they were burnt in the hearth to eke out the peat and continued in use until the end of the eighteenth century. It seems probable that heathy turf was dried and used as fuel at Howe. As shown in Table 15.2 burnt peat was also present in three of the hearths; the samples were pre-sieved so any remaining peat ash would have been removed before they were examined for plant remains.

Table 15.3 shows floor contexts with similar plant remains present although burnt peat was only noted from one context. It may be that the presumed heathy turf had other uses or that hearth material may have become scattered.

COMPARISONS WITH PRESENT DAY HEATHS

This was a strong case for the use of turf as fuel. Did such species-rich vegetation still exist on Orkney and was it feasible to collect it for fuel? My husband, Jim Dickson, suggested that this was an opportunity for a visit to Orkney and to carry out some experimental archaeology! Such species-rich heaths can only develop with moderate grazing and are now rather rare on the

	257	494	605	607	784
Calluna vulgaris (ch)	+	+	+	+	+
Carex binervis (n)	–	+	+	+	–
C. hostiana (n)	–	+	–	+	+
C. nigra (n)	+	–	+	+	+
C. panicea (n)	–	–	–	+	–
C. pilulifera (n)	+	–	–	+	–
C. pulicaris (n)	+	–	–	–	–
C. serotina (n)	–	–	–	+	+
Danthonia decumbens (c)	+	–	–	–	+
Empetrum nigrum (fst)	+	–	–	+	–
Juncus squarrosus (s)	–	+	–	+	–
Luzula sp. (s)	–	–	+	+	–
Polygala cf *serpyllifolia* (s)	–	–	–	–	+
P. cf *vulgaris* (s)	–	–	–	–	+
Potentilla erecta (a)	+	–	+	+	+
Rhizome fragment	–	–	+	–	+
Burnt plant	–	+	–	–	+
Peat	–	+	–	+	+

KEY: As for Table 15.1

Table 15.2 Selected hearth contexts with 'heathy turf' taxa, Late Iron Age, Howe, Orkney.

Orkney Islands where the grassland has been improved by reseeding with clover and grasses for grazing. We were fortunate in having the help of the Orkney botanist, Miss Elaine Bullard, who took us to Ward Hill on South Ronaldsay. She showed us two areas of grazed heath. At the bottom of the hill on level ground was an area of wet heath which could only be cut with the mineral soil included and was therefore not very suitable for fuel. On a gentle slope at the top of the hill was an area of dry heath. Using a spade, it proved possible to pare off a largely organic layer of turf bound together with heather and grass roots; when dried it was about 3 cm thick. The opportunity to burn the turves on a peat fire and collect the ashes to examine is still awaited. Meanwhile small quantities of wet and dry heath have been broken up and sieved. Many *Calluna* stems and roots were present and occasional *Carex* rhizomes. Seeds were found of *Calluna*, *Erica cinerea* (bell heather), *Empetrum nigrum* (crowberry), *Carex* species and *Potentilla erecta*, with megaspores of *Selaginella*. This is a similar assemblage to the plant remains in the hearth and floor samples at Howe. Table 15.4 is a list of the flowering plants growing in the vicinity of the wet and dry heath at Ward Hill. *Calluna*, *Empetrum nigrum*, *Erica cinerea*, *Juncus squarrosus*, *Pedicularis sylvatica* (lousewort) and five *Carex* species were also recorded in hearth or floor contexts from Howe. Of the ten species of grasses only *Danthonia*

Camilla Dickson

	250	284	383	564	581	615	644	786	1098	1352
Calluna vulgaris (ch)	+	+	+	+	+	+	+	+	+	+
Carex binervis (n)	+	+	+	+	+	+	+	+	+	+
C. dioica (n)	–	–	–	–	–	–	–	–	+	–
C. flacca (n)	–	+	–	–	–	–	–	+	–	–
C. hostiana (n)	+	–	+	+	cf	+	+	–	–	+
C. nigra (n)	+	+	+	+	–	+	+	–	–	+
C. panicea (n)	–	+	–	+	–	–	–	–	–	–
C. pilulifera (n)	+	–	–	–	–	–	–	–	+	–
C. pulicaris (n)	+	–	–	–	–	+	+	+	–	+
C. serotina (n)	–	–	–	–	–	–	–	–	+	–
Danthonia decumbens (c)	–	+	–	–	–	–	–	+	–	–
Empetrum nigrum (fst)	+	+	–	–	–	+	–	+	–	+
Erica cinerea (s)	–	–	–	–	–	–	–	–	–	+
Juncus squarrosus (s)	–	–	+	–	+	–	+	+	+	+
Luzula sp. (s)	–	–	–	–	–	–	–	–	–	+
Pedicularis sylvatica (s)	–	–	–	–	–	–	–	–	–	+
Plantago lanceolata (s)	–	–	–	–	–	–	+	+	–	–
Polygala serpyllifolia (s)	cf	–	–	+	–	–	–	–	–	–
Potentilla erecta (a)	–	–	+	–	+	–	–	+	+	+
Bryophyta (1st)	–	–	–	–	–	–	–	+	–	–
Rhizome fragment	–	+	–	–	–	–	–	+	–	+
Burnt plant	–	–	–	–	–	–	–	+	+	–
Peat	–	–	–	–	–	–	–	–	–	+

KEY: As for Table 15.1

Table 15.3 Selected floor contexts with 'heathy turf' taxa, Late Iron Age, Howe, Orkney.

decumbens, with its tough resistant seeds, was preserved in the hearth and floor samples. Seeds of some grass genera such as *Festuca* (fescue) germinate quickly and are less likely to be found in the seed bank. Nevertheless the similarity between the present day heathland plants and those represented in the hearths and floor deposits is a striking one. There can be little doubt that similar heathy turf was burnt at Howe. As indicated from the lists from both Orkney and Shetland it is important to identify *Carex* nutlets to the species level in order to characterise grass and heathland vegetation used for turf. The number of species likely to be represented is not large and confident identification can be made on the nutlet morphology, especially when the cell pattern is observed (Nilsson & Hjelmquist 1967).

PAST USES FOR TURF

Where should we be looking for turf remains? The following list of uses for turf, all of which probably go back hundreds of years, have been extracted from Fenton (1978).

Walling	Composite wall: stone inside with turf core and turf or stone outer walling.
	Lower part stone, upper turf.
	Alternate courses of stone and turf.
Roofing	Turf over timber framework, usually with oat straw thatch.
	Turf over flagstones (Orkney).
Indoors	Seating: turf on low stone wall.
	Turf as back peats on the fire.
Industrial	Turf covering peat charcoal burnt in pits, to delay combustion.
Agricultural	Turf as compost on fields.

As an instance of an industrial use there is possible evidence for the use of turf in peat charcoal production at The Biggins, Papa Stour, a Late Norse site in Shetland. Unburnt seeds of plants associated with damp grassland were found with lumps of burnt peat in pits (Dickson forthcoming b).

Evidence for an agricultural use for turf may be sought in the thick dark layers around old crofts which may in part be composed of turf.

	wet heath	dry heath		wet heath	dry heath
Agrostis canina	+	+	*Juncus squarrosus*	–	+
Anthoxanthum odoratum	+	+	*Linum catharticum*	+	–
Calluna vulgaris	+	+	*Luzula multiflora*	+	+
Carex binervis	–	+	*Molinia caerulea*	+	–
C. flacca	+	+	*Nardus stricta*	+	+
C. nigra	+	+	*Narthecium ossifragum*	+	–
C. panicea	+	+	*Parnassia palustris*	+	–
C. pulicaris	+	–	*Pedicularis sylvatica*	+	+
Danthonia decumbens	+	–	*Pinguicula vulgaris*	+	–
Deschampsia flexuosa	–	+	*Plantago maritima*	+	–
Empetrum nigrum	+	+	*Poa subcaerulea*	–	+
Erica cinerea	+	+	*Potentilla erecta*	+	+
E. tetralix	+	–	*Prunella vulgaris*	+	–
Euphrasia officinalis s.l.	+	–	*Rhinanthus minor*	+	–
Festuca ovina	–	+	*Succisa pratensis*	+	–
F. viviparum	+	+	*Trichophorum cespitosum*	+	–
Holcus lanatus	+	–	*Vaccinium myrtillus*	–	+

Table 15.4 Flowering plants in the vicinity of wet and dry heath samples taken at Ward Hill, South Ronaldsay, Orkney, 1984. Identifications by E H Bullard and J H Dickson

CONCLUSION

Although turf cut from mineral soil and used largely for walling and roofing may decay away leaving mainly mineral components, seeds and other macroscopic plant remains can survive as evinced at Kebister. The more organic heathy turf used as fuel to supplement peats may survive especially in hearths as heather charcoal and burnt seeds of heath and grassland plants along with burnt peat as seen at Howe.

We should be looking for macroscopic plant evidence for the use of turf in all areas where wood and other building materials and fuel were in short supply.

ACKNOWLEDGEMENTS

I am grateful to Miss E H Bullard and to my husband J H Dickson for helpful discussion and assistance with field work on Orkney. This paper was funded by and based on work funded by Historic Scotland.

REFERENCES

Clapham, A R, Tutin, T G & Moore, D M 1987 *Flora of the British Isles*, 3rd edition. Cambridge: CUP.

Dickson, C 1994 'Plant remains', *in* Ballin Smith, B (ed), *Howe – four millennia of Orkney prehistory excavations 1978–1982*, Society of Antiquaries of Scotland Monograph No 9.

Dickson, C forthcoming (a) 'Macroplant report', *in* Owen, O A & Lowe, C E, *Excavation at Kebister, Shetland: the archaeology of a multiperiod settlement site*, AOC (Scotland) Ltd Monograph.

Dickson, C forthcoming (b) 'Plant remains', *in* Crawford, B & Ballin Smith, B, *Excavations at The Biggins, Papa Stour, Shetland.*

Fenton, A 1978 *The Northern Isles: Orkney and Shetland.* Edinburgh: John Donald.

Nilsson, O & Hjelmquist, H 1967 'Studies on the nutlet structure of South Scandinavian Species of Carex', *Bot Notiser*, 120, 460–85.

Smith, A J E 1978 *The Moss Flora of Britain and Ireland.* Cambridge: CUP.

Spence, D H N 1979 *Shetland's living landscape.* Shetland: The Thule Press.

16. Dark Age agricultural practices and environmental change: evidence from Tentsmuir, Fife, eastern Scotland

Graeme Whittington and John McManus

Abstract

Fire has been commonly used to increase the low agricultural potential of marginal land. Deflation of a sand dune system at Tentsmuir, Fife, has revealed an extensive burned layer which provides a calibrated radiocarbon date of AD 424–616. Pollen, particle-size and particle-texture analyses indicate that this layer provides an early example of burning of heathland to improve pasture. However, this burning was apparently not always well-controlled and a severe burning event led to the disruption of the heathland cover and to localised erosion. Paradoxically, the degree of erosion was moderated by the effects of climatic deterioration. Heightened storminess led to the burial of the burnt-over heath by fine, organic-rich windblown deposits from adjacent arable land, encouraging the growth of grasses. This process, eventually, made the former heather management system redundant.

INTRODUCTION

The utilisation of fire as an adjunct to agricultural practices is a world-wide phenomenon but its ultimate aim can be variable. Today the most common use of fire is in the clearance of land for arable agriculture. The current firing in the Amazon region has made the practice notorious, while the swidden and slash and burn systems of the tropical regions have generated, in general, a large body of writing (see, for example, Harris 1972). In Europe the use of fire in clearing woodland has been discussed and reviewed frequently (eg Tolonen 1983) since Iversen (1941) produced his landnam model. As befits an hypothesis emanating from north-west Europe, the majority of convincing work on this topic has come from scholars in that area (eg Huttunen 1980; Tolonen 1985; Vuorela 1986; Sarmaja-Korjonen 1992). For some other areas, the use of this clearing technique has been challenged, as for England, where Rackham (1986) has claimed that burning a felled woodland cover would have been inefficient and impracticable. Fire has also been used to increase the low agricultural potential of marginal land. Perhaps the most celebrated arable form of this is embodied in the *citemene* system of north-east Zambia (Allan 1964; Strømgaard 1985) where weak, nutrient-deficient soils are enriched with ash obtained from the cyclic burning of a lopped or felled arboreal cover. The heathlands of Atlantic Europe also come into this category in a pastoral sense. An overview of burning in such areas is provided by de Smit (1979) while the work of Norwegian scholars (eg Kaland 1986; Øvstedal 1985) gives particular examples. For Britain the practice has been discussed by Gimingham (1971 & 1972).

THE INVESTIGATION SITE

On the western shore of the North Sea in Scotland, there lies between the estuaries of the rivers Tay and Eden an area of windblown sand, covering 1800 ha (Figure 16.1). This area, called Tentsmuir, is well known for its Meso-lithic site at Morton (Coles 1971) and today is largely given over to coniferous plantations and an airfield. One part (National Grid Reference NT 3472 7232) is still occupied by a visible dune system (the Earlshall Muir Site of Special Scientific Interest). It has a vegetation cover which displays a major difference in the distribution of ericaceous species (Figure 16.2). Nearer the sea, where the water table is high, the dune system is uneroded, less

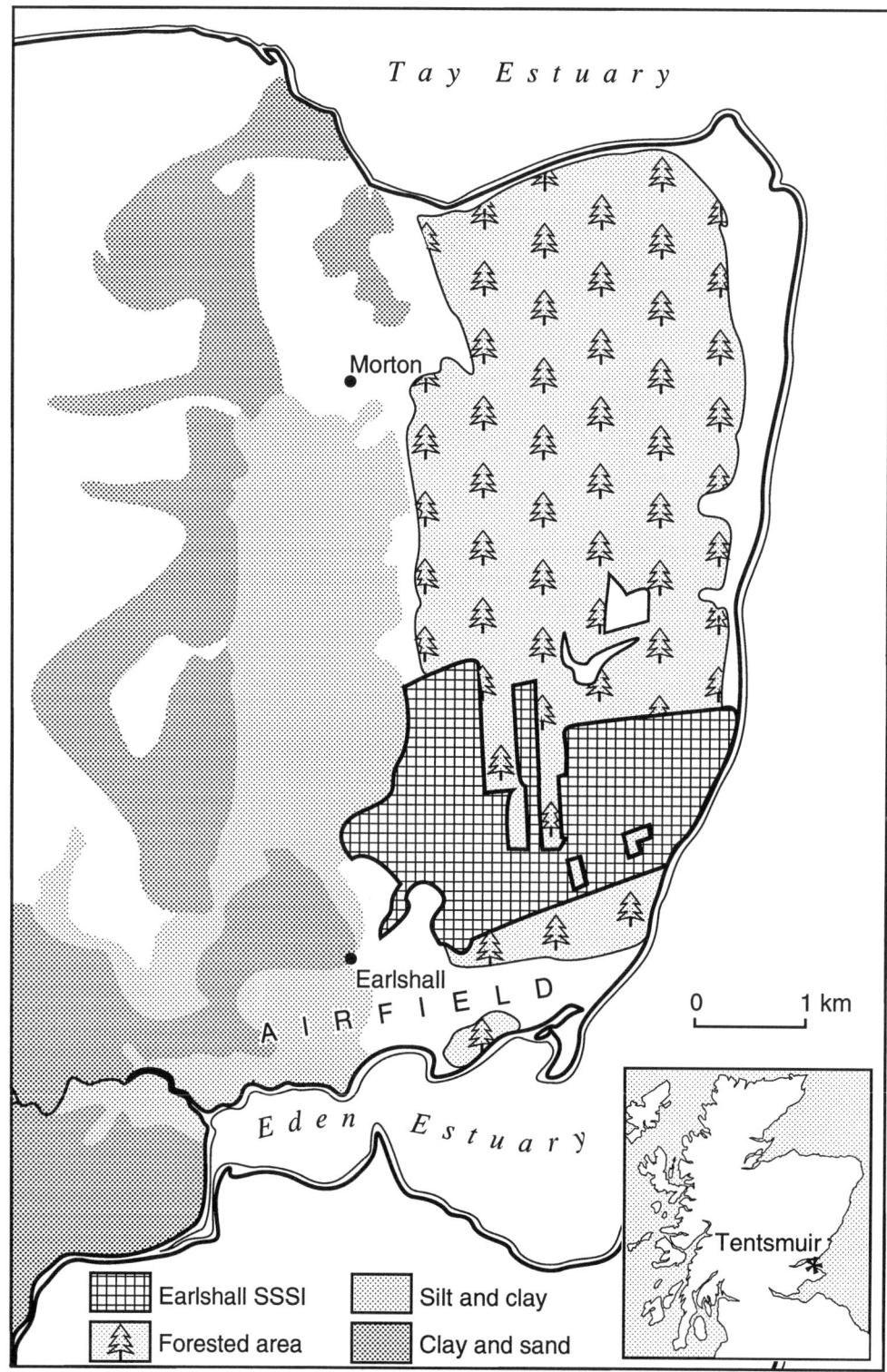

Figure 16.1 The location of Tentsmuir within Scotland, and the study area of Earlshall Muir in its local context.

accidented and supports *Erica cinerea* on the dune crests while the more extensive lower slopes and slack zones are dominated by *Calluna vulgaris* and *Erica tetralix*. Further inland the dune system achieves a greater height, exhibits extensive deflation hollows, has a lower water table and supports a flora dominated by Gramineae and *Salix repens* and from which *Calluna* is largely absent. As a result of

the deflation, a conspicuous band of burned material has been exposed, lying below the present surface at a depth varying between 1.5 and 2.0 m. It can be traced in the unforested zone over an area of more than one km². This burned layer would seem to be related to a former land surface where the vegetation had undergone prolonged and systematic firing. The altitude of the burned zone

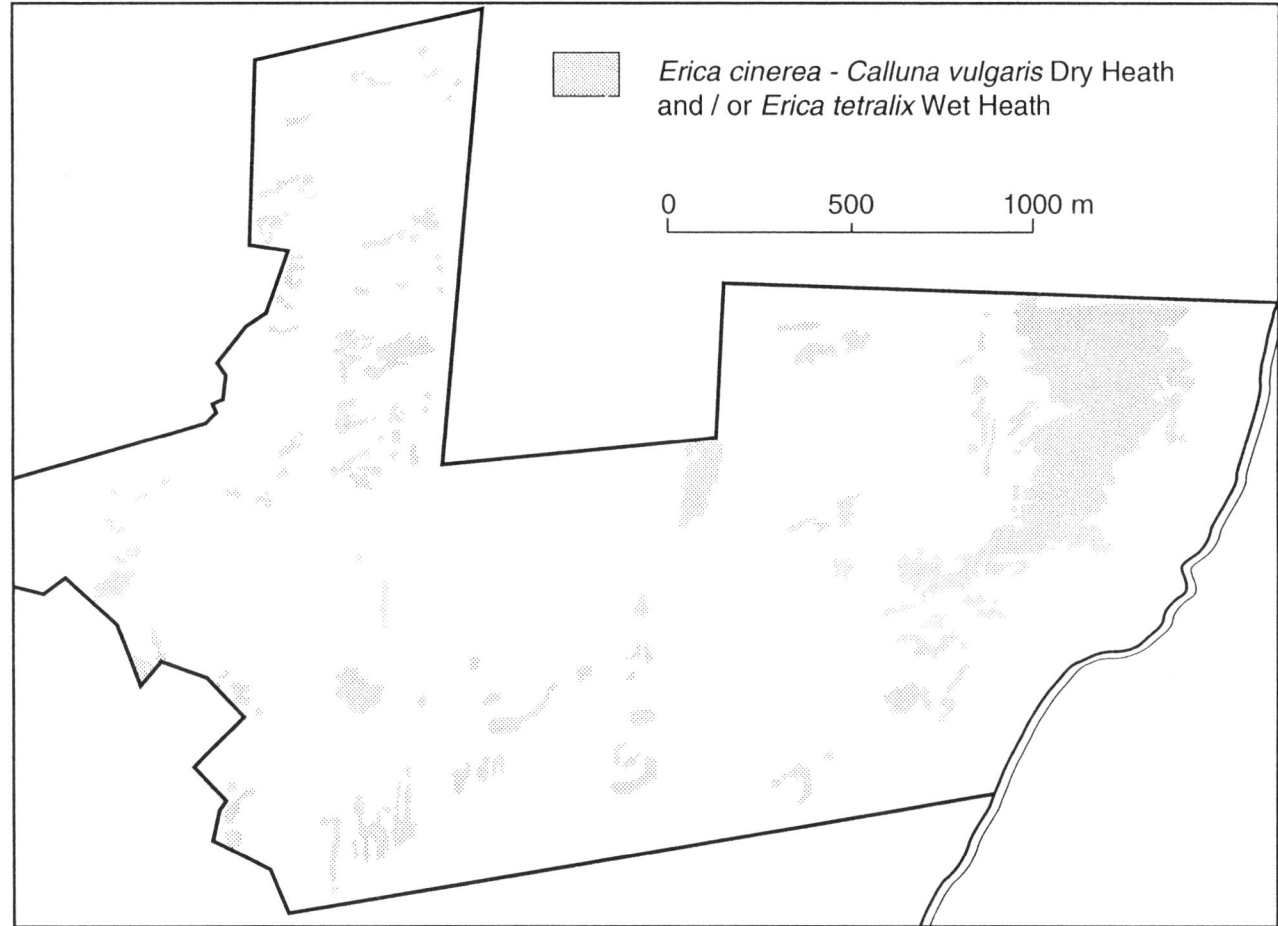

Figure 16.2 The present-day distribution of ericaceous species on the Earlshall Muir SSSI area of Tentsmuir.

approximates to that of the current land surface of the dune system to the east, where *Calluna* is an important constituent of the vegetation cover. Thus it is feasible to hypothesise that the buried former land surface once supported a *Calluna* cover which was burned in the prosecution of an agricultural practice. As Kaland (1986, 36) has said, 'The systematic burning of heath is meaningless if it is not deliberately done to improve the quality of the pasture.'

METHODS

Material was collected from a 20 cm extent of a freshly cleaned exposure, which straddled the burned layer, in overlapping aluminium boxes of 10 x 10 x 5 cm dimensions. In the laboratory sub-samples of 0.5 cm vertical thickness were taken. These were divided into two groups. The first was prepared for palynological investigation using the standard treatments prescribed by Faegri *et al* (1989). Five hundred land pollen grains were counted at each level. Pollen concentrations were calculated by means of the addition of known quantities of the spores of *Lycopodium clavatum* (Stockmarr 1971). During the pollen counting,

material identified as charcoal (Patterson *et al* 1987) was also quantified. This was achieved using, as a standard, a unit of size measuring 400 μm^2. The results of the pollen and charcoal counts are given in Figures 16.3 and 16.4.

The second group of samples was used to provide information on particle-size and particle-nature. The particle-size results are shown in Figure 16.5.

A sample of 2 mm thickness over an extended horizontal distance was obtained from the most dense concentration of charcoal in the section. This was submitted for radiocarbon analysis.

RESULTS

The pollen analysis

Throughout the sampled section, the percentage pollen record is dominated by *Calluna vulgaris* and Gramineae. These two taxa do, however, display a varying relationship and others have a part to play. At the base of the pollen profile the spectrum is dominated by *Calluna* (40% Total Land Pollen or TLP) and that taxon shows a continuous rise over the succeeding 8 cm where it reaches its maxi-

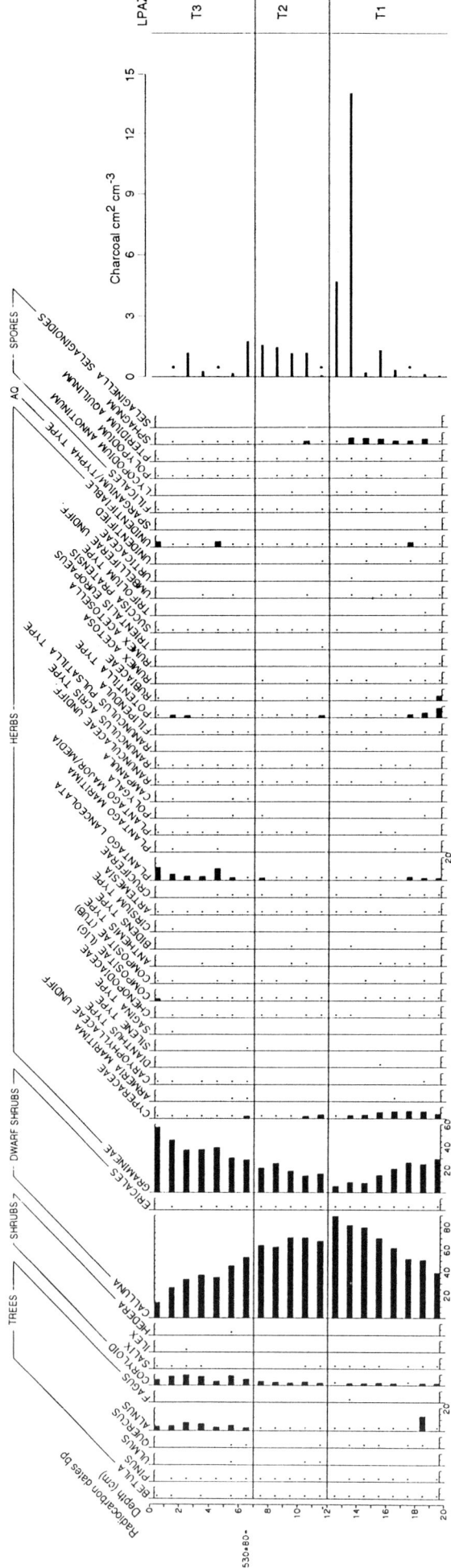

Figure 16.3 Pollen percentage and charcoal concentration diagrams.

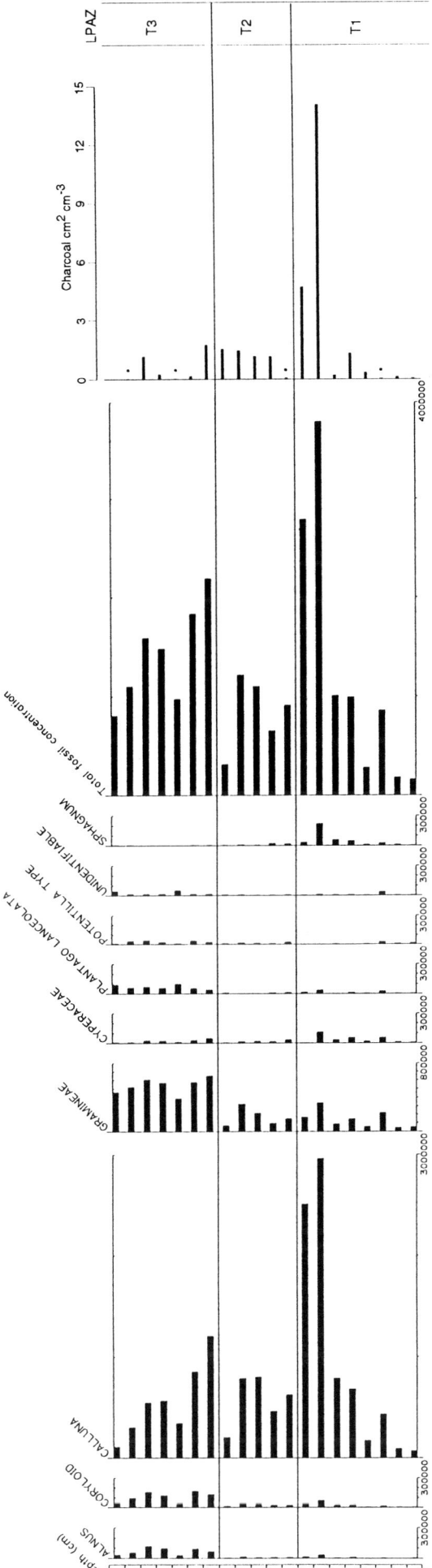

Figure 16.4 Pollen and charcoal concentration diagrams. Note the variable scales on the pollen diagram.

Cumulative percentage

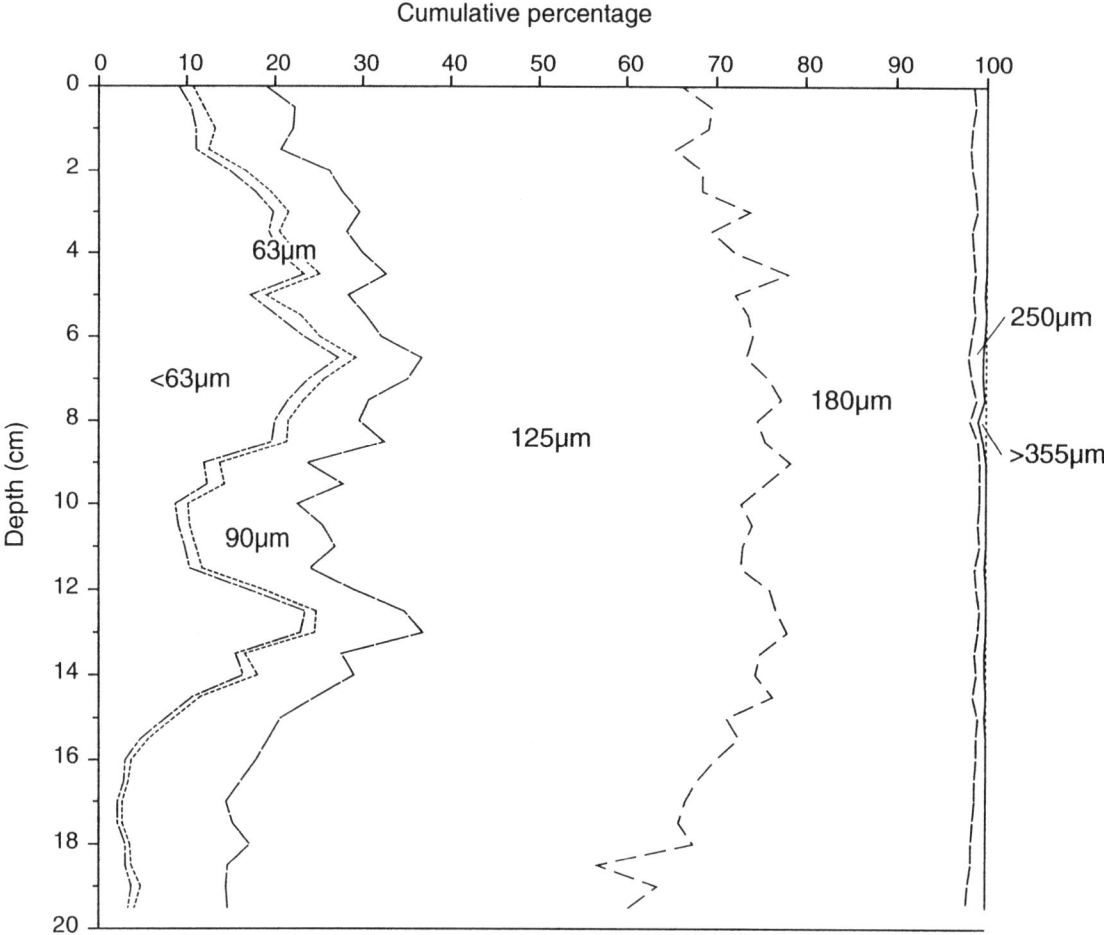

Figure 16.5 Particle-size analyses.

mum of 90% TLP. As the *Calluna* expands three other taxa go into decline; Gramineae falls from 30% to 6%, Cyperaceae from 13% to 0.5% and *Potentilla* type from 9% to zero. The *Calluna* maximum is reached at 12–13 cm and from then on falls steadily to its lowest level of 14%. It continues to be the dominant taxon even after its peak has been reached but at 5 cm in the profile it has been replaced as the dominant by Gramineae. As it rises three other taxa show increasing percentages. *Alnus*, Coryloid (cf *Corylus avellana*) and *Plantago lanceolata* all reach levels of over 5% with *Plantago* exceeding 10%.

Throughout the whole depth of the profile, arboreal pollen, apart from *Alnus*, is recorded only at percentages of less than 1% and is restricted to *Betula*, *Pinus*, *Quercus* and *Ulmus*. The herbaceous taxa are not only limited in number (despite scanning of the slides as a means of obviating the blanketing effects of large pollen producers such as *Calluna* and Gramineae) but also occur at very low percentages. Those that are present do agree, both in their sparseness and type, with the pollen taxa noted by Kaland (1986, 31) in his studies of heath burning in Norway.

The pollen diagram suggests that not only was the immediate locale one that had an extremely limited arboreal

cover but that even the wider region was likely to be treeless. *Pinus* is remarkably low given its propensity to aerial transport and even the lower pollen producing *Quercus* and *Ulmus* are poorly represented. Within the pollen site's catchment, *Alnus* and probably *Corylus* did occur and their representation increases in the profile above 7 cm.

The base of the pollen profile appears to indicate that a coastal heath was extant, becoming increasingly dominated by *Calluna* which eventually led to the decline of other taxa. At 12 cm in the profile *Calluna* had become totally dominant but from then on the area began to undergo an ecological change which brought about the development of coastal links in which Gramineae became increasingly important along with *Plantago lanceolata*. Taken overall the pollen record parallels very closely that reported by Edwards (1978) from the Dundrum sand dune system in the Murlough National Nature Reserve in Northern Ireland.

Analysis of particle-size and particle-nature

The particle-size analysis was carried out by sieving. The majority of the sediments at the sampling site consists of particles over 90 μm in size, putting them into the fine-sand class. The silt-size particles, <63 μm, never exceed

25% and are the most variable in their contribution to the profile. Individual grains from the profile were examined, for surface textures and shape, by scanning electron microscopy. Three classes were found: first, well rounded, often nearly spherical grains with smooth faces; second, angular to sub-angular grains with stepped faces and little abrasion; third, elongate and flaky grains of feldspar and mica, often bounded by cleavage fractures. The first class dominates in the lower part of the section, below 15 cm, and is derived from a marine source. The second class is of fluvio-glacial origin and dominates above 10 cm. The third class is poorly represented but more common above 12 cm.

The radiocarbon dating

The material for radiocarbon analysis was submitted to the laboratories of Teledyne Isotopes. The date obtained is 1530 ± 80 bp (I-16,338) which gives a calibrated value in the range of AD 424–616 (Stuiver & Becker 1986), thus lying within the Dark Age.

THE NATURE OF HEATHLAND AND ITS MANAGEMENT BY FIRE

In his study of *Calluna* in sand dunes in South Sweden, Wallén (1980) has shown that under undisturbed conditions the age structure of that taxon can remain constant. This is achieved by the change of parts of the basal stems through the production of adventitious roots, so converting above-ground biomass to below-ground biomass. Under such conditions, *Calluna* is able to exclude other competing species. Such a steady state, however, can be disrupted if there is a variable accumulation of organic material, an improvement of the nutritional status or a disturbance in local conditions; each of these will reduce the development of adventitious rooting. The effect of firing will be to produce all of these conditions and thus it has the potential for removing the dominance of *Calluna* in dune systems.

Herbivores such as cattle, red deer and sheep can obtain sustenance from *Calluna*. Shoots of all ages are equally digestible (Milne 1974) but a preference is shown for the long and short shoots of the current year's growth and the short shoots of the previous year (Miller 1979). Such palatable material is stimulated if the heather is periodically burned. Chapman (1979) has suggested that the benefits accruing from such burning, as far as nutritional effects are concerned, may last only some three or four years. At its optimum use, the periodic burning of *Calluna* is a labour-efficient system of pasture management.

Only with the combined use of burning and grazing, however, is the pasture's nutritional value maintained. As Gimingham (1971) has shown there are also considerable dangers to the heather pasture if firing is undertaken injudiciously. On poorly drained ground, such as that which exists at the coastal site of Tentsmuir, burning can induce a total replacement of heather by other species. If a 'hot burn' is allowed to develop the *Calluna* may be completely destroyed. A greater frequency of firing, even with a 'cool burn', can also create a situation where the total ecosystem becomes more prone to erosive damage.

DISCUSSION

An examination of Figure 16.3 indicates that the domination of the dune vegetation by *Calluna* increased to its maximum at 13 cm but thereafter declined as it was successively replaced by Gramineae. Thus there are signs that the steady state suggested by Wallén (1980) may have been approaching but, at the stated depth, one or more of the disrupting factors he identified came into operation. Figure 16.3 shows that the peak in *Calluna* pollen coincides with the highest occurrence of charcoal. This suggests that the disruption factor involved was fire and, following Kaland's observation, that it was undertaken to improve pasture. The time over which this burning was occurring cannot be determined but it is clear from the charcoal record that it was severe at a depth of 13 cm, after which there was a decline. It does appear that either a 'hot burn' occurred at 13 cm or that the effects of over frequent firing came into play, and that this led to severe erosion and the eventual interment of the contemporary land surface. Thereafter, the pollen diagram shows that a different vegetation cover emerged on the sediments which built up on the former land surface. Even after that, the charcoal record suggests that some burning did still occur.

Reference to Figure 16.4, which presents pollen concentrations, shows that the picture presented by the percentage diagram was not such a simple event. The pollen concentrations show that both *Calluna* and Gramineae fall to lower levels after the major burn at 12–14 cm. Above that point, charcoal becomes a less marked feature in the stratigraphy, although there is a continuing phase of substantial charcoal concentration accumulation up to 6 cm. It is only at that point that *Calluna* concentrations decline and those of Gramineae show a marked dominance (from a low of 7×10^5 to 65×10^5 grains cm^{-3} wet sediment).

Thus the question arises as to what occurred from agricultural and environmental standpoints in this area after the major burn. The preservation in the stratigraphy of the major charcoal band suggests that the erosion which followed the burning and subsequent local vegetation disruption was not total. The continuation of *Calluna* as a significant vegetation component, as demonstrated by the concentration diagram, tends to confirm that suggestion. The presence of charcoal also indicates that *Calluna* burning was persisted with until such time as the input of new sediments into the area and persistent burning created conditions which favoured grasses at the expense of heather. The existence of the present current lower altitude dunes to the east indicates that further supplies of sand were available.

The date obtained for the creation of the main burned layer is of considerable interest. Investigation of blanket mires has shown that there is evidence for a climatic deterioration in the mid-first millennium AD (Blackford & Chambers 1991), during which time the *Calluna* burning was occurring. An increase in storminess then, coupled with the availability of sand on the shores of Tentsmuir, at that time much closer to the investigation site, would have allowed a continued influx of sand on to the severely burned land surface and the eventual development of a land surface which was more suited to the needs of Gramineae than *Calluna*. This, eventually, would have made the former heather management system redundant.

Sectioning of the dunes where the burned layer occurs reveals that it is not horizontally continuous. There is evidence for the widespread but localised development of deflation hollows in the now fossilised land surface. This suggests that the severe burning, occurring around 1530 bp, did cause localised but frequent destabilisation of the land surface and that the burned layer was then progressively buried by material derived from the deflation hollows. This would have led to the redistribution of previously stratified charcoal and pollen, which may go some way to account for the level of charcoal and *Calluna* concentrations in the material which overlies the burned layer. This redistribution would have been encouraged by increased storminess and any continuing use of the area for the pasturing of cattle. With sand from the deflation hollows accreting to the land surface over an extended time period, the *Calluna* would also have been able to survive until such time as the depth of accumulation and the continued use of fire, as shown by the secondary charcoal concentration up to 5–6 cm, led to the increasing dominance by invading grass. This is a process which Gimingham (1971) has identified in heathlands where burning has proceeded to a point where the heather can no longer withstand competition.

The depth of accumulated sand above the burned layer is, however, too great to be accounted for purely by the redistribution of sand from local deflation hollows; this suggests that there could have been an input to the area from the coastal zone. The particle-size and SEM analyses are of interest here. They shows that silt-size particles not only increase again in the levels immediately above the burned layer, before declining as fine and coarse sand-size particles become increasingly dominant (Figure 16.5), but that there is a change in the origin of the particles. The area to the immediate west contrasts strongly with Tentsmuir. It consists of a belt of fine sands and silts (Figure 16.1) on a raised beach which today supports high grade arable land. It is prone to wind erosion under stormy conditions and annually deposits silt-size and fine sand-size particles on Tentsmuir. If this land was cultivated at this time, any increase in storminess would have led to a transfer of material from the fields into the dune zone. It is attractive to see the heathland managed as a zone of pasture and exploited in conjunction with arable farming on the silts as an integrated management system. Place-name evidence shows that this area was occupied by Pictish people in the Dark Age and, although little is known of their agricultural practices, they did settle in areas with high agricultural potential (Whittington 1977).

Blackford & Chambers (1991) have shown that the mid-first millennium AD was a period of increased rainfall. Despite the accompanying heightened storminess, it might be thought that this greater degree of wetness might have inhibited aeolian soil movement. This is unlikely because the Tentsmuir area lies in the rain shadow of the Scottish mountain mass. Current rainfall totals do not exceed 800 mm, while the months of March and April are not only among the driest but experience strong winds. It is at this time that the present transfer of soil occurs. Of further interest, in this context, is the increase in the pollen of *Alnus* and Coryloid in the upper part of the pollen spectra. These taxa release their pollen in March and April and in this case it is most likely to have been transported to Tentsmuir by winds with a westerly component.

CONCLUSIONS

Pollen, particle-size and particle-texture analyses of a fortuitously revealed burned zone in fossil dunes have allowed two conclusions to be reached. First, it would seem that a dated occurrence for the early utilization of burning to improve pasture in a system of animal husbandry has been obtained. Second, the severity of the burning could well have brought about severe and widespread local site degradation due to the sealing of the soil's surface by charcoal particles. This has been shown to reduce water infiltration by up to 75% (Mallik *et al* 1984) and on an unvegetated surface that would precipitate rainwash erosion. That situation, in any widespread sense, was, paradoxically, prevented by the onset of the climatic deterioration of the mid-first millennium AD which led to the burial of the burnt-over heath by fine, organic-rich windblown deposits from adjacent arable land, encouraging the growth of grasses. As Wilson (1991) has commented for a site at Portstewart in Northern Ireland '... detailed investigations of small exposures in coastal aeolian sands can reveal much valuable information about local landscape evolution ...' and also add to our understanding of the early use of marginal land.

ACKNOWLEDGEMENTS

Thanks are due to Rona Charles and Stewart Pritchard of the Fife Sub-Regional Office of Scottish Natural Heritage for providing information on the Earlshall Muir SSSI; to Janet Mykura who drew the diagrams; to Colin Cameron for the data on particle size and Donald Herd for that from the SEM; to Pat Doody for permission to base Figure 16.2 on a map in *Focus on Nature Conservation*, 1985, 13, 140.

REFERENCES

Allan, W 1964 *The african husbandman*. Edinburgh: Oliver and Boyd, 66–76.

Blackford, J J & Chambers, F M 1991 'Proxy records of climate from blanket mires: evidence for a Dark Age (1400 BP) climatic deterioration in the British Isles', *The Holocene*, 1 (1), 63–67.

Chapman, S B 1967 'Nutrient budgets for a dry heath ecosystem in the south of England', *Journal of Ecology*, 55, 677–689.

Coles, J M 1971 'The early settlement of Scotland: excavations at Morton, Fife', *Proceedings of the Prehistoric Society*, 37, 284–366.

Edwards, K J 1978 'Palynology of buried soil horizons', *in* Nairn, R G W (ed), *Murlough National Nature Reserve, (National Trust) scientific report 1978*. Dundrum: Murlough National Nature Reserve, 20–21.

Faegri, K, Kaland, P E & Krzywinski (eds) 1989 *Textbook of pollen analysis*. 4th edition. Chichester: Wiley.

Gimingham, C H 1971 'British heathland ecosystems: the outcome of many years of management by fire', *Proc 10th annual tall timbers fire ecology conference*, 293–321.

Gimingham, C H 1972 *Ecology of heathlands*. London: Chapman and Hall.

Harris, D R 1972 'Swidden systems and population', *in* Ucko, P J, Tringham, R & Dimbleby, G W (eds), *Man, settlement and urbanism*. London: Duckworth, 245–262.

Huttunen, P 1980 'Early land use, especially slash-and-burn cultivation in the commune of Lammi, southern Finland, interpreted mainly using pollen and charcoal analyses', *Acta Bot Fenn*, 113, 1–45.

Iversen, J 1941 'Landnam i Danmarks Stenalder', *Danm Geol Unders* II RK, 66, København.

Kaland, P E 1986 'The origin and management of Norwegian coastal heaths as reflected by pollen analysis', *in* Behre, K-E (ed), *Anthropogenic indicators in pollen diagrams*. Rotterdam: Balkema, 19–36.

Mallik, A U, Gimingham, C H & Rahman, A A 1984 'Ecological effects of heather burning. 1. Water infiltration, moisture retention and porosity of surface soil', *Journal of Ecology*, 72, 767–776.

Miller, G R 1979 'Quantity and quality of the annual production of shoots and flowers by *Calluna vulgaris* in north-east Scotland', *Journal of Ecology*, 67, 109–129.

Milne, J A 1974 'The effects of season and age on the nutritive value of heather (*Calluna vulgaris* (L.) Hull) to sheep', *Journal of Agricultural Science*, 83, 281–288.

Øvstedal, D D 1985 'The vegetation of Lindås and Austrheim, Western Norway', *Phytocoenologia*, 13, 323–449.

Patterson, W A III, Edwards, K J & Maguire, D N 1987 'Microscopic charcoal as a fossil indicator of fire', *Quaternary Science Reviews*, 6, 3–23.

Rackham, O 1986 *The history of the countryside*. London: Dent.

Sarmaja-Korjonen K 1992 'Fine interval pollen and charcoal analyses from S. Finland', *Abstracts: 8th International Palynological Congress*, Aix-en-Provence, 132.

Smit, J T de 1979 'Origin and destruction of northwest European heath vegetation', *in* Wilmanns, O and Tüxen, R (eds), *Werden und Vergehen von Planzgesellschaft*. Stolzenau, 411–435.

Stockmarr, J 1971 'Tablets with spores used in absolute pollen analysis', *Pollen et Spores*, 13, 615–621.

Strømgaard, P 1985 'A subsistence economy under pressure: the Bemba of Northern Zambia', *Africa*, 5, 39–59.

Stuiver, M & Becker, B 1986 'High precision decadel calibration of the Radiocarbon timescale, AD 1950–2500 BC', *Radiocarbon*, 28 (2B), 863–910.

Tolonen, K 1983 'The post-glacial fire record', *in* Wein R W & Maclean, D A (eds), *The role of fire in northern circumpolar ecosystems*. Chichester: John Wiley & Sons Ltd, 21–44.

Tolonen, M 1985 'Palaeoecological record of local fire history from a peat deposit in SW Finland', *Ann Bot Fenn*, 22, 15–29.

Vuorela, I 1986 'Palynological and historical evidence of slash-and-burn cultivation in South Finland', *in* Behre, K-E (ed), *Anthropogenic indicators in pollen diagrams*. Rotterdam: Balkema, 53–64.

Wallén, B 1980 'Structure and dynamics of *Calluna vulgaris* on sand dunes in South Sweden', *Oikos*, 35, 20–30.

Whittington, G 1977 'Placenames and the settlement pattern of Dark Age Scotland', *Proceedings of the Society of Antiquaries of Scotland*, 106, 99–110.

Wilson, P 1991 'Buried soils and coastal aeolian sands at Portstewart, Co. Londonderry, Northern Ireland', *Scottish Geographical Magazine*, 107, 198–202.

17. Roman Egypt – provisioning the settlements of the Eastern Desert, with particular reference to the quarry settlement of Mons Claudianus

Sheila Hamilton-Dyer

Abstract

There are some sixty Roman forts and other settlements in the Eastern Desert of Egypt (Figure 17.1). These range from large quarry settlements, such as Mons Claudianus, to small way-stations en route between the Nile and the Red Sea. This mountainous region of Egypt has been arid for several thousand years. Average rainfall is only 3 mm annually and the Romans used wells up to 50 m deep for their water supply. Desert adapted plants and animals such as *Zilla spinosa* and goats support the pastoral nomads, but the environment cannot supply the needs of large numbers of people in permanent settlements. Almost all provisions, particularly grain, flour, oil and wine were brought to these settlements from the Nile by pack animals. These were principally donkeys, the predominant faunal remains at Mons Claudianus, together with camels. According to Strabo the journey between the Nile and the Red Sea took five to seven days, the caravans resting at way-stations. The large animal enclosures of these testify to the enormous supply operation being carried out.

INTRODUCTION

During the past several thousand years life in Egypt has been almost entirely dependent on the Nile, as rainfall in this region is very low and unpredictable. With irrigation the rich floodplain soils can produce all manner of foods but beyond the influence of the Nile the land is desert. The demarcation between the two zones is remarkably abrupt. Few would wish to enter this harsh environment but the Eastern Desert contains mineral rich rocks and valuable decorative building stone. It also provides a link between the Nile and the Red Sea and beyond to India and Ethiopia.

In this region between the Nile and the Red Sea a mountain range of pre-Cambrian plutonic and volcanic rocks runs north-south, rising steeply to 2,000 m from a narrow coastal plain. Deeply cut by wadis, dry water-courses, the mountains run down to a limestone plateau 800 m above the Nile which is 75 m above sea level at Qena.

The whole of Egypt is classed as arid and hyperarid, with less than 300 mm mean annual rainfall, the amount required for wheat cultivation. In the Eastern Desert the mean annual precipitation is less than 5 mm (Butzer & Hansen 1968, 423). When rain does fall it is often local-ised, causing a flush of new growth in one place while nearby areas may have none for decades. The rain can be heavy and with virtually no vegetation run-off can be dramatically destructive.

In the mountains there are some small permanent seeps and springs. These, together with rain filled rock basins of limited capacity and duration, are well known to the present inhabitants, the semi-nomadic pastoralist Bedou-in. They rely on flocks of goats, sheep and some camels and donkeys, trading these and desert plants for other goods in the Nile Valley and the coastal settlements. Their population density is very low at about one person per 90 km² (Hobbs 1990, 2).

Scarcity of water is not the only problem for life in this area, the temperature range is large, typical of desert regions. In summer the temperature often reaches 35°C and has exceeded 45°C. In winter temperatures frequently drop to freezing at night, rapidly rising again to 20°C and higher, during the morning.

Few animals and plants tolerate these extreme conditions. Slow growing perennial plants such as the *Acacia* tree with deep roots and leathery leaves grow along wadi edges where there is permanent water deep underground. Other plants like the ubiquitous *Zilla spinosa* thorn bush are specially adapted to grow quickly when water is available and then conserve it for months or even years. Animals which need little or no free water include ibex, dorcas gazelle, foxes and various desert birds, rodents and reptiles.

In prehistoric times this area was more wooded and supported a greater variety of plants and animals than today. The change seems to have already occurred by 2000 BC (Butzer 1961; Butzer & Hansen 1968; Kassas & Girgis 1964). The Romans probably found the climate exactly as today but with more vegetation. Their extensive use of acacia and tamarisk particularly for charcoal production probably caused considerable habitat damage. Huge quantities of charcoal and slag can be seen littering the smithy huts, continuing a process started by earlier visitors using charcoal in gold separation. Some of the Roman forts and stations are built on the tailings of earlier gold diggings, re-using ore crushing mills in the walls. Near Qena the wadi is littered with mounds formed by material gathering round the root systems of Tamarisk, *Tamarix aphylla*, now long dead. The desert plants take a long time to recover from such pressure. A similar episode occurred early this century when whole groves of acacia were cut for fodder and charcoal. There is still little sign of regeneration and the Bedouin now take great care to conserve their perennial plants, having been made painfully aware of the fragility of the ecosystem (Hobbs 1990, 100, 109).

ANCIENT ROUTES AND SETTLEMENTS

Four principal ancient highways (Figure 17.1) cross the desert from Coptos (modern Qift) and Kaine (Qena) on the Nile to the coast following the major wadis, a fifth runs east from Apollonopolis and joins the Coptos to Berenice road with a further possible southern route from Syene (Aswan). These routes were established after 2000 BC for trade with lands across the Red Sea and down the coast. The well, 24 m deep, halfway along Wadi Hammamat was dug under the orders of Rameses IV using 8000 labourers (Robinson 1935). Strabo describes a journey from Coptos to Myos Hormos, a Red Sea port (modern Quseir al Qadim, Peacock 1993), as taking six or seven days stopping at watering stations en route. Roman forts and other settlements abound along and near these routes; over sixty have been identified. Each of the major forts and watering stations has its own well inside or outside the enclosure, often with associated cisterns (hydreumata). Some of these wells are very deep; one Roman example at Qattar still contains water at a depth of 50 m (Hobbs 1990, 47). The quarry settlements needed much water, not only for themselves and their animals, but also for the blacksmiths tempering baths and for splitting and cleaning the stone.

PROVISIONING: EVIDENCE FROM EXCAVATIONS AND WRITTEN RECORDS

Excavations at the major quarry settlement of Mons Claudianus have revealed a great deal of information about the quarrying operations, administration and provisioning of this large complex (Bingen 1987; 1990). The stone here is an attractive black and white speckled quartz diorite and, like the pink granite of Aswan, is unusual in being sufficiently uniform for the extraction of enormous blocks. The other major quarry site, Mons Porphyrites, is the unique source for Imperial Porphyry, a purple stone which takes a high polish. At Mons Claudianus midden material outside the settlement, and in abandoned rooms, is several metres thick. These deposits often contain exceptionally well preserved textiles, leather, and plant and animal remains. In addition to this unique resource there are several thousand ostraca, documents written on pot-sherds. These range in subject from personal letters and military documents to simple bread and water receipts (Bingen *et al* 1992). One document reveals the large number of people at Mons Claudianus by listing the several hundred quarry workers and military personnel present on a particular day.

The main period of occupation was 2nd century AD, Trajanic to Antonine, but operations were started earlier and carried on into the 3rd century AD.

Transport

The considerable demand for animal feed for the transport of stone is revealed in a papyrus written in December AD 118 (Peña 1989). This letter requests that all barley in the nome, or administrative area, 115 Roman miles (1 Roman mile = 1.48 km) below Kaine (Qena) should be sent as soon as possible up the Nile to Kaine. The urgency of the request can be gauged by the translated statement, '... for we have a great number of animals for the purpose of bringing down a fifty-foot column, and already we are nearly out of barley ...' and also by the repeated desire that this operation should be carried out swiftly.

This column may have been intended for the temple of the deified Trajan in Rome. Broken and unfinished examples, still lying in the Mons Claudianus quarries, can be estimated as weighing in the region of 100 tonnes. These columns would therefore need a draught team of some size. It is not specified in the text whether men or animals were being used although it has been assumed to be camels and oxen (Peña 1989; Meridith 1952). Recent evidence, however, from the documents and animal remains found at Mons Claudianus strongly suggests that donkeys and camels were the only animals used for transport. The conditions are so severe that cattle, which need large quantities of water, would have been useless. This is supported by a lack of cattle bones and nowhere in the documents is there any mention of cattle.

Camels are of course superbly adapted for arid, sandy regions, and can carry twice the amount of a donkey. Camel

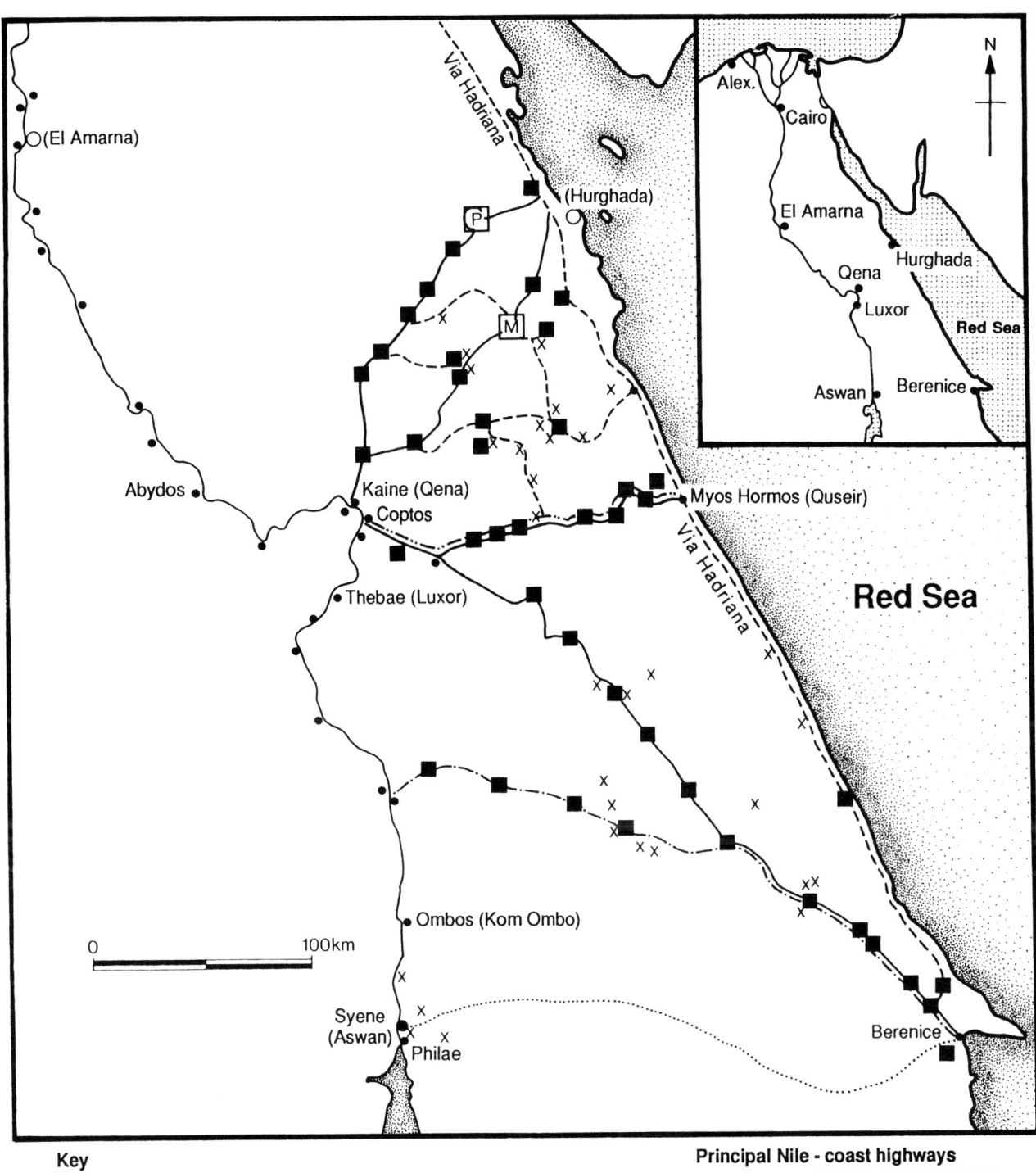

Key

[M]	the MONS CLAUDIANUS complex
[P]	the MONS PORPHYRITES complex
■	forts, way stations
O	towns - modern with no ancient name
●	towns - present in Roman period
x	mines, quarries

Principal Nile - coast highways

————	4 routes from Coptos & Kaine
—··—··—	Wadi Hammamat
—·—·—	Apollonopolis
·············	possible route from Syene

Figure 17.1 Roman remains in the Eastern Desert of Egypt (after Meredith 1958).

caravans were already using the Nile-Red Sea routes in the first and second centuries BC and became increasingly important throughout the Roman occupation for trade and military purposes (Bulliet 1975, 117). No doubt they are the fast carriers *phoros dromicon* of the documents. They are, however, at a disadvantage in rocky areas. They could not have been used in the quarry complexes where it is already known that donkeys were used. Many ancient Egyptian reliefs and inscriptions indicate the extensive use of donkey caravans for all types of supply and expedition.

The British Army manual (War Office 1923) specifies a maximum pack load for a donkey of 100 lb (50 kg), an amount which seems to correlate with ancient records. Harnessed in long teams, donkeys were probably used to pull at least some of the stone wagons in the same way as the mule teams Martial describes used to haul marble and also, in the recent past, the mules used as wagon teams in the American West. Manpower could have been used for the largest columns with donkeys as logistic support. Camels were probably not used as their anatomy causes harnessing problems (Bulliet 1975). Horses seem to have been rare high status animals in Egypt. The cavalry unit had horses and remains at Mons Claudianus, though rare, are present.

The assumption that it is the donkey and not the stronger mule that was used is based partly on the lack of any positively identified remains of the latter, although it is difficult to separate them. It is also donkeys and not mules that are mentioned on the ostraca. Documentary evidence indicates that mules have always been rare in Egypt, probably as a result of the rarity of horses, as mules are hybrids of male donkey and female horse. They are also sterile and therefore not self sustaining as is the donkey (Nibbi 1979). Elsewhere in the Roman empire mules were in common use. They were present in Roman Egypt albeit in low numbers, as there is a record of the sale of a mule wagon (Johnson 1936, 407).

The wagons used to carry the columns and other large loads of stone must have been substantial; in some places along the routes wheel tracks have been observed with two gauges, one of seven feet six inches and a larger one of nine or eleven feet (Murray 1925; Tregenza 1955). Most of these are now obliterated by the passage of army and other modern vehicles.

Among the ostraca are mentioned the arrival of a twelve wheeled wagon, a four wheeled cart and also two loads by fast carrier.

As the local environment could not support large numbers of people and animals, an extensive supply line had to be organised. This was probably under military control; pass-tablets are mentioned for travel between the stations. There also seems a need for protection of the routes as one letter (ostracon inventory number 4888) from a son to his father asks for work elsewhere as he has not eaten for two days from fear of the barbarians. The government organised camel troops to keep the nomads under control (Robinson 1935).

A feature of the two northern routes, past Mons Claudianus and Mons Porphyrites, are the extensive animal lines outside the enclosures such as those at El Heita and Qreiya. These often have various troughs and conduits associated with hydreumata, plaster lined cisterns. The plan of these animal lines is in the form of a rectangle, sometimes as large as the station enclosure itself, with internal partitions dividing the space into three long areas and providing troughs/byres along the length of the walls. There is currently insufficient evidence to show whether or not these structures were roofed. The numerous stone piers of the 'granary' adjacent to the animal lines at Mons Claudianus indicate that this building was roofed, not with the large slabs found in the fort but probably with tamarisk branches or similar material.

That these enormous structures are absent along the other desert routes attests to the quantity of beasts required for the provisioning of the quarry settlements and the transport of stone. Each station is approximately 25 km from the next, which coincides with the distance a loaded donkey is expected to travel per day.

Dates on papyri and ostraca indicate year-round presence but transport of columns to the Nile may have been reserved for the cooler winter months (see Peña 1989).

Food supplies

Many of the ostraca are entolai, documents relating to the activities of the cibariatores, the quartermasters. These are usually for supplies such as wine, oil, dates and other foods and equipment, paid for by cash or on account. The style of the ostraca is very similar to that of the Vindolanda tablets, with which they are contemporary.

Not only were cattle absent from harness, the meat is also absent with no mention of cattle or beef in the documents and a total lack of bones on site, although it is acknowledged that cured meat such as sausages would leave no physical evidence.

Although figures have yet to be finalised it is clear that donkey bones form the bulk of the mammal remains with ample evidence of butchery. Remains of pig take second place with camel, goat, sheep and other animals trailing behind. Very large amounts of fish bones and skin are also present, mostly of Red Sea species. Of 8000 bones examined so far, 2000 are fish. Equid bones are also in the region of 2000 fragments with pig around 600. Camel numbers 140, about the same amount as goat. Much material has not been positively identified to species but the large amount of chopped rib and vertebra fragments are almost certainly donkey. Bones of wild animals are few but hunting, a pastime mentioned in the documents, produced a few remains of ibex, dorcas gazelle and desert partridge.

Requests for donkeys are frequent in the documents; one, subject to final interpretation, seems to ask for animals still shedding teeth, presumably old enough to carry a pack but still tender for slaughter on arrival. One letter informs the recipient that a donkey has died in harness and men have been sent to collect it, a worthless task unless the intention was to eat it.

Like today, the donkey did not normally figure in Egyptian food but there is no reason other than cultural for not eating donkey or other equids. Horse meat, for example, can be leaner and more tender than beef if the animals are young. Including donkey meat as part of military rations is mentioned by Xenophon. Camel meat is also a good substitute for beef, but the price in the 2nd century AD was twice that of an ox and four times higher than a donkey (Johnson 1936, 230–232). The camel also has a low rate of breeding success. These factors probably explain the relatively small quantities of camel bone found; the animals were more valuable alive. It would seem, therefore, that donkey meat replaced the usual beef provision as a cheap and readily available alternative under difficult supply conditions.

Although the Mons Claudianus complex is currently the only site in the desert to be excavated, surface examination of the midden material at other sites appears to be similar with remains of donkey, pig, Red Sea fish and shells prominent. It is therefore extremely interesting to note that the few faunal remains examined from the Roman levels at Quseir al-Qadim on the coast have included a few cattle bones; 9 out of 701 mammal bones, amongst an assemblage dominated by fish (amount not stated) and sheep/goat (163) remains. Only one bone was identified as donkey (Wattenmaker 1982). Pig bones were also found. These are common at Mons Claudianus where they were kept, or at least delivered, alive; a receipt recovered from deposits near the well asks for water for piglets. The rarity of adult and neonatal pig remains implies that these were weaners brought alive to the site to be kept and fattened until required. There was certainly no shortage of rubbish for them to root through. Waste parts of the body are included in the cattle bones from Quseir implying slaughter on site rather than the import of preserved joints. In view of the transport difficulties where did these animals come from? The coast is as desolate as inland areas and could not have supported cattle. Wadi Hammamat does have more frequent way-stations than the northern quarry routes, possibly enabling the droving of limited numbers of cattle to provide fresh meat. Such a journey must have taken more than a week to cover the 150 km. With less than 1000 animal bone fragments from Quseir compared with eight times that amount from Mons Claudianus the total absence of cattle at the latter site remains a little puzzling. The presence of personal letters requesting various items for individual use precludes total military control over supplies. Perhaps the quality and amount of meat supplied was enough to satisfy all inhabitants. It would be interesting to determine whether sites nearer the Nile valley have cattle remains.

Depending on the amount allocated the protein supply seems good with donkey and fish the major source of animal protein, much of this probably fresh. The fish remains are overwhelmingly Red Sea species such as Scaridae (parrotfish), Serranidae (groupers) and Lethrinidae (emperors). The few Nile fish recovered are mainly Siluridae (catfish). The high quality tilapia and Nile perch are absent, yet these are similar in culinary qualities to the sea fish found. A full discussion of the fish remains is the subject of another paper (Hamilton-Dyer 1994); briefly it would seem that the inhabitants preferred fresh fish, and/or sea fish. The journey from the Nile would have taken too long for the transport of fresh river fish, other than the catfish which will live for several days if kept damp.

Of the other food supplies, recent excavations have shown that bread seems to have been baked on site, probably using both high grade flour from the Nile valley and other types milled on site. Cereals identified in the deposits include two types of wheat and also barley (this probably as animal fodder but could also have been for beer; malt is mentioned in one document). Other species identified include at least four pulses, several nuts, fruits – particularly dates, garlic and spices (van der Veen 1990). The presence of wine, oil and fish pickle/paste is evident from both documents and jar inscriptions. Fresh vegetables may have been more difficult to obtain, although various salad items are mentioned in documents as being grown in a nearby garden. Experiments have shown this to be possible with plenty of irrigation, but tender plants can be damaged by overnight frosts in the winter.

The ceramics for the transport and cooking of these foods are mostly of Nile produced amphorae together with Egyptian faience, oil lamps and Aswan fine wares. The numbers of imported vessels are very small but include items from as far as Italy, Spain, Gaul, Palestine, Tripolitania and Rhodes (Tomber 1989). The numerous amphorae supplied a ready source for the manufacture of ostraca. Vessels have often been reworked probably indicating the difficulties of supply. Some personnel were of sufficiently high status to have not only fine ceramics brought to site but also glass vessels, some so thin it is remarkable that they survived the journey at all.

CONCLUSION

The wealth of evidence pertaining to the settlements in the Eastern Desert indicates successful occupation of the area for some time. For this to be viable in such a marginal area the settlements had to be almost entirely artificially supported. When the Romans left this area the settlements were abandoned. The buildings have been largely ignored by the nomads and remain standing empty to this day; mute testament to the considerable, but ephemeral, influence of the Empire on the Eastern Desert.

ACKNOWLEDGEMENT

Assistance with travel expenses was given by the British Academy.

REFERENCES

Ancient sources

Martial, *Epigrams* V, 22, 7, 8)
Pliny, VIII, 170
Strabo, XVII, I, 45
Xenophon, *Anabasis* II, 1, 6

Other sources

Bingen, J 1987 'Première campagne de fouille au Mons Claudianus', *Cairo: Bulletin de l'Institut Français d'Archéologie Orientale*, 87, 45–52.
Bingen, J 1990 'Quatrième campagne de fouille au Mons Claudianus', *Cairo: Bulletin de l'Institut Français d'Archéologie Orientale*, 90, 65–81
Bingen, J, Bulow-Jacobsen, A, Cockle, W, E, H, Cuvigny, H, Rubinstein, L & Van Ringen, W 1992 *'Mons Claudianus Ostraca Graeca et Latina I, O. Claud. 1 à 190'*, Cairo: Institut Français d'Archéologie Orientale Documents de Fouilles, XXIX.
Bulliet, R W 1975 *The camel and the wheel.* Cambridge, MA: Harvard University Press.
Butzer, K W 1961 'Climatic change in arid regions since the Pliocene', *in* Stamp, L D (ed), *A history of land-use in arid regions.* Paris: UNESCO Arid Zone Research, 17, 31–56.
Butzer, K & Hansen, C 1968 *Desert and river in Nubia.* Wisconsin: University of Wisconsin Press.
Hamilton-Dyer, S 1994 'Preliminary report on the fish remains from Mons Claudianus, Egypt', *in* Heinrich, D (ed), *Archaeo-Icthyological Studies* (Papers presented at the 6th meeting of the ICAZ Fish Remains Working Group). *Offa*, Sonderdruck 51, 275–278.
Hobbs, J J 1990 *Bedouin life in the Egyptian wilderness.* Cairo: American University in Cairo Press.
Johnson, A C 1936 'Roman Egypt in the reign of Diocletian', *in*

Frank, T (ed), *Economic survey of Ancient Rome, Volume III.* Baltimore.
Kassas, M & Girgis, W A 1964 'Habitat and plant communities in the Egyptian desert. V. The limestone plateau', *Journal of Ecology*, 52, 107–119.
Meridith, D 1952 'The Roman remains in the Eastern Desert of Egypt', *Journal of Egyptian Archaeology*, 38, 94–111.
Meridith, D (ed) 1958 *Tabulae Imperii Romani, Coptos sheet NG 36.* London: Society of Antiquaries of London.
Murray, G W 1925 'Roman roads and stations in the Eastern Desert of Egypt', *Journal of Egyptian Archaeology*, 11, 138–150.
Nibbi, A 1979 'Some remarks on ass and horse in ancient Egypt and the absence of the mule', *Zeitschrift für Agyptische Sprache und Altertumskunde* 106, 148–168.
Peacock, D P S 1993 'The site of Myos Hormos; a view from space', *Journal of Roman Archaeology* 6, 226–232.
Peña, J T 1989 'Papyrus Giss.69: evidence for the supplying of stone transport operations in Roman Egypt and the production of fifty-foot monolithic column shafts', *Journal of Roman Archaeology*, 2, 126–132.
Robinson, A E 1935 'Desiccation or destruction: Notes on the increase of desert areas in the Nile Valley', *Sudan Notes and Records*, 18, 119–130.
Tomber, R 1989 'Mons Claudianus', *Cairo: Bulletin de Liason XIII, Institut Français d'Archéologie Orientale*, 35–37.
Tomber, R 1990 'Mons Claudianus', *Cairo: Bulletin de Liason XIV, Institut Français d'Archéologie Orientale*, 26–27.
Tregenza, L A 1955 *The Red Sea Mountains of Egypt.* London: Oxford University Press.
Veen, M van der 1990 'Archaeobotany', *in* Peacock, D P S & Maxfield, V A, *Archaeological reports from Mons Claudianus 1990.* Southampton: Unpublished interim report.
War Office, Veterinary Department 1923 *Animal management.* (British) Army Orders July 1923.
Wattenmaker, P 1982 'Fauna', *in* Whitcomb, D S & Johnson, J H, *Quseir al-Qadim 1980 Preliminary Report*, Malibu: ARCE Reports 7, 347–353.

18. An exploratory survey of the water supply structures on the Syrian and Egyptian pilgrim routes to Mecca and Medinah

Ibrahim M Al-Resseeni, Lawrence A S Butler, Pauline E Kneale and Adrian T McDonald

Abstract

Mecca and Medinah were both ancient sites on trade routes through Saudi Arabia. Mecca, a religious site from the time of the prophet Abraham, increased in importance after Mohammed made it an annual place of pilgrimage. Medina became his religious and political capital. Since the seventh century tens of thousands of pilgrims have walked to Mecca through inhospitable semi-arid desert. Supply of water for washing, drinking and cooking for such large numbers required considerable organisation. The caliphs provided facilities for pilgrim caravans and it is the structures that remain from the periods of provision which are the subject of investigation. Facilities for water supplies and storage include water tanks, cisterns, channels, canals and wells. This paper summarises a field survey of structures at over 130 sites on the two principal routes from Syria and Egypt. A typology based on site size and function is created and applied. Larger sites have multiple functions.

INTRODUCTION

Settlement in marginal semi-arid and desert lands is usually at sites where agriculture can be practised on a temporary or permanent basis, at transhumance sites, or at centres for markets and trade. Often these factors are found in combination. Less usual are settlements in response to cultural or religious pressures.

The rise of Islam and its edict that every Muslim must perform the pilgrimage (*Al Hajj*) to Mecca once in their life time if able, physically and financially, has led to settlement at sites in Saudi Arabia that would not normally be viable, or worthy of investment to develop water supplies and irrigation canals.

Mecca and Medinah were pagan sites for worship and also trade centres on four active trade routes through Saudi Arabia, prior to the time of Mohammed. After the rise of Islam these caravan trade routes continued but developed an additional pilgrimage function and consequently two further routes, from Basra and Oman, were established (Figure 18.1). Water is not only a physical necessity but is also needed for purification and bathing rituals. Prior to air and rail transport, pilgrims travelled in large caravan groups from Africa, Europe, India and Asia. In early

Islamic times, AD 600–800, the number of pilgrims was small and water demands were limited. Water could be provided by wells or springs along the caravan routes. As the number of pilgrims increased, difficulties spurred the Caliphs to seek more sophisticated water resources and since relatively early Islamic times built *birak* – water tanks; *ahwadh* – cisterns; *qanawat* – canals or channels; and *abyar* – wells, in order to make the Al-Hajj journey more comfortable. At various times through to the end of Ottoman rule castles were built to defend water supplies and safeguard pilgrims.

Although pilgrim numbers fluctuated according to political and military circumstances, the general growth in numbers was tremendous, and despite the early civil engineering works of the caliphs, many travellers faced water shortage, dehydration and death.

In general, the pilgrim stations lie 30 to 60 km apart. These distances depend on water resource availability and on the terrain which limits the speed of travel between stations.

This study reports upon a rapid field survey of all the sites identified on the Syrian and Egyptian routes within Saudi Arabia. These were the most heavily trafficked by

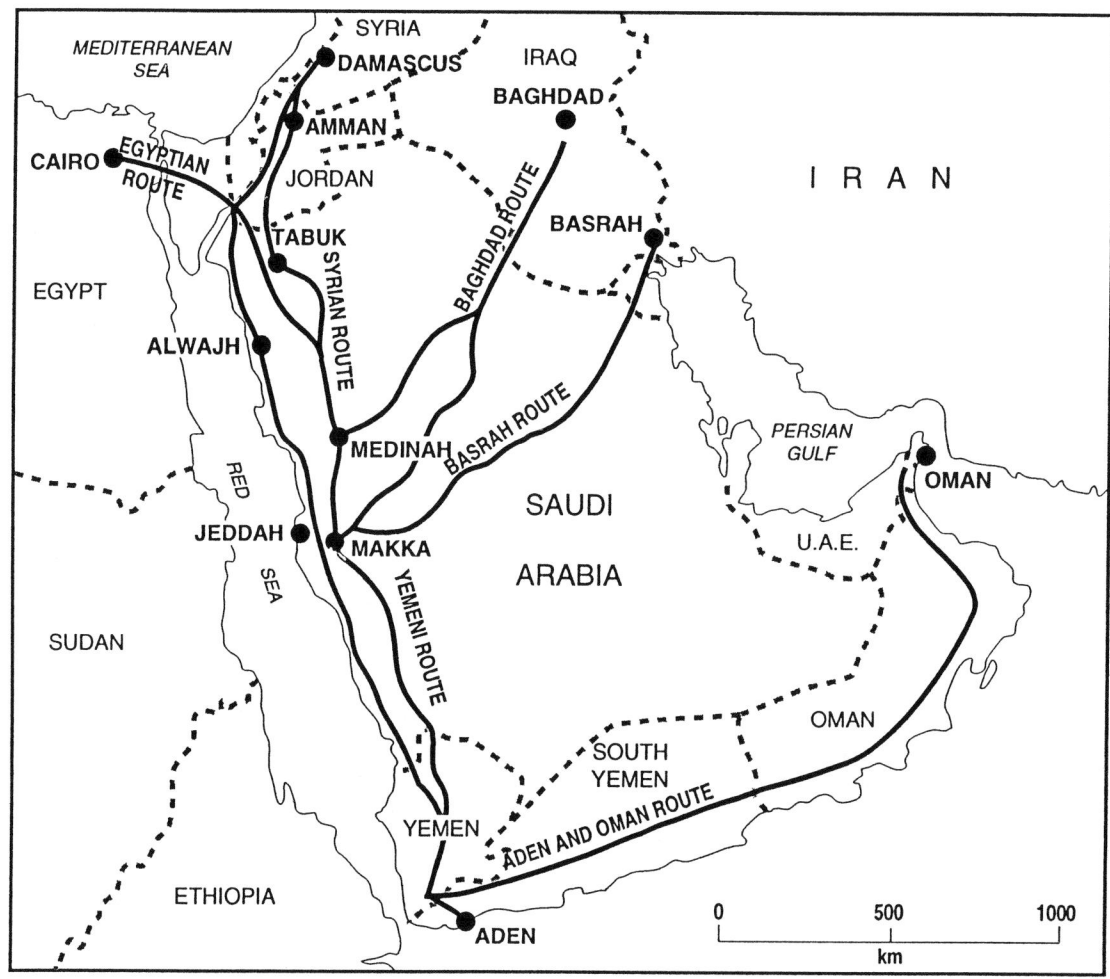

Figure 18.1 The pilgrim routes in Saudi Arabia.

pilgrims but this is the first attempt to make a systematic archaeological record of the stopping points and water resource structures (Al-Resseeni 1992).

This paper comments on the settlements and provides:

* documentation of the sites on the basis of the literature and field evidence.
* the sources of water supply, filling mechanisms for storage tanks, and makes comment about the engineering required at the sites.
* considers the archaeological variations evident in both design and construction, and makes some deductions about the age of the structures.

This initiates a base-line survey of water resource archaeology of the pilgrim routes which can then be used to comment on:

* the ecological state of these fragile sites.
* the relative efficiency of supply vis a vis evaporation, percolation and other losses.
* whether water was a limiting factor on the numbers who travelled.

Hydrology

In Saudi Arabia rainfall is low and infrequent and there are no perennial rivers or lakes. The annual average rainfall for the area under consideration is 20–30 mm while potential evaporation rates may be 3000 mm per year. Such rainfalls as do occur are generally high intensity events. Infiltration capacities have been shown to be particularly low (Berkowicz *et al* 1992) due to the very low hydraulic conductivity of dry soils exacerbated by the presence of hydrophobic biofilms and physical and chemical sealing of the ground surface. Rainfall in excess of infiltration capacity runs off as sheet flow. As the strength of flow increases, debris is picked up and over a long time erosion results in wadi formation. Overland flows, however, can be diverted into cisterns and tanks for storage and this project has mapped a number of long low diversion walls at wadi sites that serve this function.

Water that infiltrates will either become throughflow or percolate to groundwater where it may be extracted by wells, adits or springs. However in the last few hundred years the increased demand for water has led to over exploitation of this resource (Al-Ibrahim 1991). Currently

75–85% of Saudi Arabia's water is taken from ground water, far in excess of natural recharge. In Wadi Fatima, 354 of 360 natural springs have dried up and the water table has dropped 10 m in the Riyadh region in the last 30 years (Al-Ibrahim 1991). Groundwaters are renewable only if the long term extraction rate does not exceed the replenishment rate and if the rate of transport from the site of replenishment exceeds the rate of utilisation.

It is evident from this study that water table levels were higher in the past. Springs and wells formerly used by pilgrims are now dry and adits no longer flow.

TYPOLOGY OF SETTLEMENTS

In excess of 130 sites have been identified from the literature of earlier geographers and pilgrims. They require classification by size and by function to aid discussion. The size categories are large, medium, small and lost. The function groupings are simple, religious, junction and seaport (Figure 18.2).

Large stations

The major or large stations cover 2–3 km² or more. Al-Jumum, Site 5, covers an area of about 10–15 km² close to Mecca. Here caravans rested after the journey and pilgrims camped for several weeks while visiting Mecca. It has all the types of water resource structures, water tanks, wells, canals, and many remains of building foundations. It is still a permanent settlement. The larger sites on the routes typically had other functions, especially trade, and are some of the most difficult to examine because of modern building.

Al-Muazzam (Site 108, Figure 18.3) has a well which is defended by a castle, and storage tanks just outside the castle walls. This tank has a capacity of 7500 m³ or 7.5 million litres. It shows features common to many tanks; a stairway for access, presumably to facilitate cleaning out windblown sand and water borne sediments, and a gypsum plaster lining of the tank bed and walls to minimise seepage losses. We also find the remains of diversion walls built strategically across the wadi floor to divert overland flow into the tank. Al-Dar Al-Hamra, Site 103, shows similar features (Figure 18.4).

Medium-sized route stations

The ten medium route stations offer rather more than the simple 'overnight stop' facilities of the small stations. There are frequently a number of water resource structures. Al-Rauha, Site 35, has evidence of a water tank and several wells and canals. This site has a mosque rebuilt on much older foundations and wells, with both these and the mosque being still in use. That Al-Hajj, Site 126, is another station of this type, with a castle for defence.

Small route stations

These have water resource structures but no evidence of permanent settlements. Many sites are simply isolated

wells of variable capacity. Some with narrow diameters and rough undressed stone linings may be very old, sites originally used by Bedouin. Others have dressed stone, are much larger, 2–3 m in diameter, and may have been enlarged to supply pilgrims. They are sites where pilgrims might camp for a short time, overnight, watering and perhaps resting the camels.

There are 75 small route stations on the two routes, for example Birkat Antar (Site 91) on the Egyptian coastal route and Maqna, Site 119, a seaport site with a spring water source.

Lost route stations

These are documented by early Islamic geographers and travellers but could not be located. Waddan was mentioned as an important pilgrim station where water is abundant. It is located near Masturah (Site 16) where there is a wadi called Whadden but no evidence in memory or on the ground of settlement or wells.

Simple route stations

These have no apparent function other than as an overnight stop and watering sites. These may be small as at Al-Hafirah, Site 48, which has a well. This site continued as a settlement serving pilgrims from the time that a railway station was built here on the Damascus to Mecca line which carried pilgrims until it fell into disuse after 1918. Medium sized simple sites can be exemplified by Shajwa, Site 56. This group includes the majority of the sites mapped.

Religious route stations

These are known to be places where pilgrims performed religious rites as part of Al Hajj. They tend to be close to the main sites at Mecca and Medinah. Al-Juhfah, Site 11, is a good example with a well protected by a castle. Pilgrims would stop here and wash, as part of the ritual of cleansing, before travelling the final hundred miles to Mecca.

Seaport route stations

These stations on the Red Sea were embarkation and disembarkation points for pilgrims using the sea routes. Yanbu Al-Bahr, Site 41, was a small seaport which has since developed into a very large and important industrial city. By contrast Al-Jar, Site 25, the seaport for Medinah, is now in the 'lost' category. The nearest village is 10 km to the south.

Junction or traverse way stations

These can be exemplified by Ainunah (Site 118) on the Egyptian route. Pilgrims who wish to visit Medinah before making Al-Hajj go from Ainunah south-east to meet the Syrian pilgrim route at Al-Suqya (Al-Khushaybah), Site 84, and continue their journey to Medinah.

Renamed sites

In some cases sites appear to have changed their names

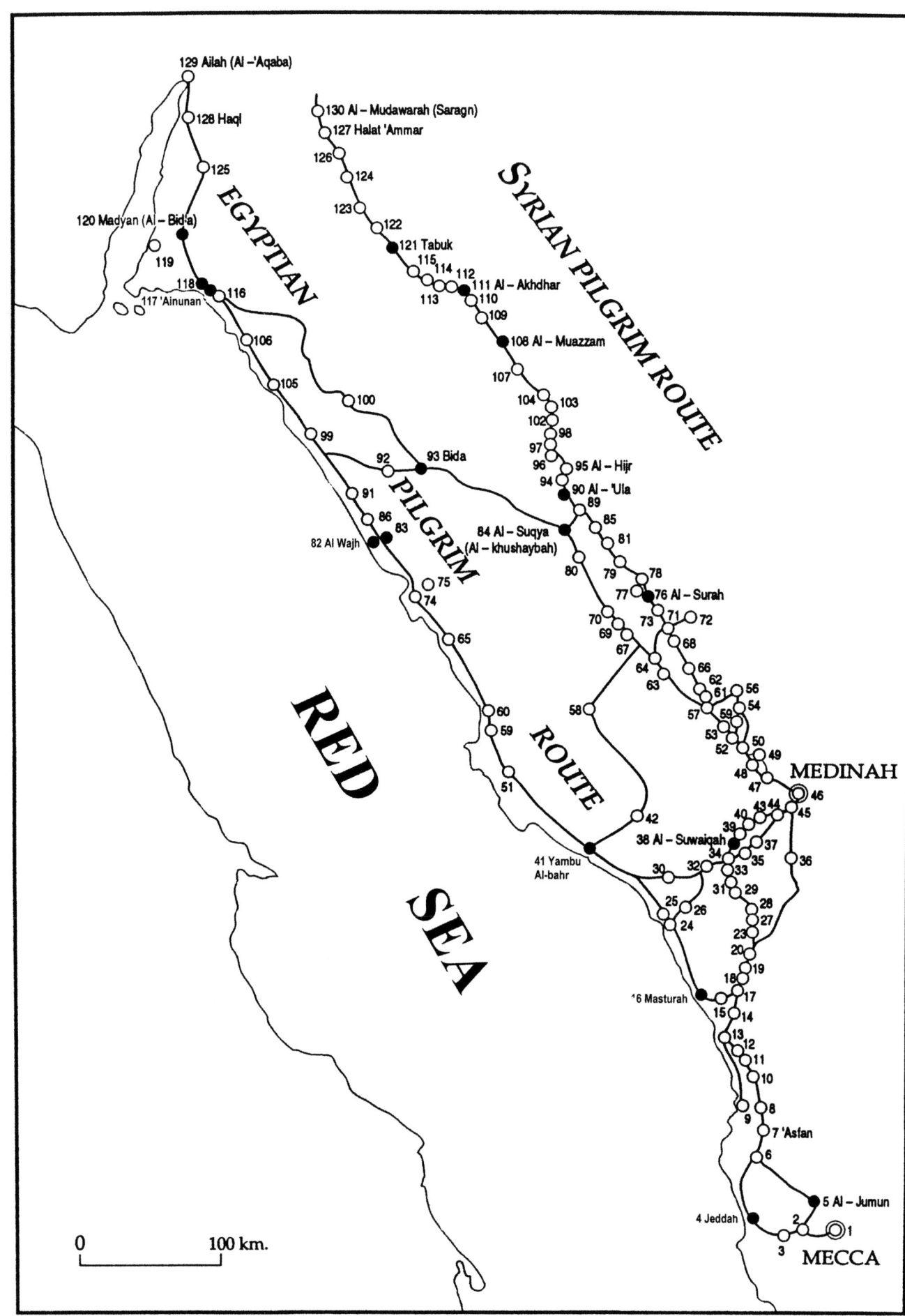

Figure 18.2 Locations of settlements on the Syrian and Egyptian routes.

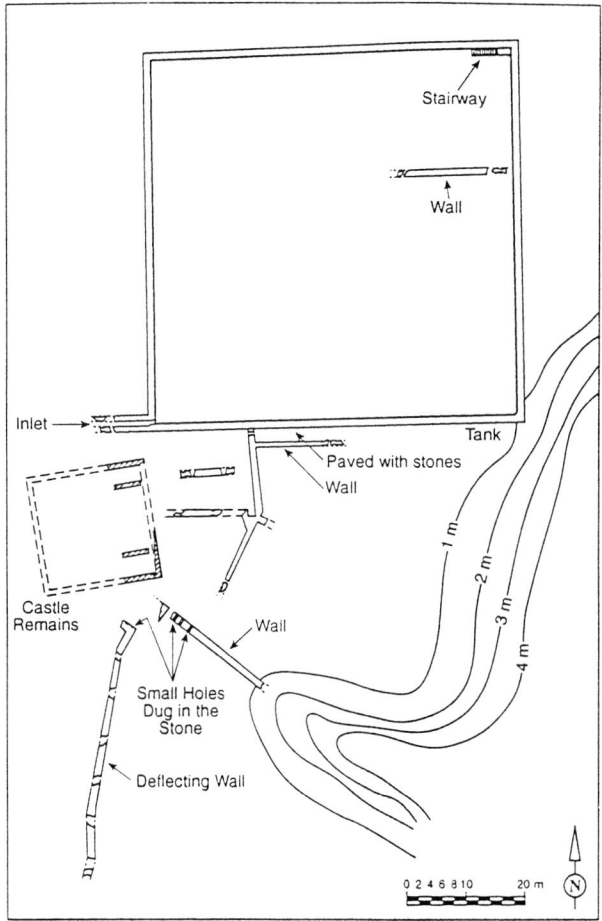

Figure 18.3 Plan of the water tank at Al-Muazzam.

and in other cases the modern settlement has moved from the ancient location to a new one, perhaps promoted by changes in contemporary road patterns. For instance, Al-Suqya (Site 84) was a well-known pilgrim station where the Syrian and the Egyptian pilgrim caravans gathered before continuing their journey to Medinah. Now it is called Al-Khushaybah. Thu Khushub (Site 49) which is a pilgrim station situated on the Syrian pilgrim route is now called Al-Mendassih.

Al-Suqya-Umm Al-Birak (Site 20) was a large and well known pilgrim station situated between Medinah and Mecca. The village of that name is now located about 10 km from the original site.

Where names and locations of pilgrim stations appear to have changed, we interviewed local residents to attempt to locate pre-existing sites. The disappearance of pilgrim stations could be attributed to natural factors such as storm flows, where the structures were reported in wadis and have been swept away, or simply to surface sand movement covering the site. Further wind or water-promoted movement of the surface may again reveal some residual structures.

Unknown sites

In this group we have found a simple site called Al-Hafiyah-Al-'Amair (Site 43). As far as is known it has not been mentioned by any writer. Bedouin who are familiar with this area said that this site was a pilgrim station and showed us the route through which the pilgrim caravan passed. The site consists of both a tank and remains of a canal, both of distinctive construction, in an area where there are many remains of canals that might indicate that the area had been cultivated by irrigation. The tank is a rectangle about 30 m x 24 m uniquely constructed with two stone walls set with mortar, infilled with rubble. The tank has two stairways and is fed by a canal about 40 cm wide and 20 cm deep, built of stones and plastered with gypsum. Two channels branch off from the canal in the form of a cross. The main channel runs for about 3 km and then disappears at the foot of the mountains. Manholes are located at varying distances and connect to the main channel. This channel was probably the means by which the tank was fed with water from springs which have subsequently dried up.

Tank design

The two storage tanks described so far are rectangular but a superb 3 m deep and 100 m diameter circular tank was found at Al-Fayjah on the Al Jumun site. It has an intriguing central structure which might have allowed filling from an artesian groundwater supply and may also have served to aerate the water. Certainly this tank in the centre of a wadi with a capacity of 23500 m^3 would have taken much organised filling (Figure 18.5).

Looking at this site and comparing it with Marr Al Zahran in the same area, where the narrow canals are still maintained and working, we find in the latter case an irrigated and cultivated surface which supports a small settlement.

Bir Al-S'aidini, Site 120, has a watertank, a well and the remains of a further well and canal. The watertank is rectangular, measuring 27.60 m x 18.20 m and built with locally available materials, sedimentary stones plastered with gypsum and mud. This tank is the earliest one found where mud was used as plaster.

Most of the materials used in tank construction were local. Amongst the stones used were granite, other volcanic rocks, limestones and sandstones. Gypsum was used extensively as plaster to seal against leakage and strengthen the structure. Sand, earth, gravel and small stones have been used as infill material. The structures on the pilgrim routes are quite varied in shape, in size and capacity. The tanks appear as perfect squares, rectangles, circles and occasionally as complex irregular rhomboids. Only two circular tanks are to be found along the pilgrim routes and they are at Marr Al-Zahran, Al-Jumun, near Mecca. By contrast the tanks at Tabuk seem to have been built in sequence and enlarged (Figure 18.6).

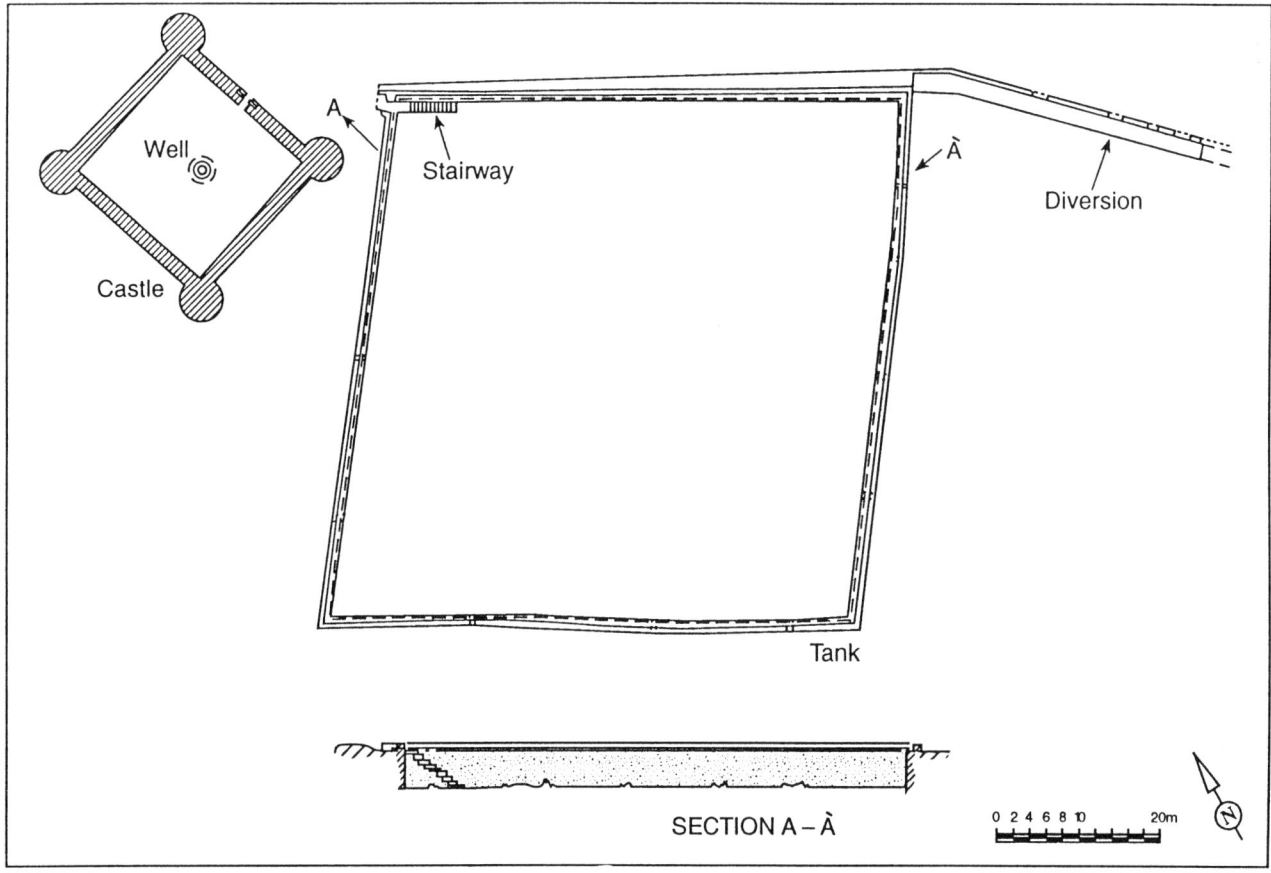

Figure 18.4 Plan of the water tank at Al-Dar Al-Hamra.

To engineer tanks of this size indicates organised supervision of the needs of pilgrims and planned building. It will take excavation to see whether the sites illustrated here were built in a single stage or developed on sites that previously had smaller facilities. The complexities of the structures provide clear evidence of the skills and abilities of the engineers and labourers who constructed them. The remains of foundations at sites vary in character but settlements must have been associated with the medium and larger sites from an early time. It required effort and organisation to ensure that ditches, canals and tanks were maintained and free flowing. Pilgrims required food, fruit and vegetables so there was a market for produce which must have encouraged settlers to extend irrigation systems and cultivate land. Outside of the annual pilgrimage period these were still trade routes with customers passing regularly along the routes, so settlers could trade on a year round basis.

Dating and design characteristics

Both water resource structures and the remains of buildings, particularly castles or fortresses, are constructed from local materials, and usually plastered with gypsum. However, the structural relationships between the Ottoman castles and the wells which are located inside these castles and fortresses are much clearer than the relationships between the castles or fortresses and the water resource structures which are located outside them. This indicates that the castle and water structures within the courtyard were probably built at the same time, while water structures outside were constructed at different times. The relationships are well illustrated at Al-Mu'azzam (Site 108; Figure 18.3) where the castle and its well dates to 1622 but the tank, of different construction, is most probably earlier in date.

Contemporary construction of castle, well and tank is shown at That Al-Hajj (Site 126) and at Al Wajh (Zurib) Site 82. In the absence of dated inscriptions, dating by plan and other features is uncertain. Generally, however, the water tanks are older than associated castles. There are a few places where the style of architecture makes dating more secure. At Al-Sayalah (Site 39) there is a cistern with a style of pointed arch also seen in Darb Zubaydah. A circular water tank at Al-Jumum (Site 5) has a style and structure similar to Birkat Al-Rabathah at Darb Zubaydah. Bir Abbas (Site 33), in its style and structure, looks like Al-Abyar in Darb Zubaydah particularly at the Al-Juf area. They can be firmly dated to the Abbasid period (AD 750–1258).

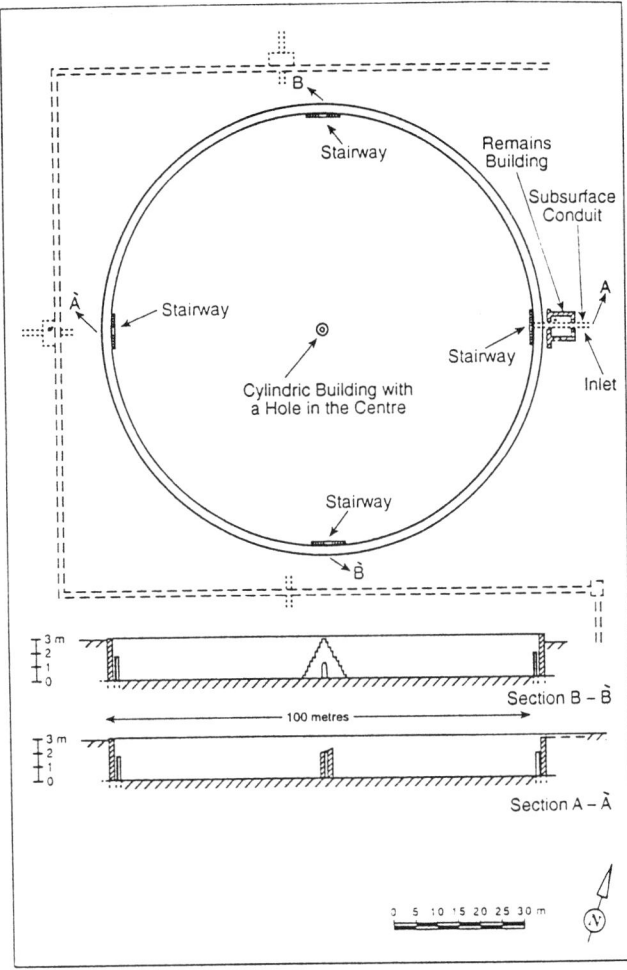

Figure 18.5 Plan of the water tank at Al-Fayjah, Al Jumun.

There are two paved trails on the Syrian and the Egyptian pilgrim routes: one is at Al-Hamra (Site 54) and the other at Riya Harsha (Site 14). These paved mountain ways are similar to those found in Darb Zubaydah, the pilgrim route from Iraq to Mecca, at Al-Rasifa and also suggest a date in the Abbasid period.

Well design

Wells are very variable in design and construction. At Bir Al-S'aidini (Site 120), the upper shaft of the well was dug in a circular shape while in the middle the shaft is roughly semi-oval. It probably existed prior to Islamic times. A hundred metres away is a circular well dug into rock. About 15 m north-west of this site is a well with a much larger circular mouth, about 6 m in diameter. Some examples of well plans are summarised in Figure 18.7 but as yet no common features in design are identified.

Labour requirements and journey times

It is impossible to identify directly the number of labourers and engineers involved in the construction of these struc-

tures. We do not know whether these were local workers or slave labour. In order to estimate the relative amount of labour involved, the weight of construction material has been determined for a variety of structures. Stone was generally quarried within 1–2 km of the site. In most structures the stones had to be trimmed to fit the design, particularly the faces to be plastered. Estimates of masonry weights involved are presented in Figure 18.8. Before construction could start partial excavation of the site was necessary and the weight of spoil material removed would have been substantial (Figure 18.8). In all cases the estimates are calculated as the minimum figure.

Because there are so many variables in wall building and tank construction the only valid comparison of effort is in the weight of spoil shifted during the period of construction. We have assumed that one cubic metre of spoil would require approximately 2.24 man hours of effort. Therefore, in Al-Jumum the removal of the spoil would take 158,565 man hours. It would, therefore, take ten men nearly 320 days at 5 hours per day, but if the spoil was to be moved any distance then these times would be greatly extended.

Water requirements

Water consumption by pilgrims or camels depends on many factors such as age and health of an individual and the time of year. In the summer heat a person needs about 5–6 litres for drinking and 4–8 litres for washing daily. In summer a person cannot go for more than 1–2 days without water, while in winter it could be 3 days. The camel can drink between 28–48 litres at a time. The performance of the pilgrimage 'Al-Hajj' was restricted to the twelfth month in the Islamic lunar calendar. It was necessary to perform rituals at Mecca between the 8th and the 13th day. However, because of the discrepancy between the Islamic year and the solar year, the period of 'Al-Hajj' varies from winter to summer and back to winter over a 32 year cycle.

It is difficult to identify precise numbers of pilgrims who made Al-Hajj before this century because there were no annual records. However, there are some sources which provide an estimate of the number of pilgrims. About 300 Muslims went on the first Hajj in AD 620 (Ibn Kathir 1966). The number of pilgrims who accompanied the prophet Mohammed in his last pilgrimage was between 90,000 and 114,000 (Al-Jaziri 1983). Clearly the size of pilgrim groups grew rapidly. The numbers of pilgrims may be affected by the political, economic and military situation of the Islamic countries, so that the number of the pilgrims varies from year to year. The pilgrimage season in 1279 was very large '... with 40,000 Egyptians alone and the same number of Iraqis and Syrians' (King 1977).

Asking how many people a tank like that at Al Jumun could support, with its capacity of at least 23 million litres, is another approach. Assuming a requirement of 10 litres per head per day for travellers without animals, this tank

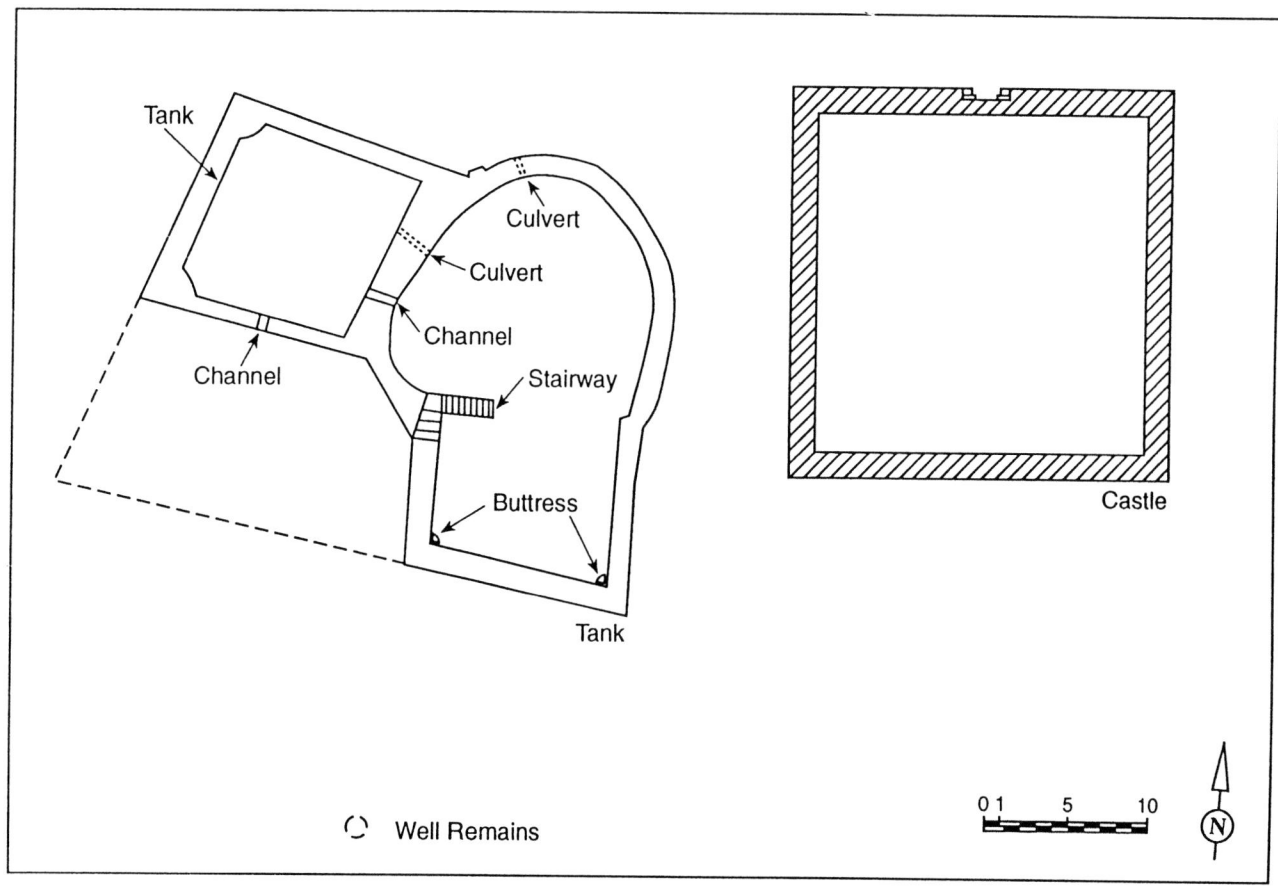

Figure 18.6 Plan of the water tanks at Tabuk.

could supply over 70,000 people for a 30 day period. In contrast, at Tabuk the combined capacity is 1.25 million litres and so if pilgrims were to rest for 3 days the pilgrim capacity would be about half that of Al-Jumum. If, however, some assumptions about canals and transit water are entered these estimates change dramatically. If each group of 3 pilgrims accompany a camel and take one day's travel supply of water, the *per capita* intake at the station becomes 43 litres. Thus the pilgrim capacity of Al-Jumum drops to 17,500 for the 30 days.

ARCHAEOLOGICAL INTERPRETATIONS

All the pottery found during field work was collected from the surface and all belongs to the Islamic period, apparently the Abbasid period. Unfortunately no coins were found during the survey. This lack of datable evidence makes establishing dates and the interpretation of the remains more difficult. Much archaeological work depends upon analogies with other sites but in Arabia generally, and Saudi Arabia in particular, there has been relatively little archaeological excavation and the problem of interpretation and dating these remains will exist until excavation takes place.

In considering dates generally, it is logical and poten-

tially likely that the majority of the route stations and the remains all belong to Islamic times. Some, however, probably belong to the early Islamic period whilst others might be more recent. The exceptions are some stations such as Al-Hijr, Madain Saleh, Al-'Ula and Tabuk, which existed in pre-Islamic times.

Dates for the castles are easier to estimate. Wallin (1979) concludes castles on the Egyptian and Syrian routes are similar in construction, although differing in size. They were probably built by sixteenth century Turkish Sultans to protect and supply the pilgrims, and to guard the wells they generally enclose. That at Al-Hajj, Site 126, is referred to by a number of Arabic writers. Ibn Batutah (nd), writing in 1327, mentions it as an important pilgrim station only and does not mention its castle. Al-Mawsawi (1965) when he passed through in 1629 states that the pilgrim station has both a castle and a water tank, suggesting that the castle was constructed between 1327 and 1631 and the inscription above that castle indicates That Al-Hajj was built in 1563. The coincidence of the written date with the inscription date gives confidence that the well associated with the castle and the route station may have been constructed between the same dates. The castle itself was renewed in 1849 and so the photographic and cartographic evidence associated with this site may be misleading under superficial examination.

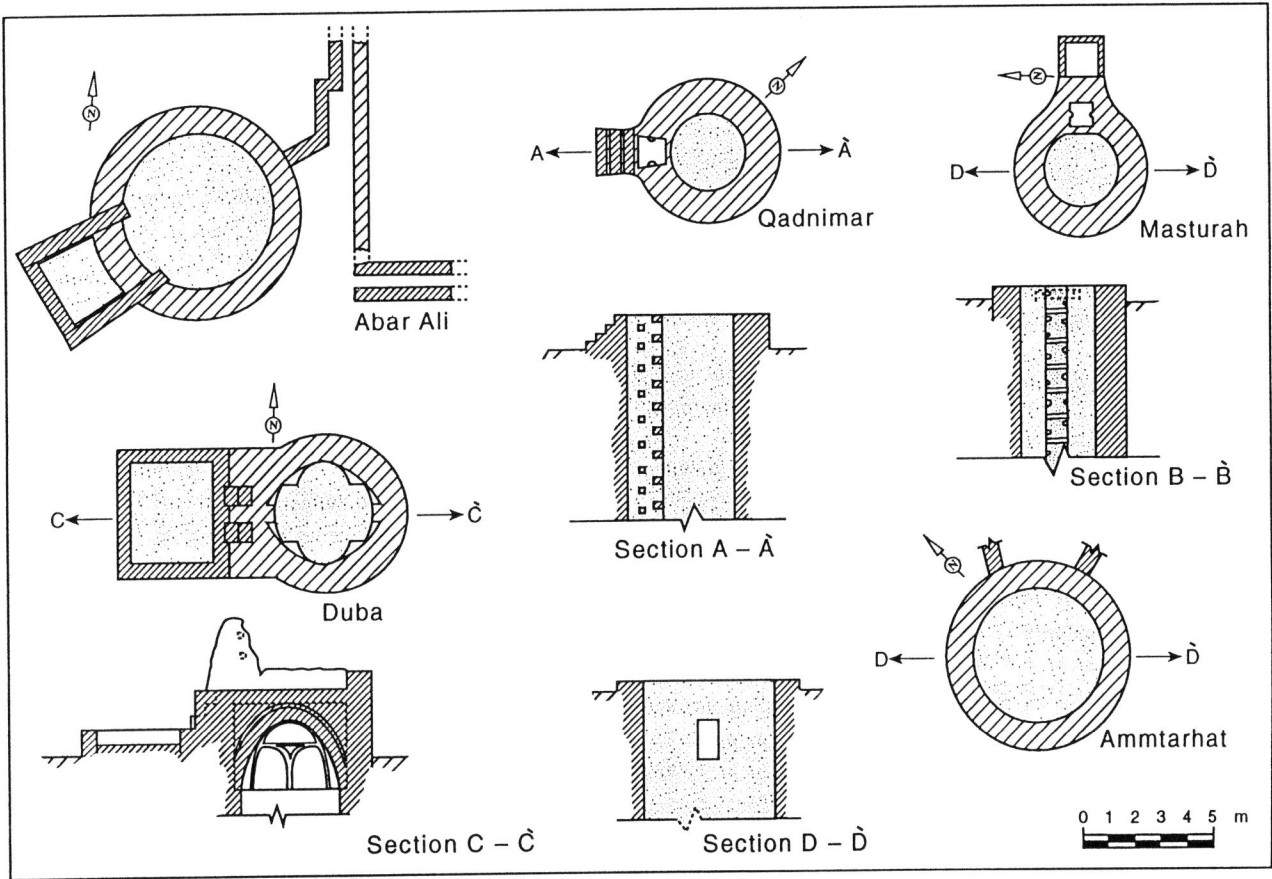

Figure 18.7 Plans and sections of wells.

Al-Akhdhar, Site 111, is a wadi visited by the prophet Mohammed on his way to Tabuk in AD 630. It has a castle, Sulimaniyah, which was built in 1532 according to the inscription and it has springs which feed three wells. Clearly this is a station which has a very early documented history of existence, and the castle was added at a much later date. Castles alone are likely to provide dating information indicating the latest possible date of creation of the water resource structures.

Al-Muwelih, Site 106, on the Egyptian coastal route is mentioned by a number of Muslim and Arab writers. Al-Jaziri, in the sixteenth century AD, says that its castle was built in 1560. He also mentions that the Sultan ordered some wells to be dug to provide the pilgrims with water. There are still some wells, in very good condition, at this site but it is difficult determine their date without further excavation.

Al-Wajh, Site 82, is a coastal caravan stop. There are a number of wells and a castle as well as a water tank connected to the castle. This castle called Qalah zurib was, according to the inscription, constructed in 1616 or 1617.

Directions for future work

Although a great deal of work has been conducted during the field investigations it is still very preliminary. It is very difficult to specify particular sites or routes to be the subject for further work because the area has received so little archaeological attention in the past and virtually the whole area needs to be explored. Nevertheless, there are some route stations which are more important than others and more helpful in terms of archaeology.

Al Jumun, Site 5, is very important because it is close to Mecca, covers a large area and contains a number of different sites with various archaeological remains including water resource structures, buildings, castles and foundations of other unidentified buildings.

Al-Swaiqah is mentioned by a number of early Islamic writers as a pilgrim station. Bedouins familiar with this area claimed that the village of Al-Swaiqah was both large and well populated but only the ruins of water resource structures which include tanks, wells and channels are still visible, with no evidence of settlement.

The traverse roads connecting the main pilgrim routes need more survey. They were the first to be abandoned, and may be the least disturbed. On the route from 'Ainunah inland to Al-Suqya-Al-khushayhah are sites which have disappeared and changed name. The road between Yanbu Al-Bahr and Medinah was a traditional road now abandoned because the modern road takes a different route and again this may have preserved some ancient features.

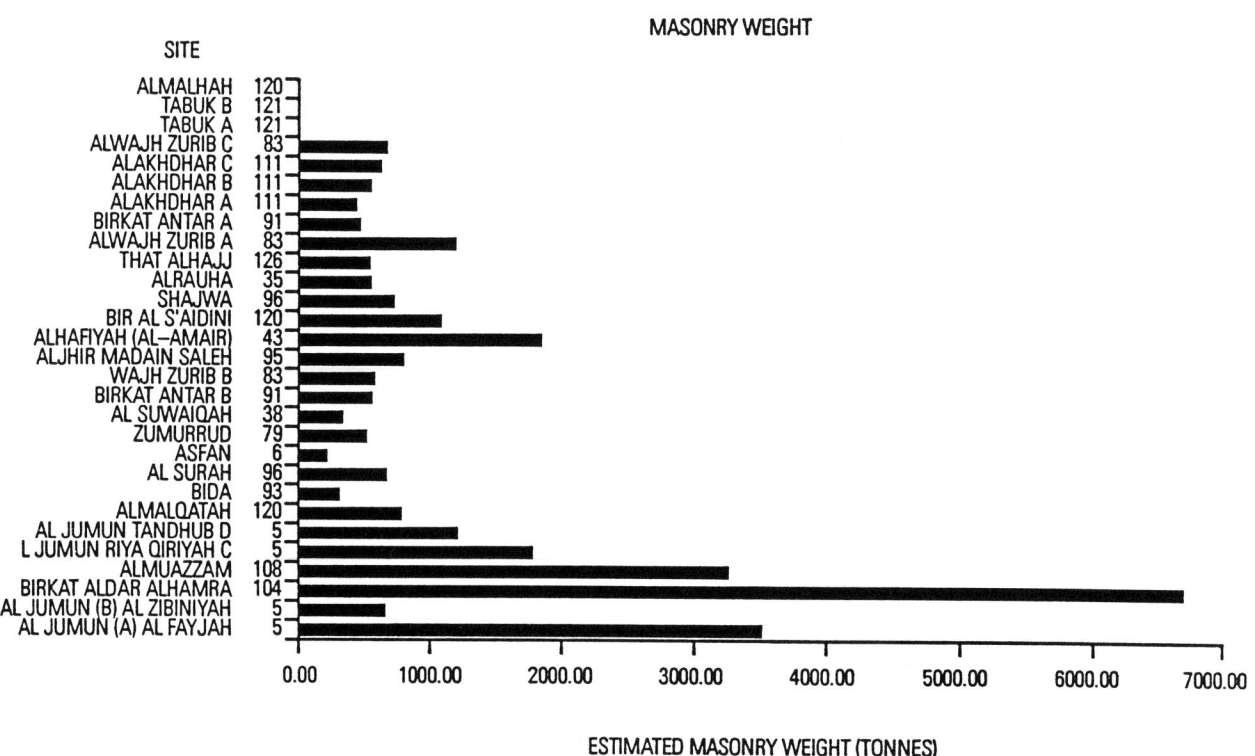

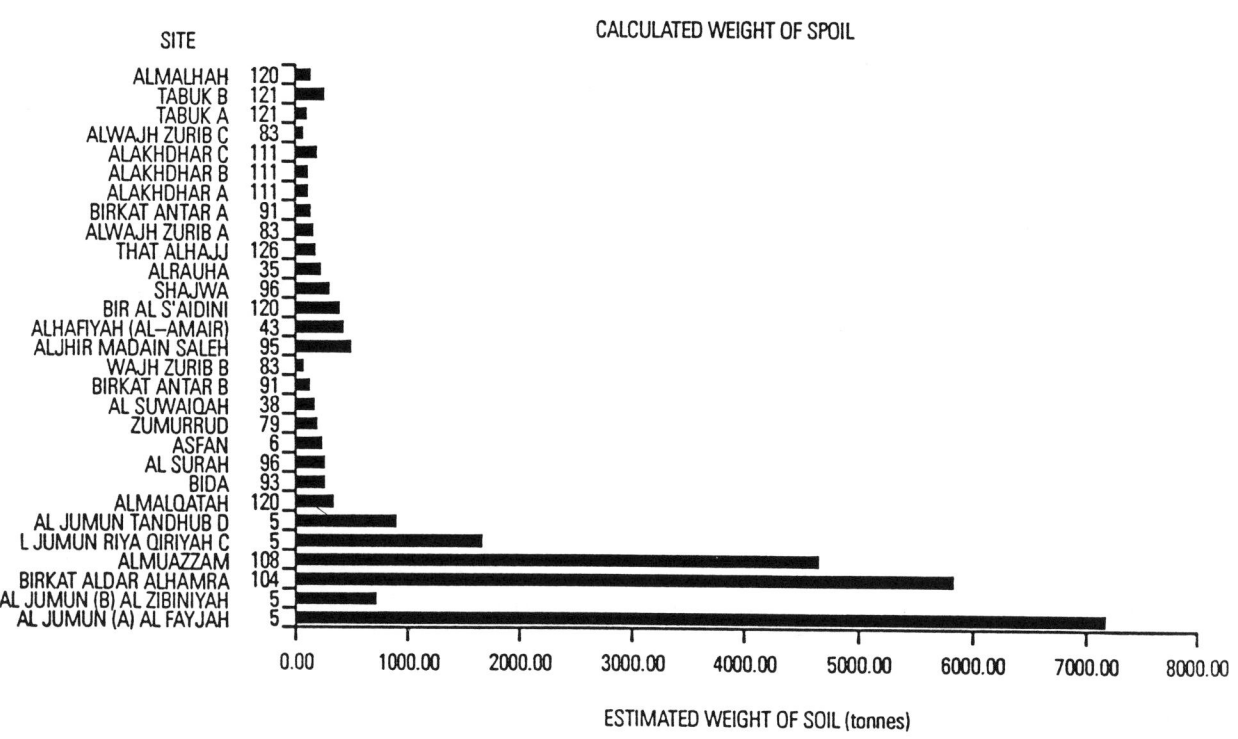

Figure 18.8 Estimates of spoil and masonry weights at various sites.

CONCLUSION AND RECOMMENDATIONS

The survey reveals a diversity of water resource structures including tanks, wells, canals, cisterns and springs. Most water resource structures appear to be built near the banks of wadis, particularly the tanks and the cisterns. The wells are usually found at the bottom of the wadis or very close to the base of the wadis. Here ground water is easier to reach. The tank and the cisterns appear to be fed from a variety of sources such as springs, wells or from trapped rainwater during rain periods. We believe that during Al-Hajj, water would have been brought to some cisterns in water skins.

The water resource structures are diverse in their shape and size. The locations of structures have been carefully selected and in general they are built from locally available materials, stone and gypsum. Most of these features are well constructed and have survived for long periods of time. They are almost entirely hand-worked and hand-built. The style of construction of cisterns, wells and tanks is characterised by the use of two faces of stones, almost always trimmed. In some cases there is evidence of settling ponds, filters and deflecting walls, both to capture the water from nearby wadis and wells and to some extent to clean the water of particulate matter. Some of the water-trapping channels (today usually known as catchwaters) run for many kilometres.

The importance of water resources is emphasized by the presence of fortresses to protect the water resources. Castles are built adjacent to the water tanks often with the security of a well within their perimeter.

It may be the case that wells were the earlier form of water resource structure. However, with the increasing number of pilgrims using the Syrian and Egyptian pilgrim routes, it may have become necessary to build tanks in order to store water from a wider area and over a longer time period. This would also have made water available to greater numbers of pilgrims. The fortresses may have performed a policing as well as a military role.

On the Syrian route the Muslim caliphs started to construct water tanks, basins, wells, and guest houses from the AD 700s. After the crusades the Mamluk caliphs built wells and basins with fortresses and citadels for protection and in some cases castles as at, for example, That Al-Hajj, dating from the 12–15th centuries.

From AD 1516 to 1918 the Ottoman rulers increased the facilities especially following the spread of Islam to the Balkans, Bulgaria and Serbia, increasing pilgrim numbers. In the twentieth century the development of the Damascus to Mecca railway led to new structures, with water holding facilities, rest houses and forts to protect the line. The railway was damaged in the first world war and was never re-opened fully.

On the Egyptian route the collection focus of African pilgrims was Al Fustat, the old capital of Egypt. Here foot travellers accumulated in caravans, either crossing the Sinai peninsula to Aqaba and then walked down the coast to Mecca, or, if more wealthy, travelling by sea to Jeddah. The route through Sinai had to be abandoned as unsafe during the two hundred years of the crusades and some of the stopping places disappeared or reappeared with new names. The last great caravan with 1170 pilgrims is recorded leaving Cairo in 1883 with pilgrims from Egypt, Syria, central and north Africa, Spain and Morocco. By the 1950s camel caravans were disappearing completely and at about this time air travel became a real alternative; Jeddah airport was built to receive thousands of pilgrims annually.

The Syrian and Egyptian pilgrim caravan routes, both coastal and inland, share some features:

* both were used pre-Islam for trade caravans.
* both were used for military purposes when the Islamic armies conquered Bilad Al-Sham, Syria and Egypt.
* both became pilgrim routes after the rise of Islam so that they have conventional and beaten roads.
* both are provided with water resource structures and other facilities such as castles, fortresses and rest houses.
* in general, water resource structures become more numerous when the pilgrim routes approach the holy cities Mecca and Medinah.

They differ in that:

* the Syrian route mostly passes through desert with few permanent settlements. The Egyptian caravan coastal route follows the Red Sea coast passing through a number of populated seaports and villages.
* the Syrian route has more and larger water resource structures, more wells and tanks, than the Egyptian route, reflecting the larger numbers travelling.
* in 1908 the Al-Hijaz railway between Damascus and Medinah was opened. The train replaced camels as the principal form of transport and the Damascus to Medinah route became the more popular.

This initial survey suggests future work to:

* obtain a clearer picture of the history and volume of traffic movement through the sites.
* remove ambiguities relating to date by detailed archaeological investigations at specific pilgrim stations.
* investigate the history and archaeology of the castles and fortresses.
* locate the 'lost' sites.
* identify the hydrological causes of dried up sites.
* explore the role of the traverse routes.
* identify and conserve the sites vulnerable to damage.
* look further at the labour input and skills of the workers who constructed tanks, wells and fortresses in these harsh locations.

REFERENCES

Al-Ibrahim, A A 1991 'Excessive use of groundwater resources in Saudi Arabia: impacts and policy options', *Ambio,* 20, 1, 34–37.

Al-Jaziri, A B M 1983 *Durar Al-Farai'd Al-Munadhamah, Fi Akhbar Al-Hajj was Tariq MAKKA Al-Muadhamah.* 3 Vols. Hamad Al-Jasir (ed), Riyadh, Saudi Arabia: Dal al Yamamah.

Al-Mawsawi, M B A 1965 (2nd edition) *Rihlah Al Shita Wal Saif.* Beirut, Lebanon: Matbaat at Maktab al Islam.

Al-Resseeni, I 1992 *The water resources structures on the Syrian and Egyptian pilgrim routes to Makka and Medinah.* PhD thesis, University of Leeds.

Berkowicz, S M, Lawford, R G & Yaalon, D 1992 'Climatic and geomorphic factors influencing environmental aridity in Israel's arid zone', *in* Atkinson, K A & McDonald, A T (eds), *Environmental management in Canada.* University of Leeds, 27–68.

Ibn Batutah, M I' A, nd *Rihlat Ibn Batutah Dar Al-Kitab Al-Lobnani.* Beirut, Lebanon: Maktabat Al-Madrasah.

Ibn Kathir, A A 1966 (1st edition) *Al-Bidayah Wa Al-Nihayah.* Volume 5. Riyadh, Saudi Arabia: Dar al Yamamah.

King, R 1977 'The pilgrimage to MAKKA: some geographical and historical aspects', *Erdkunda,* 26, 61–71.

Wallin, G A, 1979 *Travels in Arabia (1845 and 1848).* Oleander: Falcon.

19. Marginality, multiple estates and environmental change: the case of Lindisfarne

Tony Brown, Sarah Crane, Deirdre O'Sullivan, Kevin Walsh and Rob Young

Abstract

This contribution considers the environmental and socio-economic context of recent archaeological work on the island of Lindisfarne. Whilst it can be argued that the island is both physically and socio-politically marginal today this was not the case in the ninth century AD when a farmstead, which was probably part of a large multiple estate, existed on the north side of the island. Environmental work has shown that this farmstead was then close to agriculturally useable land and there was a wider range of natural resources available than today. It is also believed that the Lough on the island is not natural but was dug in the late seventh century or early eighth century AD for the monks of the priory. The farmstead was abandoned for socio-political reasons almost certainly well-before the environment became marginal due to sand dune migration.

The authors believe that this study reinforces the contention that environmental and socio-political marginality cannot be studied alone and both need to be explained in the context of social and economic factors of dependency and independency.

INTRODUCTION

This paper is the product of a collaborative programme of research on the settlement, land-use history and archaeology of the Northumbrian island of Lindisfarne. In it we will discuss the context of one site, the ninth-century settlement at Green Shiel on the north side of Lindisfarne. This is an area which is totally unpopulated today and in the popular perception is a highly 'marginal' location. We hope to show that in the ninth century the situation was much more complex. To fully understand the nature of the site and its implications for ninth century settlement and the development of modern attitudes towards its 'marginality', it must be studied in terms of contemporary, ie ninth-century, economic and social structure, as well as the continuous processes of landscape and environmental change which impinge upon it at the present time.

THE SITE

This is not the place to describe the results of the excavations at Green Shiel in detail; excavations began in 1983 and are expected to continue until *circa* 1997. A number of interim reports have appeared (Beavitt *et al* 1985; 1986; 1988; 1989; O'Sullivan & Young 1991a; 1991b). The site consists of a group of five buildings and associated yards, now located in an area dominated by sand dunes. The occupation is dated by nineteen coin finds to the middle decades of the ninth century (*circa* AD 835–871). At least two of the structures were probably dwelling houses, divided laterally into areas for animal and human shelter. The buildings are usually over 20 m long but crudely made with wide, irregularly-coursed walls. About two-thirds of the site has been excavated but apart from the coins there are few artefacts, suggesting either poverty of material culture, or a short period of occupation, or both. Two distinct phases of building were evidenced in two of the structures. Cattle dominate the animal bone assemblage from the site and an analysis of material from Building C indicates that some 45% of the total number of animals recorded were calves under two years of age at death. More cows than bulls are present in the analysed sample. The age distribution and the preponderance of female animals over males suggests a cattle economy based on dairying. To the south is an area of broad ridge and furrow ploughing which may be contemporary with the site and which was certainly not covered by sand until the sixteenth century.

THE MEDIEVAL CONTEXT

The later medieval and, to an extent, modern pattern of nucleated villages and open-field farming in north-east England is now generally believed to date from the post-conquest period, although there is little direct evidence for this as there is of course no Domesday Book for Northumberland. Models for the structure of earlier medieval settlement in this area in the ninth and tenth centuries usually adopt the framework of the multiple estate, as initially identified by Jolliffe (1926) and elaborated by Glanville Jones (1976). The distinctive features of these estates are partly size – they are substantially larger than the much smaller manors which are often viewed as the norm in feudal England – but also diversity – they usually covered a range of upland and lowland environments, each contributing to different aspects of the estate's economy.

Controversy surrounds the antiquity of these estates (Gregson 1985) but it can be readily conceded that they represent an efficient way of maximising diverse economic resources over large areas. The economic burden of supporting the lord's household could be shared among different rural groups; demesne land was concentrated in the estate centre where the various vills and hamlets owed dues.

The territories later known as Islandshire and Norhamshire in Northumberland are both fairly typical examples of such estates, and are probably the core of the original endowment of the Anglo-Saxon monastery of Lindisfarne (Craster 1954; Morris 1977). However, no specific record of the extent of the earliest endowments is recorded in contemporary sources. The monastic community abandoned their base on Lindisfarne itself in AD 875, but these estates remained in the hands of the Cuthbert community at Durham throughout the middle ages, forming a discrete area, known as North Durham, which was only incorporated into the County of Northumberland in modern times.

Figure 19.1 shows the tenurial organisation of Islandshire in the early thirteenth century, when the estate centre was clearly at Fenwick (Jones 1976, 64). The various types of obligations are indicated. These large estates are generally in a state of fission at this time. The beginning of this process is again difficult to fix because of the lack of documentation.

Within an overall framework of dispersed settlement the location of individual settlement sites is problematic. Although the organisational structure was tenacious, settlements themselves seem to have been fairly mobile. With the exception of the royal site at Bamburgh, no excavated Anglo-Saxon site in the region was demonstrably continuously occupied for more than a few generations. This suggests that choice of settlement location was economically opportunistic, rather than based on long-established cultural attachment to particular places; the system of farming must also have accommodated settlement mobility and indeed, at least partly, dictated it.

Although the ninth century is often viewed as a period of political and social turmoil in England there is increasing evidence from elsewhere in Europe that population was on the increase; the relatively well-documented estates of Carolingia, for example, provide fairly clear indications of growth (Nelson 1992, 22–3). There is no documentation which can be brought to bear on this question either way for Northumberland, nor is it yet clear whether this expansion was continuous through to the better understood rise in population of the 11th to 13th centuries (Miller & Hatcher 1978, 28–33) but the possibility of steady population increase, even for those parts of Northumbria not directly effected by Scandinavian settlement, must be born in mind.

There are thus three general trends or issues which need to be kept on the agenda when considering the specific example of Green Shiel: a large estate, commanding diverse resources; settlement mobility; and, probably, rising population.

The estate structure is reflected archaeologically in tantalising ways. The resources needed to keep the Lindisfarne scriptorium in calf skin for vellum, for example, were certainly much greater than could possibly have been supplied by herds from the island pastures; likewise the wood for St Cuthbert's coffin, supplied by a prime oak from a well-managed forest, was probably not from the island itself (Cronyn & Horie 1985). Green Shiel was obviously contributing cattle and probably marine resources into the system but it was also part of a wider-based, coin-using economy.

Far from being a geographically marginal site on a peripheral island, the site at Green Shiel was probably fully integrated into a sophisticated and highly structured system of resource exploitation and social administration. What we must now ask is how has the site and the island generally come to be currently perceived as a marginal environment? A pre-requisite for answering this question is an understanding of the development of the sand dune system.

THE DEVELOPMENT OF THE MODERN DUNE SYSTEM

Today, the north shore of Lindisfarne is dominated by an extensive and well developed dune system which is now part of the Lindisfarne National Nature Reserve. It is apparent from air photographs taken during the 1950s, 1960s and the 1970s that the northernmost area of the system took on its current specific form only during the last thirty years. In 1971 the Nature Conservancy Council (now English Nature) undertook some stabilisation of the dunes which included the planting of *Ammophila arenaria* (marram grass) and consequently the system is now relatively stable, at least more so than in previous decades.

It is quite clear from historical accounts that the history of this dune system has been one dominated by periods of

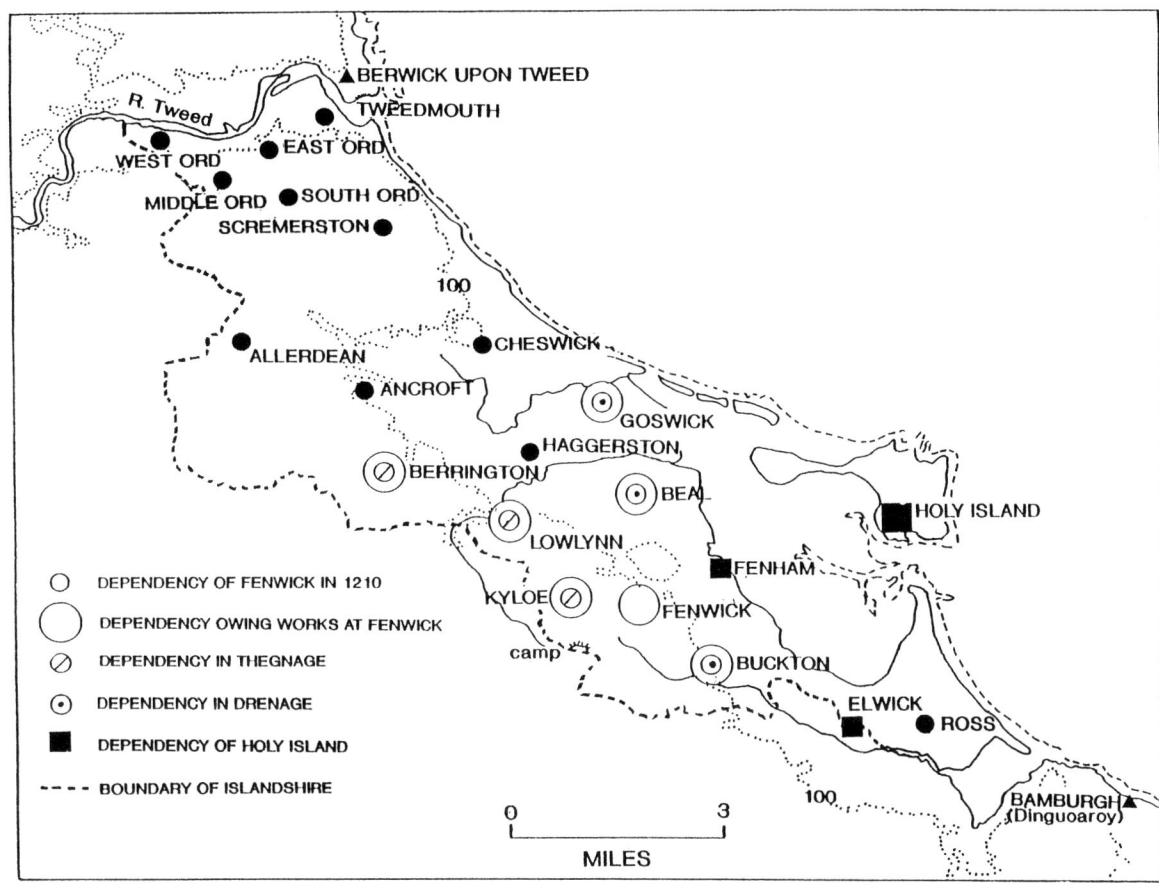

Figure 19.1 Map of the multiple estate of Islandshire in the early thirteenth century (after G J Jones).

severe instability. Buildings within the system have been recurrently buried and exposed (Darlington 1965, 90). Ecologically, the modern Lindisfarne dune system is currently one characterised by typical succession up to dune slack level (Ranwell 1972). The dune system possesses a typical flora which includes *Lotus corniculatus* (bird's-foot trefoil), *Cirsium* sp. (thistle), *Echium vulgare* (viper's bugloss), *Salix repens* (creeping willow) and *Festuca rubra* (red fescue). In order to understand the history of this environment and its early medieval characteristics we need to ascertain when and how this dune system developed, as its existence, or absence, is crucial to the Green Shiel project.

PARTICLE SIZE ANALYSIS

One of the first pieces of work undertaken as part of our reconstruction of the early medieval environment was a consideration of the geomorphology of the Green Shiel environs. The aim of the particle size analysis was to compare the modern sedimentary environment with that responsible for the sediments contemporary with and just prior to the settlement at Green Shiel.

The particle size analysis of some 60 sediment samples from in and around Green Shiel and the ridge and furrow

area has yielded groupings of the various sedimentary environments. Samples from directly beneath building walls, beaches, and dunes, show that the sediments from the buildings are better sorted and more positively skewed than those from the modern environments. All of the evidence indicates that the archaeological samples were deposited by aeolian processes less intense than those which transported the more recent sediments.

One of the most important characteristics of the relationship between wind and sand transportation is the power (cubic) relationship that exists between wind velocity and sand transport. The relationship is such that for any increase in wind velocity there is a proportionately high increase in the quantity of sand transported. Therefore low frequency, high magnitude events such as the storms of the sixteenth century, are of much greater importance than high frequency, lower magnitude events which were undoubtedly responsible for limited sand blow across the lower areas of the site throughout its history. Gottschalk believes that '... over the whole period before 1600, a rising line can be observed in the severity of the storm surges in the Netherlands, reaching its culmination in the 16th century' (Gottschalk 1975, 822).

There is no doubt that a substantial dune area had developed by the end of the sixteenth century as Speed's map of 1610 clearly indicates the existence of such an area. The

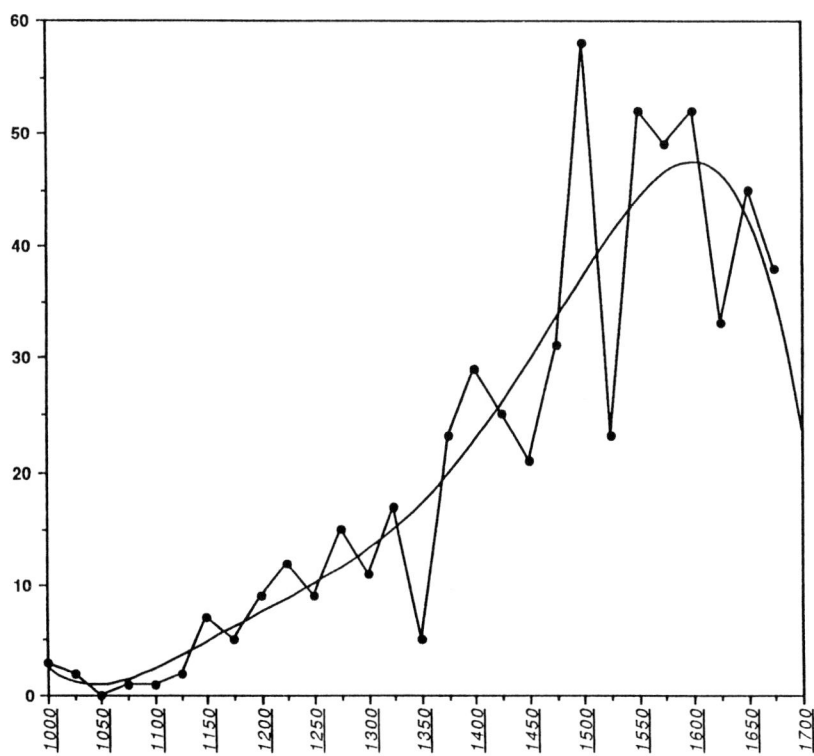

Figure 19.2 Diagram illustrating the frequency of storms in the Netherlands and northern Europe in the period AD 1000–1700 (data from Gottschalk 1971, 1975 and 1977).

most likely period for the development of the Lindisfarne dune system is the fifteenth century; Figure 19.2 shows that there is a peak of storminess in the 25 year period from 1475 to 1500, although the peak of the underlying trend is further into the sixteenth century. As Lamb comments 'Another accompaniment of some of the severe storms of the northeast Atlantic and the North Sea region in the late Middle Ages and after was the overwhelming of a number of coastal places by blown sand' (Lamb 1982, 183).

There are of course problems with this type of historical data in that the increased frequency of storm events is partly a function of the increased reporting and recording of events. However, if the trend illustrated in the graph was a function of this alone, we would expect to see a continuous upward trend into the eighteenth century; the opposite is in fact true. Therefore, there is little doubt that the data do indicate an increase in storm frequency during the late Middle Ages, although we should be aware that earlier periods are probably underreported.

It is our contention that for at least 600 years before the development of the dune system, the north side of the island may not have exhibited any features that would have obviously allowed Lindisfarne to be classified as agriculturally marginal. However, to substantiate this claim we must go further back and reconstruct the early medieval environment.

THE EARLY MEDIEVAL ENVIRONMENT

Evidence for the early medieval environment of Lindisfarne comes from a number of sources. The most important of these is the pollen diagram from the island Lough (Figure 19.5), but there is also geomorphological evidence from in and around the settlement site at Green Shiel which allows us to gain an insight into the nature of the land surface that was contemporary with the site.

Digital terrain models derived from an auger survey across Green Shiel and the adjacent ridge and furrow illustrate the topographical characteristics of the buried soil on the underlying clay till (Figure 19.3; Figure 19.7) and the modern sand dune topography (Figure 19.4). This indicates that Green Shiel was built adjacent to a stable, developed soil in an environment that was relatively unaffected by aeolian processes. The lack of cereal remains from this soil and the evidence from seeds of weeds (especially *Rumex acetosa*), as well as phytolith data, may indicate pastoral land use. This ties in with the economic evidence from the settlement. A date from this soil is an obvious requirement but an attempt to use accelerator dating was unsuccessful (ie, gave a modern date) due to reworking by ants. At the time of writing luminescence dating is under way.

As the digital terrain model shows, the edge of the clay till and the associated soil lie immediately to the

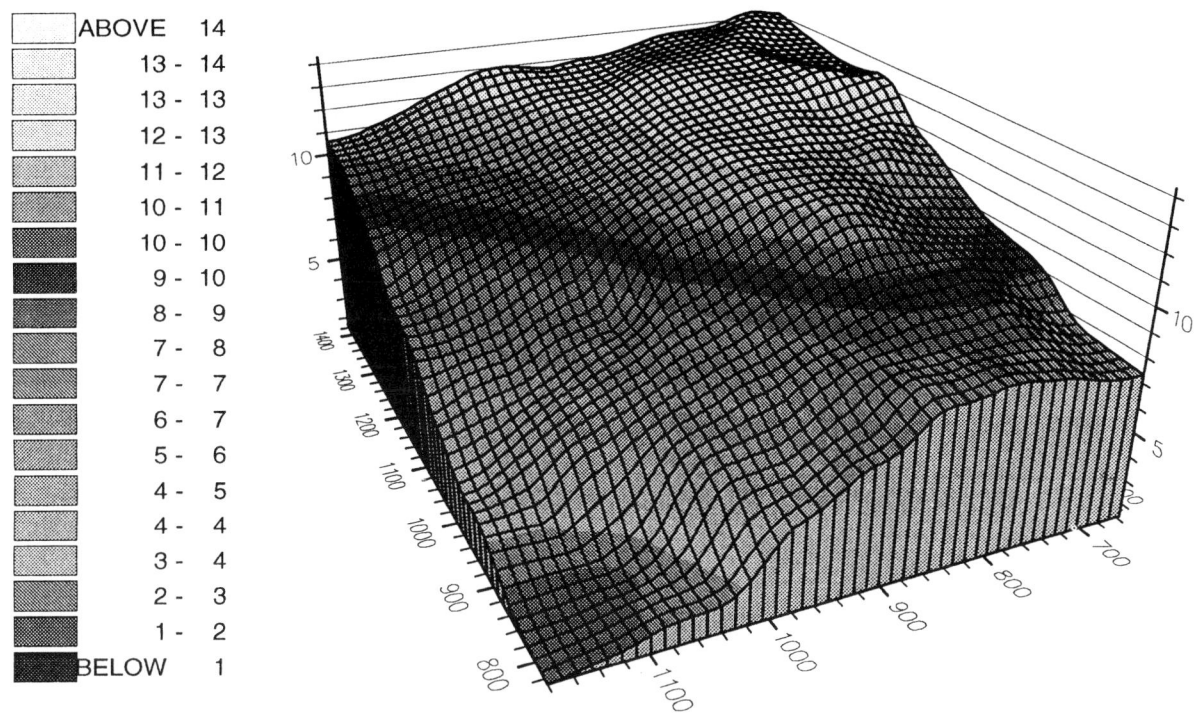

Figure 19.3 Digital terrain model of the sub-dune land surface around Green Shiel viewed from the north-west (data based on EDM survey and augering). Scales in metres from an arbitrary datum. The Green Shiel buildings are located in the bottom left corner of the square centred on x1000 y1000.

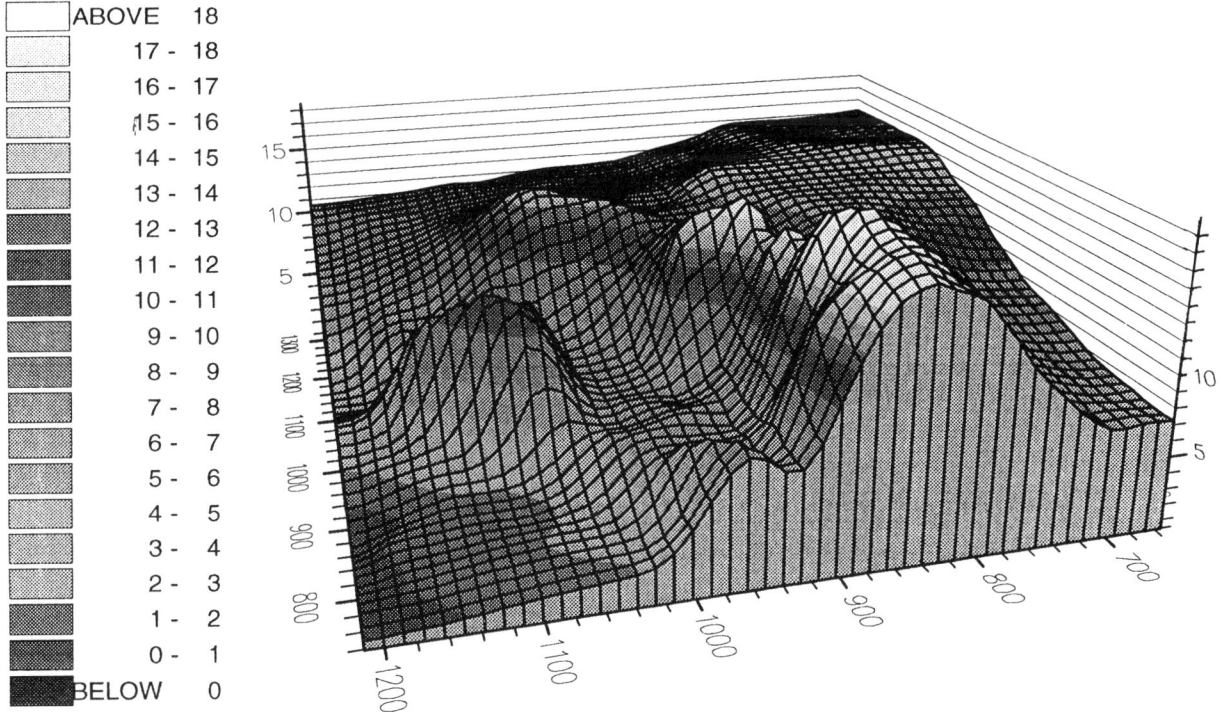

Figure 19.4 Digital terrain model of the modern dune system around Green Shiel viewed from the north-west (data from EDM survey). Scales in metres from an arbitrary datum. The Green Shiel buildings are located in the bottom left corner of the square centred on x1000 y1000.

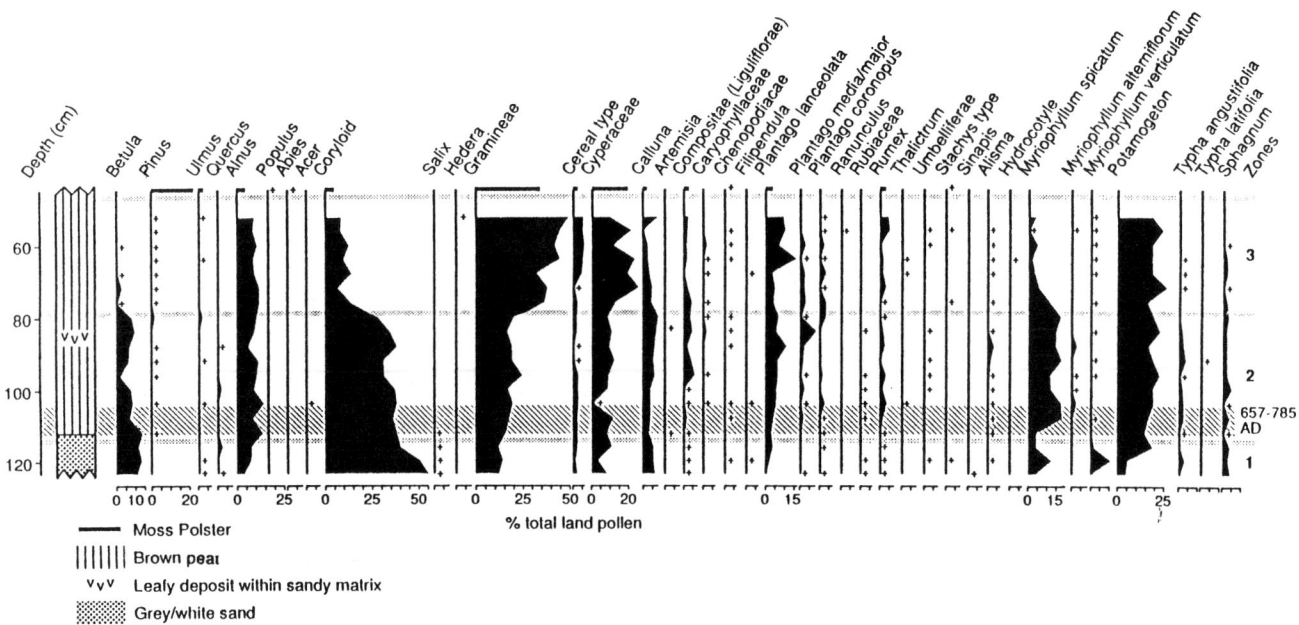

Figure 19.5 Pollen diagram from the centre of the Lough, Lindisfarne (analysis by S Crane).

south and east of Green Shiel. Today, the till cliff-line, as well as the buried soil, is covered by blown sand; in some places the buried soil lies under *circa* 70 cm of sand (Figure 19.8). This soil is approximately 15 cm thick and is a well-mixed dark grey sandy soil with an organic content of about 1% (wet oxidation method). Even if the date of this soil proves to be relatively recent, the fact that the soil exists is of some importance. The proximity of the soil to the Green Shiel site is of significance in itself; but it is also vital to ascertain whether or not it was buried by sand during the early medieval period. The width of the ridge and furrow implies that it was certainly exposed and exploited into the medieval period; this alone is evidence for the Green Shiel environs not having been exposed to the extreme sedimentation processes until the central middle ages at the earliest. Other evidence for an ameliorating climate during the early medieval period comes in a number of forms.

At Warkworth to the south of Lindisfarne, the development of a dune slack is radiocarbon-dated to the late tenth century (Frank 1982); this slack was subsequently covered by blown sand. On Lindisfarne itself the results of the biometric analyses of whelk assemblages also support this contention; specimens derived from archaeological contexts are significantly squatter than modern examples, indicating that the shore line was subjected to less intensive wave action during the early medieval period (Walsh 1988).

When all of this evidence is considered, there must be little doubt that the present dune system, which was clearly developing by the sixteenth century, did not exist during the early medieval period.

ECOLOGICAL EVIDENCE: THE LOUGH

Accepting the geomorphological characteristics of the early medieval environment on the north shore of Lindisfarne, we will now go on to consider the ecological characteristics of this area.

A core revealing 1.17 m of reed peat over sand and clay was taken from the middle of the Lough on the east side of the island. This is about 1,400 m from Green Shiel and the edge of the ridge and furrow. The pollen diagram (Figure 19.5) can be divided into three zones. The top of the diagram has very little pine and therefore pre-dates the reafforestation of the mainland which began in the Kyloe hills at the turn of this century. Reafforestation is reflected in the sample from the moss pollster.

Zone 1 is dominated by coryloid pollen, in this case thought to be mostly hazel (*Corylus*), and by birch (*Betula*), with low grasses (Gramineae) and aquatics. Willow (*Salix*) is also present. There is probably a hiatus between Zone 1 and Zone 2.

Zone 2 sees a fall and then a stabilisation of coryloid and birch and a gradual increase in grasses. The aquatics suggest the Lough had shallowed to a depth allowing colonisation by emergent aquatics including pondweed (*Potamogeton*) and milfoil (*Myriophyllum*).

The division between Zones 2 and 3 marks the most dramatic change in the diagram with a fall in coryloid from 40% Total Land Pollen (TLP) to under 10% TLP as well as a fall in birch. There is a concurrent rise in grasses and sedges (Cyperaceae). The aquatics and sedges suggest a further shallowing of the Lough and encroachment by fringing vegetation, with the disappearance of milfoil

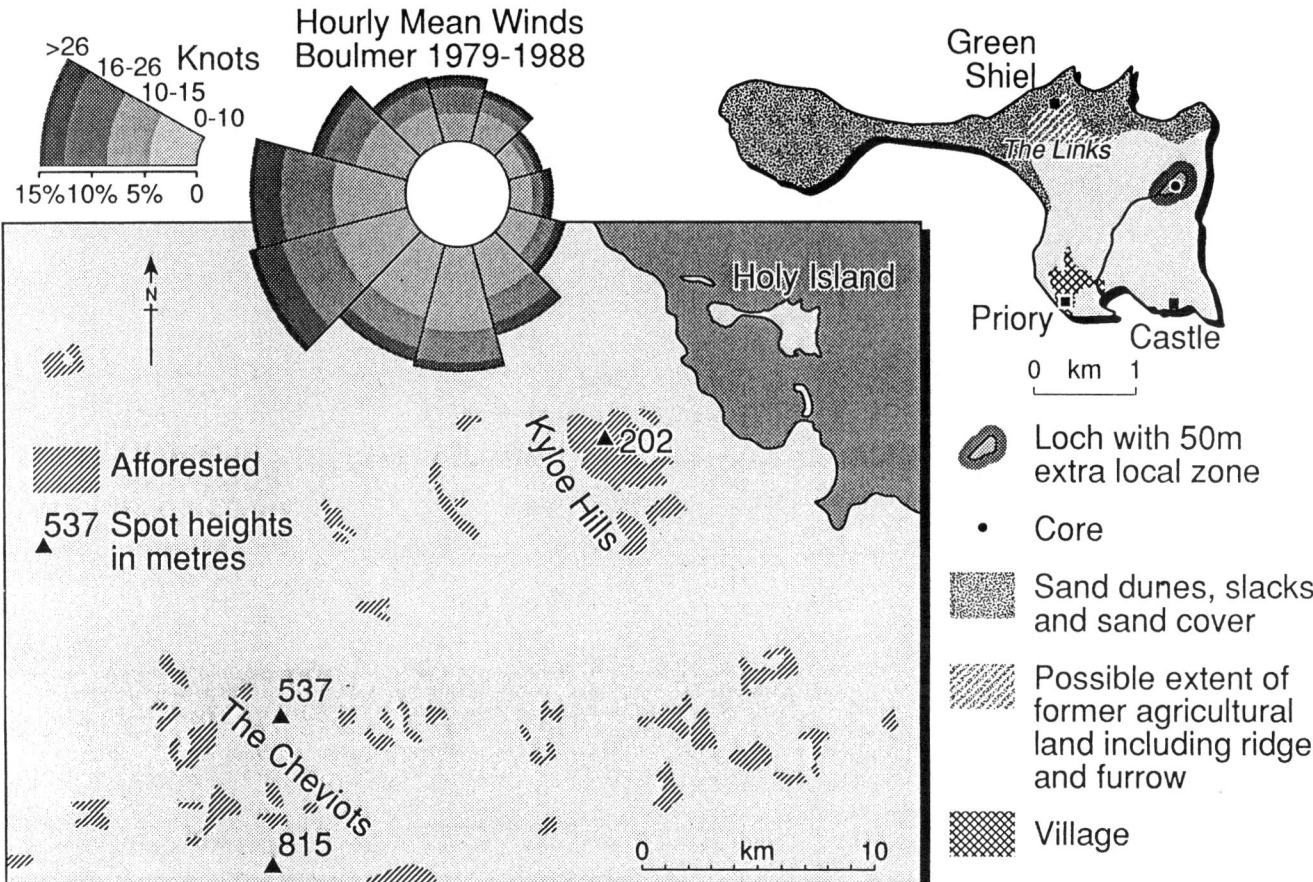

Figure 19.6 Map of the pollen source area for Lindisfarne, and (inset) wind rose diagram. (Wind data collected at RAF Boulmer, Northumberland).

probably because of seasonal drying out (a phenomenon that has been observed in recent summers). Cereal type grains increase, as do open-land species, such as the ribwort plantain (*Plantago lanceolata*) and docks (*Rumex* spp.). It is interesting that alder (*Alnus*) remains roughly constant throughout the diagram.

With a diagram from a small island close to the mainland there are three transport components: the local, from the Lough and its edge; the extra-local, from the island; and the mainland component which will contain both regional and long-distance components. Assuming that prevailing winds were similar to the present-day, then the most important areas for the mainland component would have been the coastal zone and the Kyloe and Cheviot Hills (Figure 19.6). This means that in order to interpret a diagram in terms of the island's land-use history, the mainland component has to be separated out. This can be done by comparison with mainland diagrams, if they are available, or by the identification from modern counts of a background level for types rare or absent from the island today.

Zone 1 is interpreted as reflecting a hazel/birch woodland probably to the north (ie windward) of the Lough.

The only hazel on the island today consists of a few hedges behind the dune field to the north-west. It therefore seems reasonable that the woodland (or a part of it) was between the Lough and the Green Shiel site. The first quarter of Zone 2 has been dated to 657–785 AD (UB-3585), and sees the shrinking and/or opening up of the woodland and its partial replacement by grass and scrub. One possible interpretation is that the woodland was being managed. If coppicing was undertaken, this could have supplied hazel withies as a basic building material for small wattle huts and houses; such structures are common in Britain and Ireland in the early medieval period. The change to the open, grass-dominated landscape of today must represent almost complete clearance of the woodland. The rise in cereal-type pollen (identified using standard criteria; Andersen 1979) may well be just a reflection of the opening-up of the Lough area, and not necessarily any rise in arable cultivation. The background count of coryloid and alder, both now rare on the island, clearly reflects a mainland input. Zone 3 therefore can be taken as reflecting the totally cleared landscape that exists on Lindisfarne today.

The near synchroneity of the radiocarbon date with the

Figure 19.7   Excavated section of till cliff to the south-east of Green Shiel (photo: KJW).

arrival of the early monastic community in 635 AD is remarkable. There is no obvious physical explanation for the initiation of the build-up of organic sediments in the Lough at this time, indeed sediments like this are nearly always considerably older. In addition, this period is generally considered to have been marginally drier and warmer than the present (Lamb 1982). There is no obvious natural mechanism for the removal of older lacustrine or marsh sediments, and therefore the possibility that the Lough was artificially created must be considered. The Lough may have been dug out of a dry depression and converted into a pond by controlling the outflow. The outflow has been controlled in recent times and archaeological work is planned in the outlet area. The reasons why the monks might have done this are not difficult to see, since the Lough would have provided a fresh water supply on a rather dry island and it could also have supplied the island with freshwater fish. This possibility only alters the interpretation of the pollen diagram by reinforcing the possibility of a hiatus between Zones 1 and 2 and the intensive nature of agriculture and resource use on the north of the island in the seventh to ninth centuries.

Map and documentary evidence indicate a young plantation on the north and east side of the Lough by 1854. Johnston (1875) in his notes from botanical walks on the island at this time says it included sycamore, alder, birch, black poplar, mountain ash and willow. Due to the poor condition of many of these trees, Johnston came to the conclusion that the plantation was ill-conceived. Part of the plantation was later cut down to make way for an industrial waggonway which was constructed during the latter part of the nineteenth century. Evidence of this plantation is not found in the pollen diagram (in spite of extended counting for the uppermost peat sample) which suggests that, excluding the modern moss polster sample,

the diagram pre-dates the early nineteenth century. *Acer* and *Populus* are found in the modern sample, but may be derived from the mainland.

Taken together, the settlement, the area of medieval ridge and furrow, and the pollen evidence suggest that during the early medieval period pastoral and possibly arable farming was more extensive in the north of the island. There are many barriers to the recognition of arable farming from pollen analysis of a site like this. Firstly, the site is approximately 1 km from the archaeological site and secondly the occurrence of many of the so-called arable indicators (eg, Chenopodiaceae) can be naturally high in coastal localities. Although some of the cereal-type pollen could be marram grass (*Ammophila arenaria*) the decrease in the curve in Zone 2 suggests this is probably not the case, but could well be the case for the later rise in Zone 3. The site was, however, not poor, but probably a viable farmstead with considerable grazing land and possibly some woodland resources. Other similar sites exist in the north east of England; today this area is perceived as 'marginal' but there is evidence that it supported viable farming communities in the early medieval period. The best-published example is that of Simy Folds, situated 351 m above sea-level on Holwick Fell, in Teesdale, County Durham, where both archaeological and palaeoecological data suggest mixed farming in the mid-eighth century (Coggins *et al* 1984). The ninth-century farmsteads at Gauber High Pasture, Ribblehead (King 1978) and Bryants Gill, Kentmere (Dickinson 1985) are likewise in 'marginal' locations.

DISCUSSION AND CONCLUSION

It is easy to describe Green Shiel as marginal, because it

Figure 19.8 Excavation of the sub-dune buried soil in the area of ridge and furrow (photo: KJW).

now occupies an agriculturally marginal environment in the sense that it only provides poor grazing for sheep; reconstruction of the contemporary environment suggests that this was not the case in early medieval times. A more interesting perspective on this and other sites, however, is obtained by framing the question rather differently. What did marginality imply in the ninth century? Defining a marginal area in the context of such a well-adapted system as a multiple estate is not a straightforward business.

The influential recent work by Beck (1986) can help us to conceptualise the risk environment. Physical risks (including biological risks) are always created in and operate through social systems. There have always been organisations and institutions which manage and control the risk environment, including the social distribution of risk exposure. As Lash & Wynne (1992) note in their introduction to Beck's work 'The magnitude of the physical risk is therefore a direct function of the quality of social relations and processes'. It follows that the most basic risk is that of 'Social dependency upon institutions and actors who may well be ... alien, obscure and inaccessible to most people affected by the risks in question' (Lash & Wynne 1992). In any consideration of marginality in archaeology the total risk environment should be analysed, not just the physical risks. In the case of Green Shiel this includes other external risks such as Viking raids, and probably most important of all, the relationship of the settlement with the multiple-estate and its authority. Archaeological data indicate that the settlement must have been abandoned by the late eleventh century; the total absence of medieval ceramics alone makes this certain. The coins from the site suggest occupation into the mid-ninth century. The latest dateable artefact, a spearhead, could possibly be of early tenth century date. The site was abandoned because the social organisation of settlement within the overall estate structure changed, and not in response to environmental deterioration. The re-organisation of settlement in a nucleated village around the priory is likewise a response to the Norman refoundation in the late eleventh century.

Marginality can only be a meaningful idea in relation to a particular economic and social system – places are not innately marginal even in the geographical sense, since marginality is a relative and scale-dependent concept. Places are only marginal in relation to social and economic pressures, whether or not they are impacted by physical stresses or risks. This is not to say that environmental work is secondary, as the physical risk environment must be specified if social and economic processes are to be made visible. No settlement change in response to increased stress and settlement change without a change in the stress environment are just as important, (if not more so since they are more common) as a settlement response to changes in the physical environment, even in marginal areas. As on Lindisfarne, although these stresses change, the human response is dependent on the relationship between the outer settlements and the centre of economic and political control. Green Shiel was not abandoned as a result of environmental change; Green Shiel probably became sociopolitically marginal well before it had a chance to become agriculturally marginal.

Studies of marginality cannot therefore ignore social factors through which changing stresses are perceived and acted upon. In practice this means that analysis of ecological or geomorphological factors alone cannot 'explain' marginality, but only in combination with the analysis of social and economic factors of dependency and independency.

REFERENCES

Andersen, S T 1979 'Identification of wild grass and cereal pollen', *Danm. Geol. Unders. Årbog* 1978, 69–92.

Beavitt, P D, O'Sullivan, D & Young, R 1985 *Recent fieldwork on Lindisfarne*. University of Leicester, Department of Archaeology Occasional Paper No 1.

Beavitt, P, O'Sullivan, D & Young, R 1986 *Holy Island: A guide to current archaeological research*. University of Leicester: Department of Archaeology.

Beavitt, P, O'Sullivan, D & Young, R 1988 *Archaeology on Lindisfarne*. University of Leicester: Department of Archaeology.

Beavitt, P, O'Sullivan, D & Young, R 1989 'Fieldwork on Lindisfarne, Northumberland, 1980–1988', *Northern Archaeology,* 8 (1987), 1–23.

Beck, U 1986 *Risikogesellschaft: Auf dem Weg in eine andere Moderne*. Frankfurt: Suhrkamp Verlag. (Trans M. Ritter, 1992, London: Sage publications).

Coggins, D, Fairless, K & Batey, C E 1984 'Simy Folds: an early medieval settlement in Upper Teesdale', *Medieval Archaeology* 27 (1983), 1–26.

Craster, E 1954 'The patrimony of St Cuthbert', *English Historical Review,* 69, 177–99.

Cronyn, J & Horie, V 1985 *St Cuthbert's coffin: the history, technology and conservation*. Durham: Cathedral Dean and Chapter.

Darlington, J 1965 *Lindisfarne, Northumberland: Physiography and vegetation*. (Unpublished postgraduate dissertation, University College London).

Dickinson, S 1985 'Bryant's Gill, Kentmere: another "Viking-Period" Ribblehead?', *in* Baldwin, J R & Whyte, R D (eds), *The Scandinavians in Cumbria*. Edinburgh: Scottish Society for Northern Research.

Frank, R 1982 'A holocene peat and dune-sand sequence on the coast of north-east England – a preliminary report', *Quaternary Newsletter,* 36, 24–32.

Gottschalk, M K E 1971, 1975, 1977 *Stormvloeden en rivieroverstromingen in Nederland*. 3 vols. Assen: van Gorcum and comp.

Gregson, R 1985 'The multiple estate model: some critical questions', *Journal of Historical Geography,* 11, 339–51.

Johnston, G 1875 'Our visit to Holy Island in May, 1854', *History of the Berwickshire Naturalist's Club,* 7 (1873–75), 27–52.

Jolliffe, E A 1926 'Northumbrian institutions', *English Historical Review,* 41, 1–42.

Jones, G 1976 'Historical geography and our landed heritage', *University of Leeds Review,* 19, 53–78.

King, A 1978 'Gauber High Pasture, Ribblehead – an interim report', *in* Hall, R A (ed), *Viking-Age York and the North*. London: Council for British Archaeology Research Report 27, 21–25.

Lamb, H H 1982 *Climate, history and the modern world*. London: Methuen.

Lash, S & Wynne B 1992 'Introduction' *in Risk Society: towards a new modernity*. (trans M. Ritter) London: Sage Publications, 1–9.

Miller, E & Hatcher, J 1978 *Medieval England – rural society and economic change 1086–1348*. Longman: London.

Morris, C D 1977 'Northumbria and the Viking settlement: the evidence of landholding', *Archaeologia Aeliana,* 5th Series, 5, 81–103.

Nelson, J 1992 *Charles the Bald*. Longman: London.

O'Sullivan, D & Young, R 1991a 'The early medieval settlement at Green Shiel, Northumberland', *Archaeologia Aeliana,* 5th Series, 19, 55–70.

O'Sullivan, D & Young, R 1991b 'The early medieval settlement at Green Shiel, Lindisfarne. An interim report on the excavations, 1984–91', *Archaeology North,* 2, 17–21.

Ranwell, D 1972 *Ecology of salt marshes and sand dunes*. London: Chapman and Hall.

Walsh, K 1988 *An analysis of the marine mollusca from the Green Shiel site on Lindisfarne (Holy Island)*. Unpublished postgraduate dissertation, Dept of Archaeology, University of Leicester.

20. Environmental stress in the Herefordshire and Shropshire uplands

Clare de Rouffignac

Abstract

The Marches Upland Survey (MUS) developed from an assessment of the archaeological record in the western upland areas of Herefordshire and Shropshire and has involved desk-based survey and intensive fieldwork in sample areas. The environmental component of the project has three elements: first, a desk-based survey of all available literature and unpublished data has been written up to form an assessment of the palaeoenvironmental evidence; second, the assessment of previous work has been used to choose areas suitable for environmental sampling; and third, palaeoenvironmental investigation of sampled material has been undertaken. At the preliminary stage, the project has produced some evidence for the way in which flora and fauna respond to the stresses of inhabiting the uplands. There is pollen analytical evidence for both land use and climate change, and changes in the patterns of agricultural and ritual activity are interpreted as the encroachment and abandonment of marginal areas.

INTRODUCTION

The Marches Uplands Survey was established in 1991 and aims to provide the basic information needed to protect the upland landscapes and archaeological sites of Herefordshire and Shropshire (Dinn 1991). The threats to upland archaeology in the region were identified by Rowley (1972) and stem from agricultural improvement, tree planting, mineral extraction and tourism.

The area selected for the survey consists of land above the 250 m contour line, land delimited as 'upland' following the definition set out by the Department of the Environment and the Institute of Terrestrial Ecology, and adopted by Darvill (1986). In Herefordshire and Shropshire the 'uplands' total over 900 km² and include the Black Mountains and foothills, the Wye/Arrow watershed, the Ludlow Anticline in Herefordshire, Clun Forest, the Long Mynd, Caer Caradoc, Long Mountain, and the hills west of Oswestry in Shropshire (Figure 20.1).

The study area is therefore quite extensive and in consequence a number of 1 km wide transects were selected for rapid survey, forming 13.5% of the total area (Figure 20.1). Detailed measurement and excavation of sites is not within the scope of the project, but all monuments within the transect are recorded, and fieldwalking of fields, ploughed for arable or for pasture improvement, is undertaken. The monuments recorded were very varied and included prehistoric barrows, Iron Age hillforts, prehistoric to post-medieval field systems, and eighteenth and nineteenth century lead mines and lime kilns. As well as studying standing monuments and buried archaeological sites, consideration is given to sites of proven and potential environmental importance.

Before the Marches Uplands Survey, many excavations, but little systematic survey work, had been undertaken on the English side of the Anglo-Welsh border. Variability of archaeological preservation across the border has been very difficult to assess. Excavations took place mostly on monuments such as hillforts in the years before protection by scheduling. Many of the most important sites excavated during recent years are on the Welsh side of the border. These include sites such as Collfryn, where both a prehistoric hillslope enclosure (Britnell 1989) and a fifteenth century corn drier were excavated (Britnell 1984), the Breiddin (Musson *et al* 1991) which is one of the most spectacular of the Marches hillforts, eight ring ditches at Four Crosses, Powys (Warrilow *et al* 1986), and Hen Domen, an early medieval castle excavated by Barker (Barker & Higham 1982). Environmental sampling has produced a great deal of important information on historic land use and landscape development around these sites, particularly for the Breiddin. There has also been a considerable amount of ecologically based research on land use and climatic change in the Black Mountains (Price 1981).

ENVIRONMENTAL OBJECTIVES

Three key stages were identified for the environmental work:

1. Desk-based review of all available literature and unpublished data for the survey area.
2. The assessment of previous work to choose areas suitable for environmental sampling. As the Marches Upland Survey is essentially a survey project, the areas for sampling are not dictated by existing or on-going excavation. Areas targeted for sampling include both natural sites, such as recent and prehistoric peat deposits, and archaeological sites where buried soils and anthropogenic deposits have been exposed through recent erosion.
3. The investigation of the sampled material to extend our understanding of the environmental history of the study area.

The overall objective of the project is to formulate a strategy for future research in, and suggest appropriate methodologies for, environmental archaeology in the English borderland.

Palaeoenvironmental assessment

The desk-based study to assess the published and archival material from the survey area (de Rouffignac 1992) employed the methodology of Caseldine (1990), Middleton (1990) and Keeley (1985). A model for the survival potential of environmental evidence was then created before field testing was undertaken. All the sites identified were recorded for location, geology and soils information, types of environmental remains recovered, with lists of published

and unpublished data sources. Mapping of the sites was undertaken using the Key Systems KeyARCHAEOLOGY software. This can transform database information, including site locations and symbols, into CAD (Computer Aided Design) maps suitable for screen display and printing.

Within the survey area a total of over one hundred excavations have taken place. However, a scant nineteen have yielded environmental evidence. This is mainly due to the excavations having taken place before the routine use of environmental techniques but it is nevertheless worrying that some more recent excavations have also failed to take account of the importance of environmental archaeology.

The sites within the survey area where environmental evidence has been identified from past archaeological work include Pleistocene deposits at Church Stretton (Osborne 1972), three undated prehistoric sites, one probable Neolithic site at Dorstone in Herefordshire (Pye 1969), seven Bronze Age barrow burials from both Herefordshire and Shropshire, a south Herefordshire pollen profile including material dated to the Bronze Age (Wilde 1980), an Iron Age hillfort at Croft Ambrey (Stanford 1974), a Romano-British timber fragment recovered from Church Stretton (Watson pers comm), and four medieval Marcher castles.

As there were so few previously studied sites within the survey area with recorded environmental evidence, an examination of relevant sites up to 10 km outside the survey area was also undertaken. This meant that some Welsh sites, such as Four Crosses and the Breiddin, were included for comparative purposes. A total of forty-six sites from the 10 km outside the survey area were identified as having produced environmental remains (Table 20.1).

A number of conclusions can be drawn from the assess-

Period	Site (grid reference)	Reference
Late Devensian/Flandrian	Church Stretton (SO456941, SO451932, SO446921 & SO459939)	Rowlands & Shotton 1971 Osborne 1972
Prehistoric	Walcote Hall Park (SJ350840) Cardington (SO47929499) Myndtown (SO400899)	Greig unpub Chitty unpub Chitty unpub
Neolithic	Dorstone Hill (SO326423)	Pye 1967; 1969
Bronze Age	Gannols (SO307420) Aymestrey (SO428664) Selattyn Tower (SJ25593416) Eaton Lydbury North (SO37418951) Monaughty Poeth (SO255747) Skyborry (SO26957404) Oaksfield (SO18778199) Olchon (SO27983231)	Wilde 1980 Lee, in Woodiwiss 1989 Shropshire SMR Shropshire SMR Shropshire SMR Shropshire SMR Chitty unpub Marshall 1932
Iron Age	Croft Ambrey (SO445668)	Stanford 1974
Romano-British	Church Stretton (SO46009375)	Shropshire SMR
Medieval	Brockhurst Castle (SO447925) Longtown (SO321293) Pontesbury Castle (SJ401058) Chirbury Moat (SO29609875)	Barker 1961 Taylor & Woodiwiss 1988 Barker 1966 Shropshire SMR

Table 20.1 All sites from the survey area with environmental remains.

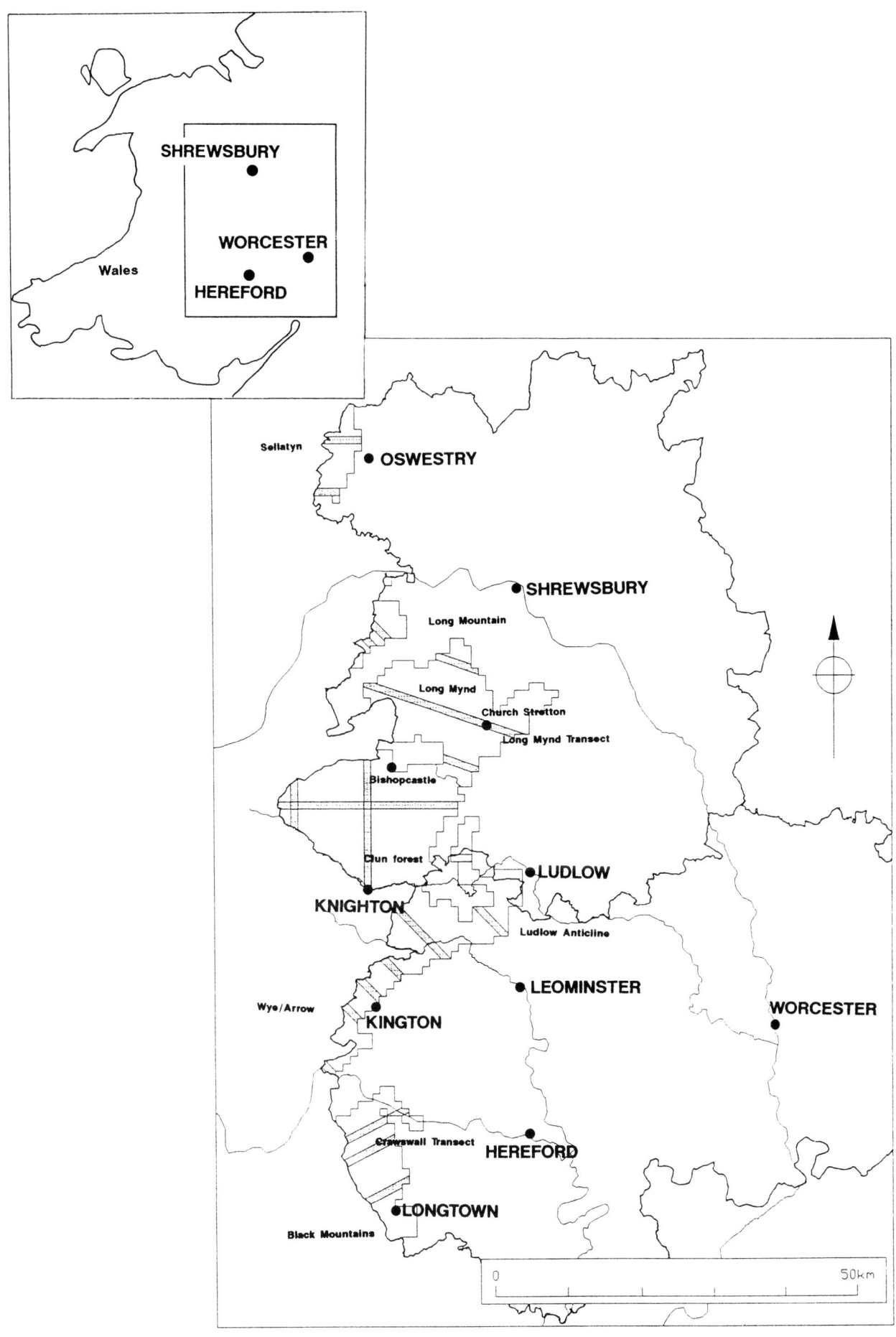

Figure 20.1 Location maps (inset) of the Shropshire, Herefordshire and Worcestershire uplands and of (main) survey areas and transects.

ment. The relative accessibility of Pleistocene sites for sampling by coring and test pits was noted in comparison to sites of more recent date. The preponderance, however, of suitable sites for sampling must be considered against the fragility of Pleistocene deposits. Such deposits are under threat in Herefordshire and Shropshire from excavation of kettleholes to create ponds for ornamental or agricultural purposes, and exploitation of peat deposits for fuel or fertilisers. Mineral extraction, for sands and gravels, will often expose and then destroy important peat deposits such as those at Wellington Quarry, Herefordshire (Roseff 1990).

There was found to be considerable potential for sampling buried soils under datable earthwork monuments. Standing monuments create problems for sampling as access for excavation is normally limited. Erosion of sites may, however, expose stratigraphic sequences suitable for sampling. A notable example is Offa's Dyke, the late 8th century bank and ditch running north-south through Shropshire, Herefordshire and Wales. The popularity of Offa's Dyke as a route for ramblers has led to widespread erosion of the turf and topsoil, exposing large areas of buried stratigraphy.

Many of the earthwork monuments in both Herefordshire and Shropshire, however, are scheduled, and it is not possible to extract environmental samples at will. Sampling from earthworks was integrated into a restoration programme by the National Trust, who were undertaking work on a number of earthworks on the Long Mynd in Shropshire. There was also a field survey transect located across the Long Mynd which included a number of earthworks *en route*.

Targeted environmental sampling

Environmental fieldwork at the Long Mynd, Shropshire
The Long Mynd is an outcrop of Silurian sandy grits, the rock formations producing large rolling hills with steep-sided valleys (Institute of Geology 1967). There are considerable numbers of earthworks on the Long Mynd, including Bronze Age round barrows, Iron Age hillforts and a number of enigmatic, undated features known as cross-dykes. The last are linear earthworks with ditches and banks, stretching across spurs, and in some places, down hillslopes.

Today, the Long Mynd is covered by rough grazing and unimproved pasture, with sheep and ponies grazing on common land. The animals are no longer taken up for the summer months to the high ground, and brought down for the winter to the lower pastures. The sheep are now kept up all winter and fed on the high ground, which leads to major erosion problems.

There are remnants of small fields and farmsteads; many of the properties are now holiday cottages. Hedge banks are still visible around the edges of the hills. *Crataegus* sp. and *Corylus* sp. are the dominant hedgerow shrubs, with the occasional surprising addition of *Laburnum* sp.

Some field systems, dating possibly to the Napoleonic Wars, are found around the northern and southern end of the Long Mynd, and have now reverted to heather cover.

The Long Mynd is a very popular recreation area. Visitor activities leading to vegetation destruction and soil erosion include car parking, rambling, pony trekking and mountain biking. The resulting erosion has produced major damage to the banks and ditches of the most prominent monuments.

As well as the effects of ramblers, one of the Bronze Age barrows (PRN SA 198; NGR SO 4210 9538) has suffered as a result of conversion into a shooting box with a concrete lining and wooden roof inserted into the central area. As a result of vandalism a decision was made by the National Trust to remove the concrete to below the top of the mound, infill the hole, and return the barrow to its original shape.

A trench was excavated adjacent to where the concrete lining was removed, and Dr Susan Limbrey of the Department of Archaeology, University of Birmingham, examined the buried soils and stratigraphy beneath the barrow. Well-defined A and B horizons were identified, with a layer of iron panning also visible. Evidence of a possible former woodland cover came from pollen samples and charcoal collected from the buried soils. The charcoal was identified as *Quercus* sp. and *Betula* sp., a far cry from today's treeless landscape on the Long Mynd.

Examination of the stratigraphy relating to one of the cross dykes at the Devil's Mouth (PRN SA 251; NGR SO 4399 9432) on the Long Mynd was less successful. However, some further charcoal samples were collected, and are awaiting identification. Accelerator dating is planned for the charcoal to determine the date of this particular cross dyke. Pollen collected from the buried soils by James Greig of the Department of Plant Biology, University of Birmingham was found to be well-preserved.

Further to the west, the cross-dyke at High Park Cottage (PRN SA 199; NGR SO 4441 9662) was found to cover a buried soil with possible evidence of ploughing at some point in the past. Charcoal samples indicated previous tree cover of *Quercus* sp. and *Corylus* sp., but dating of the samples is still awaited. Dates from charcoal recovered during excavations at the Breiddin imply clearance was taking place from around 3000 bc (Musson *et al* 1991).

The charcoal present in these buried soils is interpreted as representing the *in situ* burning of the woodland cover of the Long Mynd and suggests that the area was, in the past, covered with mixed deciduous woodland. Woodland clearance led to the soils becoming gleyed and developing iron pans. This created further problems with poor drainage and waterlogged with the result that there was a lack of regeneration of tree cover.

The soils have continued to deteriorate over the centuries to leave the Long Mynd as a landscape vulnerable to pressures of overgrazing and erosion. There are today few settlements in the area: abandoned farmhouses and hamlets are testimony to the deterioration of the soils to the point where cultivation was no longer possible or profitable.

Further work on the Long Mynd will be undertaken to examine pollen cores from suitable peat deposits away from the earthwork monuments to study the changes in the natural vegetation more closely.

Environmental fieldwork at Craswall, Herefordshire
As well as examining the problems of landscape deterioration and erosion at the Long Mynd, environmental work has been undertaken in the Craswall area of western Herefordshire where a transect for field survey was located. The Craswell transect contains a wide variety of landscapes and was largely chosen as a contrast to that crossing the Long Mynd. The transect runs from the lower slopes of the Black Mountains in the west, across small valleys and hills with rough pasture, to improved pasture on the rolling hills around Peterchurch to the east.

The stresses of human existence in the area can be seen in the landscape, particularly on the foothills of the Black Mountains where peat deposits have been identified by the Soil Survey (unpub.). The remains of small-field systems can be found in many locations, including possible relict medieval field systems around the ruins of Craswall Priory, a Grandmontine foundation dating from the twelfth century. The reasons for the abandonment can be traced to the medieval period and more recently, this century. Climatic deterioration, lack of understanding of poor soil quality, the Black Death in the fourteenth century and economic factors led to many areas being deserted. The problems of attempting to continue the open field system is seen from aerial photographs, with shrunken villages surrounded by ridge and furrow. Eventually, cultivation was abandoned, and pastoralism through the wool industry became the most important source of prosperity in the region.

The archaeological evidence for changes in agriculture are being rapidly destroyed by pasture improvement in the Craswall area. The traces of ancient fields with their boundary banks and hedges have been removed, leaving tracts of open, featureless land. The hedges which do survive contain only a few species of shrub, such as *Crataegus* sp., *Corylus* sp. and *Sambucus nigra*, and are of limited species diversity when compared to hedge banks in lowland Herefordshire (Mills 1983).

The environmental work undertaken in the Craswall area involved various techniques. Unpublished soil data collected by the Soil Survey of Great Britain has been studied to pinpoint sites for sampling. Examination of soil profiles has been undertaken to enable comparison of the effects of 'improvement' on buried soils, which may be present in areas of improved and unimproved pasture. There are no standing earthwork monuments of note in the Craswall transect and hence the possibility of recovering information from buried soils in this area is slight. Nonetheless, it is hoped that traces of buried soils may exist, particularly in areas of unimproved pasture. It is also hoped that colluvial deposits at the foot of the

hillslopes would give some indication of how farmers have managed, or failed to manage, climatic deterioration by their changes in agricultural practices.

Pollen cores are being collected from the peat deposits on the slopes of the Black Mountains to augment the full record of the environmental and climatic changes which have taken place over the centuries. These will be compared with the results from work which Bartley (1960) undertook at Rhosgoch Common over the border in the Welsh Black Mountains. Phytoliths and molluscan remains will also be considered when sampling is undertaken to determine evidence for ancient agriculture.

CONCLUSIONS

The paucity of previous palaeoenvironmental work in this archaeologically well known region is possibly surprising. The new evidence collected to date suggests that environmental factors have played a considerable role in the settlement of the Marches Uplands. In particular, it is suggested that human activities have served to accelerate natural processes of soil degradation and that nutrient loss, podzolisation, iron pan formation and consequent waterlogging may have been significant factors in the gradual decline in the arable use of these upland areas.

When the environmental work is completed from both Craswall and the Long Mynd, the results will form the basis for a research framework for future work in the region. This will stress the vital need for inclusion of environmental research in management proposals for earthwork monuments and for the protection of vulnerable Pleistocene deposits.

It is hoped that eventually the work of both the archaeological survey team and the environmental archaeologists should lead to a greater understanding of how humans have both failed and succeeded in their exploitation of the Marches Uplands.

ACKNOWLEDGEMENTS

The Marches Upland Survey is funded by English Heritage, with technical and practical assistance from the Royal Commission on the Historic Monuments of England, and is run by the Archaeological Service of Hereford and Worcester County Council, in collaboration with Shropshire County Council.

REFERENCES

Barker, P A 1961 'A pottery sequence from Brockhurst Castle, Church Stretton, 1959', *Transactions of the Shropshire Archaeological Society,* 57, 63–80.
Barker, P A 1966 'Pontesbury Castle mound emergency excavations 1961 and 1964', *Transactions of the Shropshire Archaeological Society,* 57, 206–23.

Barker, P A & Higham, R 1982 *Hen Domen, Montgomery: a timber castle on the English-Welsh border*. London: Royal Archaeological Institute Monograph Series, Volume 1.

Bartley, D D 1960 'Rhosgoch Common, Radnorshire: stratigraphy and pollen analysis', *New Phytologist*, 59, 238–62.

Britnell, W 1984 'A fifteenth century corn drying kiln from Collfryn, Llansantffraid Deuddwr, Powys', *Mediaeval Archaeology*, 28, 190–3.

Britnell, W 1989 'The Collfryn hillslope enclosure, Llansantffraid Deuddwr, Powys: excavations 1980–1982', *Proceedings of the Prehistoric Society*, 55, 89–135.

Caseldine, A 1990 *Environmental archaeology in Wales*. Lampeter: Cadw and Archaeology Department, St Davids University College, Lampeter.

Darvill, T C 1986 *The archaeology of the uplands: a rapid assessment of archaeological knowledge and practice*. London: RCHM(E) and CBA.

Dinn, J (ed) 1991 *The Marches Uplands Survey: research design and project proposal*. Worcester: Hereford and Worcester County Council Archaeological Service.

Institute of Geology 1967 *Church Stretton drift geology*. 1:50,000 Series, Sheet 166. London: HMSO.

Keeley, H C M (ed) 1985 *Environmental archaeology: a regional review Volume 2*. London: HBMC.

Marshall, G 1932 'Report on the discovery of two Bronze Age cists in the Olchon Valley, Herefordshire', *Transactions of Woolhope Naturalists Field Club*, 147–153.

Middleton, R (ed) 1990 *North West Wetlands Survey Annual Report*. Lancaster: Lancaster University Archaeology Unit.

Mills, N T W (ed) 1983 *Archaeological survey in the Leominster district, Herefordshire*. Worcester: Hereford and Worcester County Council Archaeology Section, unpublished typescript.

Musson, C, Britnell, W J, & Smith, A G 1991 *The Breiddin hillfort: a later prehistoric settlement in the Welsh Marches*. London: CBA Research Report 76.

Osborne, P J 1972 'Insect faunas of late Devensian and Flandrian age from Church Stretton, Salop', *Philosophical Transactions of the Royal Society of London*, 263, 327–67.

Price, M D R 1981 *Palynological studies on the Black Mountains*. Unpublished PhD thesis, University of London.

Pye, W R 1967 'Reports of the Archaeology Sectional Recorders: Neolithic', *Transactions of Woolhope Naturalists Field Club*, 39, 157.

Pye, W R 1969 'Reports of the Archaeology Section Recorders: Neolithic', *Transactions of Woolhope Naturalists Field Club* 39, 475.

Roseff, R 1990 *Untitled Wellington archive report*, Birmingham: Department of Archaeology, University of Birmingham.

de Rouffignac, C 1992 *The Marches uplands: an assessment of the palaeoenvironmental evidence*. Worcester: Hereford and Worcester County Council Archaeological Service.

Rowlands, P H & Shotton F C 1971 'Pleistocene deposits of Church Stretton (Shropshire) and its neighbourhood', *Quarterly Journal of the Geological Society of London*, 127, 599–622.

Rowley, R T 1972 *The Shropshire landscape*. London: Hodder & Stoughton.

Stanford, S C 1974 *Croft Ambrey*. Hereford: Private publication.

Taylor, G J & Woodiwiss, S G 1988 *Evaluation excavation at land adjacent to the Police Station at Longtown, HWCM 1036*. Hereford and Worcester County Council Internal Report.

Warrilow, W, Owen, G & Britnell, W 1986 'Eight ring ditches at Four Crosses, Llandysilio, Powys', *Proceedings of the Prehistoric Society*, 52, 31–87.

Wilde, A P 1980 *The mapping of small peat deposits in Herefordshire as a start to discovering the vegetational history of the area*, Unpublished BSc Thesis, University of Birmingham.

Woodiwiss, S 1989 'Salvage recording of a Beaker burial from Aymestrey (HWCM 7060)', *Transactions of the Woolhope Naturalists Field Club*, 46, 169–76.

21. The South Nesting Palaeolandscape Project, Shetland: mires, mounds and margins

Terry O'Connor

Abstract

The project began in 1991, with the aim of elucidating by survey and excavation the history and prehistory of land-use in the South Nesting area of Mainland Shetland. Two particular themes under investigation are the influence of the diverse solid geology of the area in constraining past settlement and affecting the visibility and preservation of the archaeological record, and the location of burnt mounds within both the prehistoric geographical environment and the operational environment of the contemporary populace. The paper outlines results from fieldwork in 1991 and 1992 when the project was at an early stage and therefore this is a report on work in progress rather than a final synthesis.

INTRODUCTION

The South Nesting Project began in 1991 as a three-year programme of investigation into settlement and environment in an area of Shetland already known to have a substantial archaeological resource (Dockrill 1992). This paper is by way of a progress report, written in 1992 after the second field season, and consists of an introduction to the study area, a brief account of the aims and results to date of the project, and then a more detailed look at investigations at three burnt mound sites, with discussion of their environmental setting, and of the preservation of evidence within them. A synthetic account of the project has been published elsewhere (Dockrill *et al* 1996).

THE STUDY AREA (FIGURE 21.1)

The islands of Shetland lie in the northeastern Atlantic Ocean at around latitude 60° 10' north, longitude 1° 12' west, some 200 km north of mainland Scotland. The islands are part of the Palaeozoic igneous and metamorphic complex which forms parts of Scotland and Scandinavia, with two major faults trending roughly north-south to give a topography dominated by hill and valley systems on that alignment. Pleistocene ice cover has only slightly modified the structural topography, and Holocene base level changes have resulted in the flooding of the lower parts of many major valleys to produce the characteristic 'voes'. The overall result is a modern landscape of steep but rounded hills interdigitating with long voes, and an immature coastline with spits, tombolos and bay-head bars.

The climate today is essentially cool and maritime, with mean January and June temperatures of 3.5°C and 12°C, mean annual rainfall of about 1000 mm, and high mean and maximum wind speeds at all times of year (Berry & Johnstone 1980, 21–6). An average year brings 236 hours of gale force wind: 1992 was first-footed by a gale gusting over 200 km per hour on exposed headlands.

Shetland can fairly be described as marginal for cereal-based agriculture. The modern vegetation is dominated by *Calluna*, *Carex* and Poaceae communities growing on blanket peat, with only small areas of deep mineral soils (Spence 1979). Prehistoric settlement is well attested throughout Shetland, with Neolithic agriculture apparent at Scord of Brouster (Whittle 1986) and Gruting. Given the environmental setting, agriculture must always have been susceptible to quite minor medium-term climatic fluctuations, or even to sustained periods of extreme weather within a period of generally stable climate. In such circumstances, the sea would be a valuable buffer, supplying a critical reserve of diverse resources.

The South Nesting area provided an opportunity to

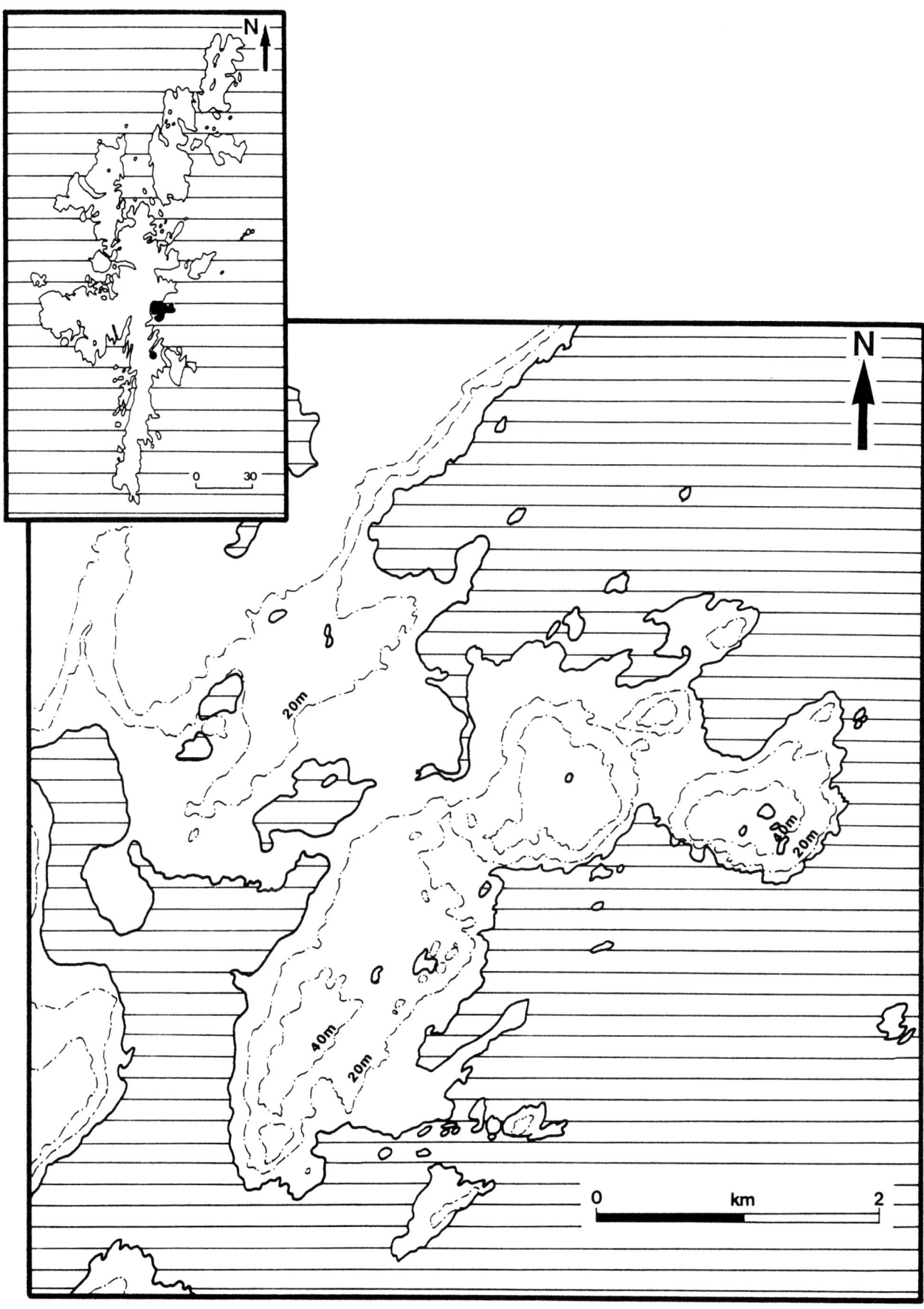

Figure 21.1 South Nesting, showing coastline, main freshwater bodies, 20 m and 40 m contours, and (inset) location.

examine the archaeology of a part of Shetland in detail. There was a substantial archaeological resource recorded, modern agricultural activity was sufficiently intensive to represent something of a threat to the visibility of that resource and the geology and topography of the area offered a contrast between areas today regarded as more or less favourable to cultivation.

To start with the foundations, the solid geology of South Nesting is structurally simple but lithologically complex, and is dominated by three southwest-northeast trending bands of rock (Mykura 1976). The most northerly band is of psammites and gneisses of the Colla Firth Group (part of the Whiteness Division of the East Mainland Succession). These rocks underlie the extensive area of hilly rough grazing which lies between South Nesting and North Nesting. The central zone of South Nesting is underlain by rocks of the Girlsta Limestone Group. This is a limestone *sensu lato*; a hard calcite marble interspersed with bands of white mica, giving generally lower relief with areas of relatively well-drained mineral soils. To the south, the Girlsta Limestone gives way to psammites and granulites of the Wadbister Ness Group, forming an area of steep, crinkly hills which run steeply to the sea along the southeast coast of South Nesting. The geology thus gives a sandwich, with comparatively fertile and tractable soils as the filling, and base-poor peat over the hills to either side.

The modern settlement of South Nesting very clearly reflects the geology, with settlement concentrated on the Girlsta Limestone belt (Figure 21.2). The crofting landscape mapped in 1878 shows much the same concentration of infield and houses, and it seems likely that South Nesting has always presented settlers with an area of comparatively good land flanked by more marginal hills, although whether the division was always seen in those terms is another question entirely.

THE PROJECT: AIMS AND METHODS

The overall aim of the project is to discover and record the development of the landscape of South Nesting throughout its history and prehistory of settlement. More precisely, one of the attractions of the area was the presence of several burnt mounds, offering the opportunity to extend research begun in Orkney (Hunter & Dockrill 1990; Dockrill 1991) by examining these enigmatic, baffling, and not infrequently frustrating monuments within their wider landscape, in an area where prehistoric occupation and field systems could also be examined, and where the whole landscape could be viewed against a background of diverse geology.

The methodology is a straightforward implementation of that used elsewhere in landscape archaeology, modified to suit local conditions and available resources. The study area was systematically field-walked by small groups of people at 10 m spacing, marking and roughly plotting on a sketch-map any major or minor features of the landscape thought to be of human manufacture. Apart from the

orthostats and earthworks to be expected of prehistoric settlement, this included ships timbers set in stone walls, abrupt changes in vegetation, and abandoned buildings. The sites located during field-walking were then re-examined critically, during which a small proportion were rejected as of non-human origins. The remainder were accurately plotted onto an overall survey, and gazetteer descriptions were drawn up. For the 1991 season, 260 sites were recorded, with many more in 1992. The more substantial, clearly archaeological, sites were then subject to a detailed earthwork survey and geophysical survey.

Limited excavation was carried out in 1991 and 1992, in order to investigate sites about which specific questions have been posed, or where erosion was clearly causing destruction of archaeological deposits. In 1991–2, excavation was undertaken on six putative boundary structures, three burnt mounds, and a prehistoric house. In addition, sub-surface deposits around sites were investigated by augering and small soil-pits.

Digging small sections of boundaries has proved valuable in two respects: demonstrating the differing internal construction of boundaries which appear similar on the surface, and locating buried surfaces with evidence of earlier cultivation. Linear features in the subsoil, interpreted as marks made by ploughing with an ard or similar implement, were located under two of the six boundaries sectioned in 1991–2, one of them close to a settlement complex (Site 114), and the other close to a partially excavated prehistoric house site (Site 229). The ard marks at Site 229 were overlain by a thick brown soil, in which the presence of abundant charcoal flecks and enhanced magnetic susceptibility indicated an origin as a 'made' soil. Evidence is thus emerging to show that the Girlsta Limestone belt was utilised for cultivation during prehistory. The use of small excavations, of just a few square metres, has proved to be a valuable means of obtaining information with little input of resources and little destruction of the surviving archaeology.

It has become apparent that the modern soils across South Nesting form a complex catena at a number of scales. The outcrops of Colla Firth and Wadbister Ness rocks are considered marginal for agriculture today, and are used almost exclusively for rough grazing. The soils in these areas are predominantly blanket peat and peaty podsols, with patches of unstable debris soils on and around outcropping rocks. On the Girlsta Limestone, the basins around some lochs, notably Trowie Loch, carry patches of mesotrophic fen peat, with the better-drained areas carrying brown soils. These pass into humus-iron podsols at one extreme of drainage and to gleys at the other extreme. Transitions from one soil type to another can be seen across very minor changes in relief. Some prehistoric field boundaries can be seen in section to bear a strip of well-eluviated podsol along the top, usually marked by a line of stunted heather, with an azonal brown soil on the sides. Excavations at Site 229 showed that the thick man-made soil associated with house structures was overlain by a

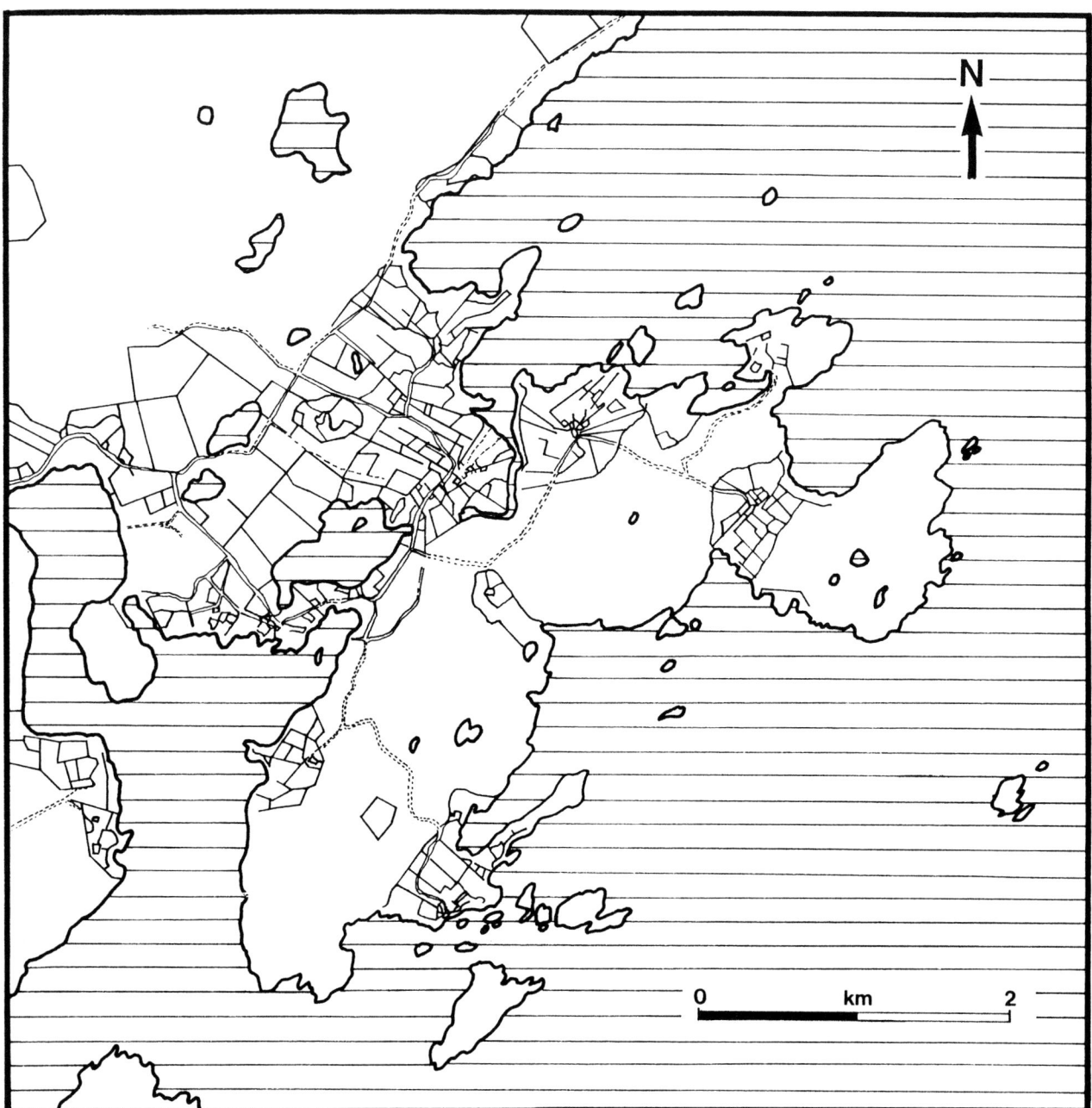

Figure 21.2 South Nesting, showing the modern field and property boundaries. Note the concentration on a central strip.

very well developed podsol, whilst nearby areas not covered by the man-made soil bore comparatively uneluviated brown soils. The Girlsta Limestone belt is obviously an area where very minor modification of drainage could make a considerable difference to predominant soil type, and thus, perhaps, to agricultural potential.

FOUR BURNT MOUNDS, THREE MIRES AND A BEACH (FIGURE 21.3)

The environmental and topographical setting of four burnt

mounds has been investigated; in three cases by excavation, and in the fourth by augering and coring. Two of the mounds lie a few metres apart and constitute the site known as Trowie Loch (Sites 251 and 252). They are located on either side of a narrow tidal channel, itself a side channel of the Vadill of Garth, close to its conjunction with Trowie Loch. Site 251 is a rather low mound, truncated on its north side by a boat naust (Site 253) and eroding onto the modern beach on its southeast side. Facing it across a narrow tidal channel is Site 252, a much larger structure, standing over 2 m high. Site 252 is undergoing erosion along its north and west edges, and is

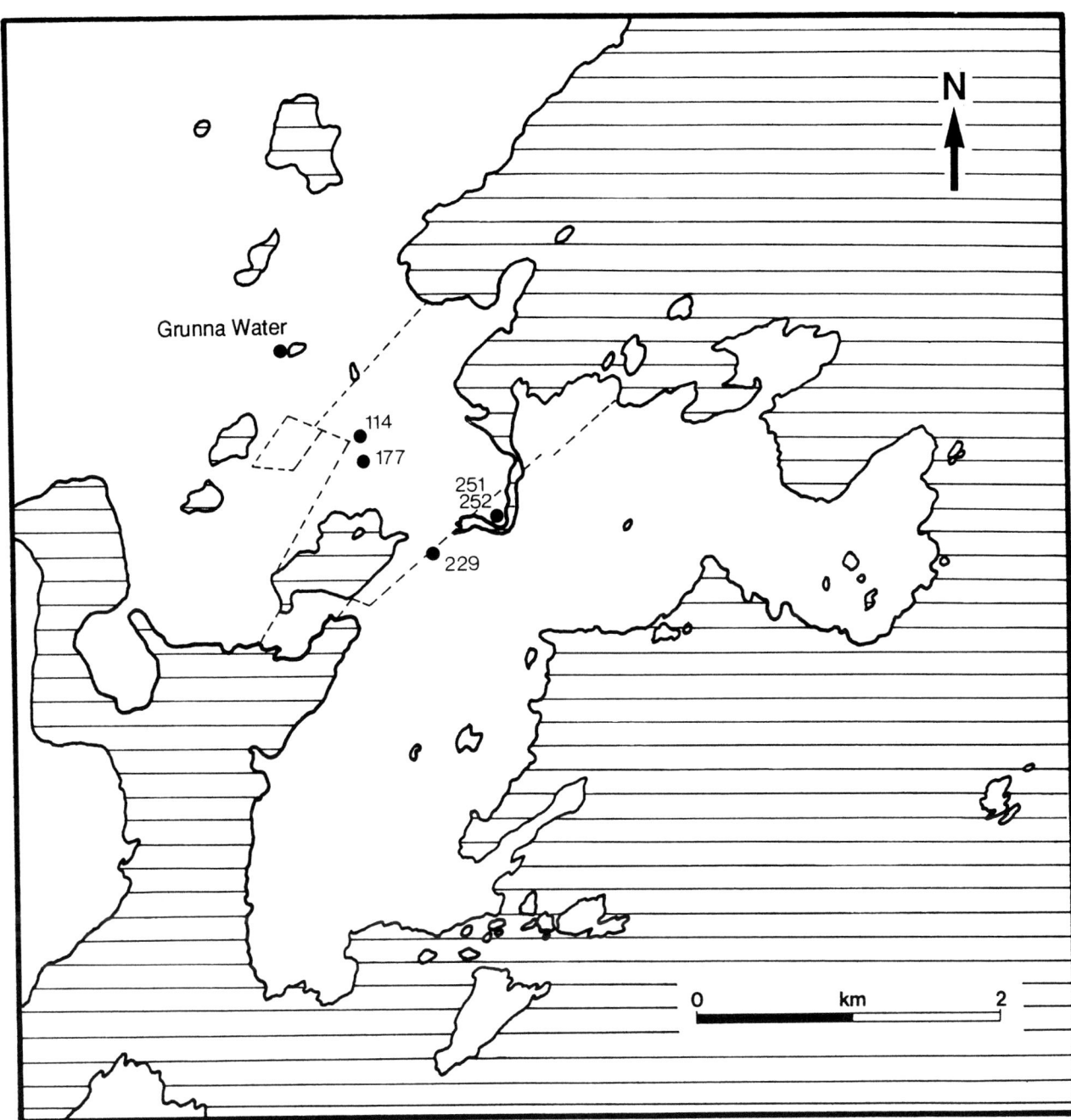

Figure 21.3 South Nesting, showing sites referred to in text in relation to major geological boundaries. Compare with Figure 21.2.

heavily mutilated by rabbits on the north side. Local oral tradition has it that the mound has been used for kelp burning. Excavation has concentrated on those parts of Sites 251 and 252 which are undergoing erosion (Bond & Dockrill 1992).

Site 177 lies on the Colla Firth-Girlsta Limestone boundary, and consists of a rather dishevelled burnt mound, burrowed by rabbits on its southwest side but otherwise in good order. A small trench was put into the northeast side of the mound in order to investigate the nature of the surface on which the mound was constructed.

A burnt mound lying just outside the study area, at Grunna Water, was investigated by augering and coring in order to address the same question as that posed at Site 177; namely, the underlying surface and deposits.

Much of the site at Trowie Loch is intertidal, and this imposed certain constraints with respect to the depth and location of trenches. The tidal channel between Sites 251 and 252 was sectioned in 1991, following a whimsical mattock blow which had revealed peat beneath the modern channel fill. This section showed the modern littoral sediments to be underlain by a brown, highly compacted

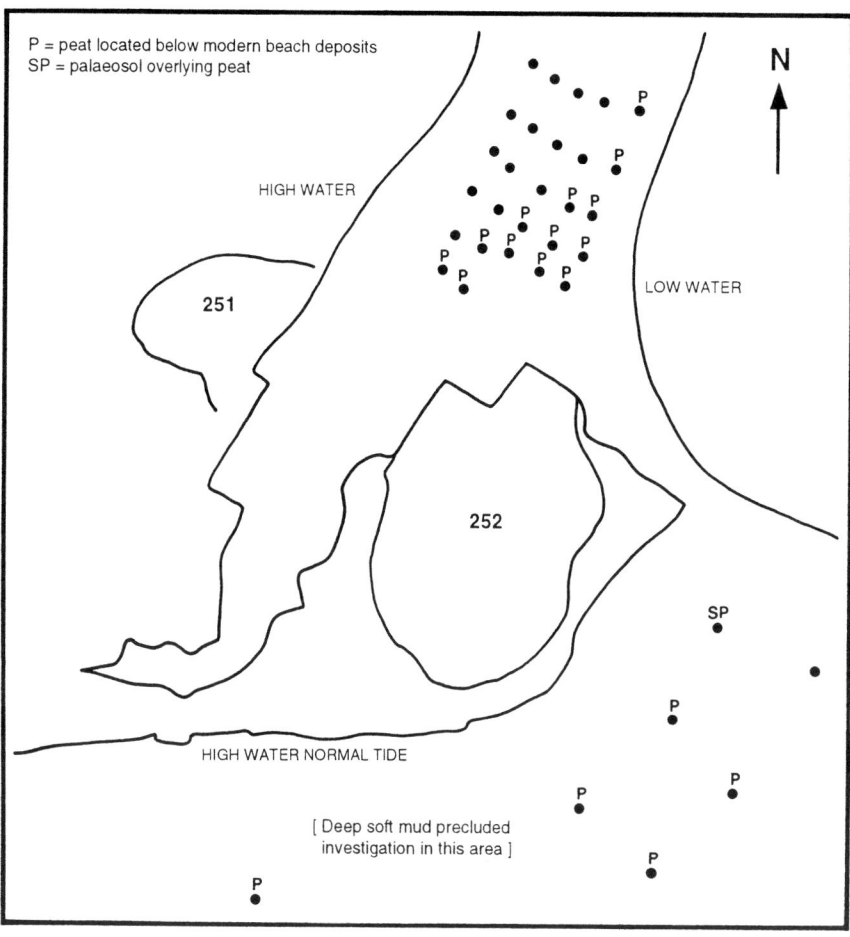

Figure 21.4 The burnt mounds at Trowie Loch (Sites 251, 252) to show location of the soil pits used to determine the extent of the buried peat.

peat which tapered away at the margins of Site 251, but clearly ran beneath Site 252. Because of tidal considerations, the section could only be kept open for a few hours, during which time the section was drawn and the peat was hastily sampled. In 1992, the section was extended eastwards into Site 252, in order to observe the contact between mound and peat under the centre of the mound, rather than at the eroded and slumped margins observed in 1991. The 1992 trench could only be maintained for a few hours, and even whilst it was open the nature of the sediments concerned and an alarming throughput of water made detailed stratigraphic recording somewhat difficult. However, the mound was seen to be underlain by over 20 cm of dark grey humic sand, which may be an immature soil developed upon the peat. Towards the centre of the mound, the grey humic sand was overlain by up to 10 cm of darker sand containing numerous fragments of root material, and this darker layer thinned away towards the edge of the mound. The grey deposit was underlain by the brown peat, with a diffuse boundary between the two, although towards the edge of the mound, the grey layer became diffuse and appeared to be intercalated within the top of the brown peat.

A similar sequence was observed in a soil pit dug at low tide some 15 m east of Site 252. This pit was one of series dug to investigate the lateral expanse of the peat beneath modern littoral deposits, and was the only one to show a sediment – the grey humic sand – interlaced between the peat and overlying littoral mud. The majority of the soil pits were dug on a grid pattern to the north of Site 252, and showed that the peat thinned very rapidly, only extending 10 m or so from the present edge of the burnt mound (Figure 21.4). Beyond this area, the modern beach was underlain by a compacted, poorly-sorted deposit of sandy clay with pebbles, showing marked reduction colouring. This deposit is interpreted as a colluvial sediment accumulated at the foot of the slope which forms the west side of the Vadill of Garth. The limited extent of the peat could be explained in terms of erosion following marine transgression except that the sequence to the east of the mound appears to show the uneroded top of the peat with its soil. If the interpretation of this sequence is correct, then the peat deposit may not have been much thicker than the 65 cm maximum observed during excavation, and did not extend beyond a relatively small area; that occupied by Site 252. The implication of this interpretation, which requires further testing by field observation, is that the placement of Site 252 may have been quite deliberately associated with the peat deposit. The sequence under the centre of the mound showed the grey sand to be overlain by a black deposit rich in plant debris. This could be interpreted as a 'turf' horizon on the top of the post-peat soil, in which case this upper component of the soil would appear to be missing towards the margins of the mound.

There is an obvious explanation to which to jump. The fuel used in the production of the burnt mound appears to have been predominantly or exclusively peat. The stones of the mound are interleaved with layers of peat ash. Perhaps the mound was simply located close to its fuel source. This is an enticing argument, but the nature of the peat deposit itself has to be considered. It is a highly compacted fibrous *Carex* and *Eriophorum* peat with appreciable inwashed mineral material, apparently formed in an oligotrophic topogenous mire. Such a peat would have made a very poor fuel, being very slow to dry to the point at which it would burn, and burning very erratically. It is possible, however, that the surviving peat represents the lower facies of a formerly thicker deposit. If an overlying thickness of more suitable peat was cut for burning, and the cut surface remained unburied for long enough to allow a soil (the grey humic sand) to begin to form, then the observed sequence could have developed, but there would be a substantial temporal gap between the top of the peat and the burnt mound. To this end, it is hoped to obtain a radiocarbon date for the top of the peat. The burnt mound is difficult to date directly, but this class of monument has a fairly well understood date-range in the Northern Isles (Buckley 1990). If a date for the peat falls several millennia too early for a burnt mound – say, 5th or 6th millennium bp – then the possibility that the peat had been cut over prior to deposition of the mound would have to be considered. Attempts to investigate the vegetational history of this small mire have been thwarted by the fact that the peat contains only very sparse, poorly preserved pollen.

Another point which needs to be made about the peat at Site 252 is that it shows a minor marine transgression to have taken place subsequent to the deposition of the mound. Perhaps this is unsurprising; sub-marine peats have been reported from a number of sites around Shetland (eg Berry & Johnstone 1980, 53). The Vadill of Garth is today a shallow, narrow inlet of the sea, which is practically dry at a point close to its mouth during spring low tides. It has not been within the resources of the project to undertake a detailed study of the past topography of this inlet, but it seems likely that at the time of deposition of Sites 251 and 252, they were located in a valley, probably with a stream draining the basin around Trowie Loch and running into the sea at a coastline located well to the north, in the modern East Voe of Skellister.

Discovery of the peat below Site 252 raised questions about the contemporary topography of other burnt mounds in the study area. To this end, a small excavation was undertaken in 1992 at Site 177; a 3 m by 1 m trench was excavated on the periphery of the mound, and was intended to sample the body of the mound and to allow examination of the deposits upon which it lay. This excavation showed the mound to have slumped appreciably. Extensive slumping had been noted at Site 252, but was explained in terms of wave and tidal movement of water gradually undermining the mound. The same explanation obviously could not be advanced for Site 177, where mound material, particularly larger stones, seems to have fallen towards the periphery by gravity, perhaps destabilised as the mound 'settled' after initial deposition.

The excavation at Site 177 was bedeviled by water. The mound was evidently holding a body of ground water above the level of the surrounding water table. When the excavation intersected the level at which water was held, a series of springs proceeded to debouch into the trench, and a sump had to be dug and continuously bailed. Water notwithstanding, the mound was shown to be set upon a deposit of up to a metre of compact sedge peat. There was no trace of an intervening soil: in fact, the junction between the peat and the overlying mound material was quite irregular, with stones sunken to some distance into the peat. Two auger transects were undertaken orthogonal to the excavation sections, and these showed the peat to thin rapidly to the south of the mound, whilst maintaining a thickness of about a metre to the east. The mound was evidently placed on the edge of a small topogenous mire. Again, the peat would have made a poor fuel, there being a substantial mineral component, which increased towards the base and to the east end of the auger transect. The similarity of this location to the location of Sites 251 and 252 is obvious. The peat was cored, and a pollen analysis is reported elsewhere (Dockrill *et al* 1996).

A burnt mound at Grunna Water was also investigated (O'Connor 1992). This mound lay just outside the study area, on the edge of the Colla Firth gneiss permeation belt (Figure 21.3). The modern topography is a small basin containing a shallow loch at its eastern end. A clear break of slope around the basin indicates appreciable drift infilling. The mound is placed about 70 m west of the loch. The surroundings of the mound were investigated by coring. The compact nature of the mound material precluded coring through it, but auger transects were undertaken as close to the margin of the mound as was practicable. Augering showed the mound to be set upon a highly minerogenic peat (or peaty mud?) probably of telmatic origin. The typical sequence around the west, south, and east sides of the mound was of about 30 cm of modern sedge peat passing sharply to this browner minerogenic peat, the whole sequence usually bottoming to rock at about 1 m. Close to the mound, it was possible to see a band of granular ashy material at the junction of the two sediments. This appeared to be material eroded from the mound, thus stratifying the mound below the modern peat but above the minerogenic brown peat. To the north of the mound, the modern peat was underlain in a small area by a mid-brown moss peat which extended to a maximum thickness of 2.8 m and could be traced over a spatial extent of only about 30 m square. Mound material could not with certainty be observed in cores of this peat, but a slight increase in magnetic susceptibility at the base of the modern peat seems to confirm that the mound is stratigraphically later than the brown peat. Thus we again have a burnt mound placed on the edge of a small peat deposit.

There is some consistency of location with the mounds described here. They have been placed in concave topography of some kind where drainage has led to the formation of a small area of peat. Whether this peat has been used for fuel is an unanswered question though peat of some kind has been the predominant fuel used in the mounds. One exception to this pattern should be mentioned. During 1992 field walking a burnt-mound was located towards the south west corner of the study area which appears from a preliminary examination to have been emplaced directly upon outcropping rock. This mound and its surroundings merit further examination. A further point which should be made is that the burnt mounds located to date in South Nesting lie in, or close to the margins of the Girlsta Limestone belt. Admittedly this is the area which was first field-walked in 1991 but field-walking in 1992 on the Wadbister Ness rocks to the south east located ample evidence of prehistoric activity, though not in the form of burnt mounds. Evidence of prehistoric settlement in the form of linear boundaries and possible house structures have been located in the 'marginal' south-eastern part of the study area but not burnt-mounds. There may therefore be a correlation between burnt-mounds and the potentially more fertile part of South Nesting, and within that area there may be an association with small contemporary peat deposits. As with all correlations cause must be carefully distinguished from effect. Considerations of geology, soil productivity and topography may have influenced the location of the mounds but that is far from saying that the locations were entirely functionally determined. Further field-work is required to complete the mapping of sites on the 'marginal' land and dating evidence will be necessary to establish the sequence of the sites and monuments recognised to date.

CONCLUSIONS

To draw detailed conclusions about prehistoric settlement in South Nesting would, at this stage, be premature. However, a picture is beginning to emerge of an area where the slight advantages offered by the better drained mineral soils overlying the Girlsta Limestone have had a marked effect on land-use. There is an obvious problem of visibility. Recent crofting and present day farming are concentrated in this central belt as well, and the archaeology of this area is much less apparent than on the marginal hill land to the north and south.

It would be easy, and probably wrong, to point to the

existence of a homestead complex similar in structure and extent to the Neolithic site at Scord of Brouster on marginal hill grazing to the south-east and to argue that either the agricultural potential of the land or the perception of that potential has radically changed. That would be to ignore the marked difference in subsequent land-use over the different parts of South Nesting, however, and equal attention must be paid to the much more fragmented prehistoric evidence in the crofted areas. Any discussion of prehistoric land-use will need to take account of the markedly varying solid geology of the area, though without adopting the associated modern perceptions of what constitutes useful and 'marginal' land.

ACKNOWLEDGEMENTS

The author is grateful to Steve Dockrill and Julie Bond for their help, advice, and comments on earlier drafts of this paper, and absolves them from responsibility for the final version.

The South Nesting Project is jointly directed by Steve Dockrill and Val Turner, and is principally funded by Shetland Amenity Trust, the British Academy, and the University of Bradford. The contributions of these and other bodies is gratefully acknowledged, as are the efforts of the numerous locals and incomers who have worked on the project to date.

REFERENCES

Berry, R J & Johnstone, J L 1980 *The natural history of Shetland.* Glasgow: William Collins and Son.

Bond, J M & Dockrill, S J 1992 'Excavations at Trowie Loch', *in* Dockrill, S J (ed), *The South Nesting Palaeolandscape Project: Report on 1991 field-work.* Bradford: University of Bradford Department of Archaeological Sciences, 59–69.

Buckley, V (ed) 1990 *Burnt offerings: Proceedings of the First International Burnt Mound Conference.* Dublin: Wordwell Academic Publications.

Dockrill, S J 1991 'Geophysical survey of burnt mounds in the Northern Isles: the magnetic response', *in* Hodder, M A & Barfield, L H (eds), *Burnt mounds and hot stone technology: Proceedings of the Second International Burnt Mound Conference.* Sandwell: Sandwell Metropolitan Borough Council, 35–39.

Dockrill, S J (ed) 1992 *The South Nesting Palaeolandscape Project: Report on 1991 Field-work.* Bradford: University of Bradford Department of Archaeological Sciences.

Dockrill, S J, Bond, J M & O'Connor, T P 1996 'Beyond the burnt mounds: the South Nesting Project', *in* Turner, V (ed), *The shaping of Shetland.* Lerwick: The Shetland Times.

Hunter, J R & Dockrill, S J 1990 'Recent research into burnt mounds on Fair Isle, Shetland, and Sanday, Orkney', *in* Buckley, V (ed), *Burnt offerings: Proceedings of the First International Burnt Mound Conference.* Dublin: Wordwell Academic Publications, 62–68.

Mykura, W 1976 *Orkney and Shetland (British Regional Geology).* Edinburgh: HMSO for British Geological Survey.

O'Connor, T P 1992 'Environmental investigations' *in* Dockrill, S

J (ed), *The South Nesting Palaeolandscape Project: Report on 1991 Field-work*. Bradford: University of Bradford Department of Archaeological Sciences, 71–90.

Spence, D 1979 *Shetland's living landscape. A study in island*

plant ecology. Sandwick: Thule Press.

Whittle, A 1986 *Scord of Brouster. An early agricultural settlement on Shetland*. Oxford: Oxford University Committee for Archaeology Monograph No. 9.

22. Identifying marginality in the first and second millennia BC in the Strath of Kildonan, Sutherland

David C Cowley

Abstract

Within the catchment of the River Helmsdale, Sutherland, prehistoric settlement and land-use remains, in which hut-circles are the common factor, have been broken down into classes representing broadly differing intensities of activity, the distributions of which highlight what may have been core and more peripheral areas. Hut-circles excavated in Scotland suggest that this form of prehistoric settlement may have undergone a cycle of expansion and contraction, a process which may be manifested in the core and more peripheral areas identified here. The identification of such broad patterns of settlement intensity across a landscape is useful in assessing the applicability of environmental sequences to archaeological survey and excavation data at local and regional scales, facilitating an examination of local diversity in responses against a back-drop of broader-brush archaeological and environmental patterns.

INTRODUCTION

To understand the nature of regional patterning, field archaeologists depend largely on the information presented by site distribution maps. In this study the archaeological record within the Strath of Kildonan has been examined to see whether some form of marginality can be identified in the survey data. The study area comprises the larger part of the catchment of the River Helmsdale, an area of just under 400 km², most of which is heather moorland, with varying depths of peat cover, as well as small pockets of improved land. The lower reaches of the Strath are defined by steep valley sides, with open, rolling land dominating the middle and upper reaches of the Strath. RCAHMS survey of 95 km² of the Strath in 1991 prompted an examination of the archaeological records held in the National Monuments Record of Scotland (NMRS) for the whole of the valley. This examination suggested that a database of a consistently high quality, although with a lower intensity of coverage than the RCAHMS survey, had been compiled by the Archaeology Division of the Ordnance Survey during the 1970s and 1980s. All sites referred to in the text have been fully referenced, and further details can be found in the NMRS, through which the Strath of Kildonan survey report (RCAHMS 1993) can also be consulted.

SURVIVAL AND RECOVERY

Biases in the survival and recovery of certain classes of monument within the study area limit the utility of their known distributions. RCAHMS survey in the Strath forms a window of systematic and detailed landscape survey which provides a yardstick for the quality of the earlier survey data in the Strath as a whole. Burnt mounds provide an illustration of the potential significance of these biases, for example the recent survey has suggested that as much as eight-fold increases in the numbers of known burnt mounds may be expected from systematic coverage of the whole Strath, limiting the utility of the existing distribution in a wider regional context. The known distribution of the hut-circle settlements within the Strath appears to be reasonably representative and they do not suffer from the problems of identification associated with burnt mounds. Their locations are generally predictable, and rapid, low-level reconnaissance is likely to achieve reasonable results. Assessment of the records relating to the hut-circle settlements indicates that, though there are gaps in the known distribution, these are only significant in detailed analysis and are largely predictable. The hut-circles, and the land-use remains with which they are found, form the largest and most reliable data-set within the Strath for an examination of whether marginality can be identified within the archaeological record.

HUT-CIRCLES AND LAND-USE REMAINS

There are vast numbers of hut-circles in Sutherland of which only a tiny proportion have been excavated or dated, limiting the potential for detailed discussion of chronology and morphology. A general approach is more appropriate to the data and, in assessing the records, three consistently recurring classes emerged, as detailed below; a basically similar classification can be found on Ordnance Survey record cards. The groups, which are illustrated on Figures 22.2–4, with their distribution shown on Figure 22.1, comprise:-

Group 1

Hut-circles, usually isolated from other monument types, with no signs of any cultivation in their vicinity.

Group 2

Hut-circles, usually in clusters of two to three and occasionally up to six, around which there are small cairns which may be interspersed with fragmentary banks but no formal fields or plots (see Figure 22.2).

Group 3

Hut-circles, occurring in clusters of at least four and up to thirteen and lying within field-systems defined by banks and/or lynchets. Examples of this type of hut-circle/field-system site are illustrated in Figures 22.3 and 22.4. At Feranach (Figure 22.3A, NC 82NW 6) four hut-circles are levelled into the slope amongst at least three plots defined by substantial lynchets up to 1.8 m in height. A small patch of what may be cord rig survives on the west side of the fields, parts of which have been slighted by post-medieval ploughing. The broch, two huts and the nineteenth-century sheep farm lie to the north-east. Cnoc Dail-Chairn (Figure 22.3B, NC 82NE 22) illustrates a similar, but more extensive, series of fields amongst which there are at least eleven hut-circles of a variety of forms and three enclosures, with two burnt mounds lying at the edges of the rough ground. At Ach na Fhionnfhuaraidh (Figure 22.3C, NC 92SW 7 & 14) lynchets and banks define subrectangular plots, at the fringes of which there is a scatter of small cairns. A broch (NC 92SW 13) and a farmstead (NC 92SW 43) are superimposed on the field-system and indicate the longevity of the occupation of the site.

There are grey areas in the allocation of sites to a specific group. In particular, those in Group 1 need to be treated with caution since hut-circles may be placed in this group due to the attrition of the site by later land-use.

In addition, the potential of peat to mask prehistoric land-use remains is illustrated by the discovery of extensive narrow-ridged cultivation in recent excavations near Lairg (McCullagh 1991, 47) and elsewhere in Scotland.

Although there is also considerable variation in the sizes and potential chronological span of sites (cf Figure 22.3, A-C), the groups do form consistent and recurrent groupings. The distinctions between the groups are also manifest in their altitudinal spread. About two-thirds of Group 3 sites are situated below 150 m OD, while the remaining third lies between 150 m and 200 m OD, most of which lie closer to 150 m than 200 m. The Group 2 sites have a more even altitudinal spread than the Group 3 sites, but with a weighting towards 150–200 m OD, with a handful above 200 m. The Group 3 sites are generally larger, include a greater variety of hut-circle forms, and are more likely to show signs of chronological depth in the remains, for example Ach na Fhionnfhuaraidh (Figure 22.3C) and Allt Bad Ra'fin (Figure 22.4, NC 92NW 3). Field boundaries associated with Group 3 sites may be of a considerable size, with lynchets up to 1.8 m in height, suggesting a long chronological span, which is also implicit in the suggestion that some Group 3 hut-circle/field-system sites represent a development of a Group 2 site with the introduction of formal boundaries into a cluster of small cairns. At Ach na Fhionnfhuaraidh (Figure 22.3C) the process of introducing formal boundaries into an area of clearance cairns may be visible in the east of the site, though there may also have been a functional difference between the area of fields and the clearance cairns.

Although the distribution of isolated hut-circles in Group 1 is most open to bias, there are a few which lie at the altitudinal maximum of the hut-circle range (250 m OD) and have not been affected by later land-use or peat-cover. Among these there may be settlements which are early, pre-dating any climatic deterioration, or sites representing peaks in a fluctuating altitudinal limit for settlement or an element in a wider settlement and land-use pattern, for example, something like medieval and post-medieval shielings. There is a very general overlap between the altitudes of the later shielings (RCAHMS 1993) and these high altitude Group 1 hut-circles, which may add weight to the suggestion of transhumance, although changes in the role of a site may have seen the hut-circles function in a variety of ways.

The distinctions between Groups 1/2 and Group 3 can be used to identify what, for want of a better term, may be called prime areas of presumably long-lived settlement (Group 3) as against more peripheral areas of less intensive and potentially intermittent settlement, represented at one level by the hut-circle/field-clearance group (Group 2) and at the extreme by the handful of isolated hut-circles (Group 1) at the altitudinal limits of the hut-circle range (250 m OD). Stevenson has suggested a similar pattern on Islay, Jura and Colonsay, where the favourable areas attracted larger settlements, such as at An Sithean, Islay (Barber & Brown 1985), while sites like the excavated

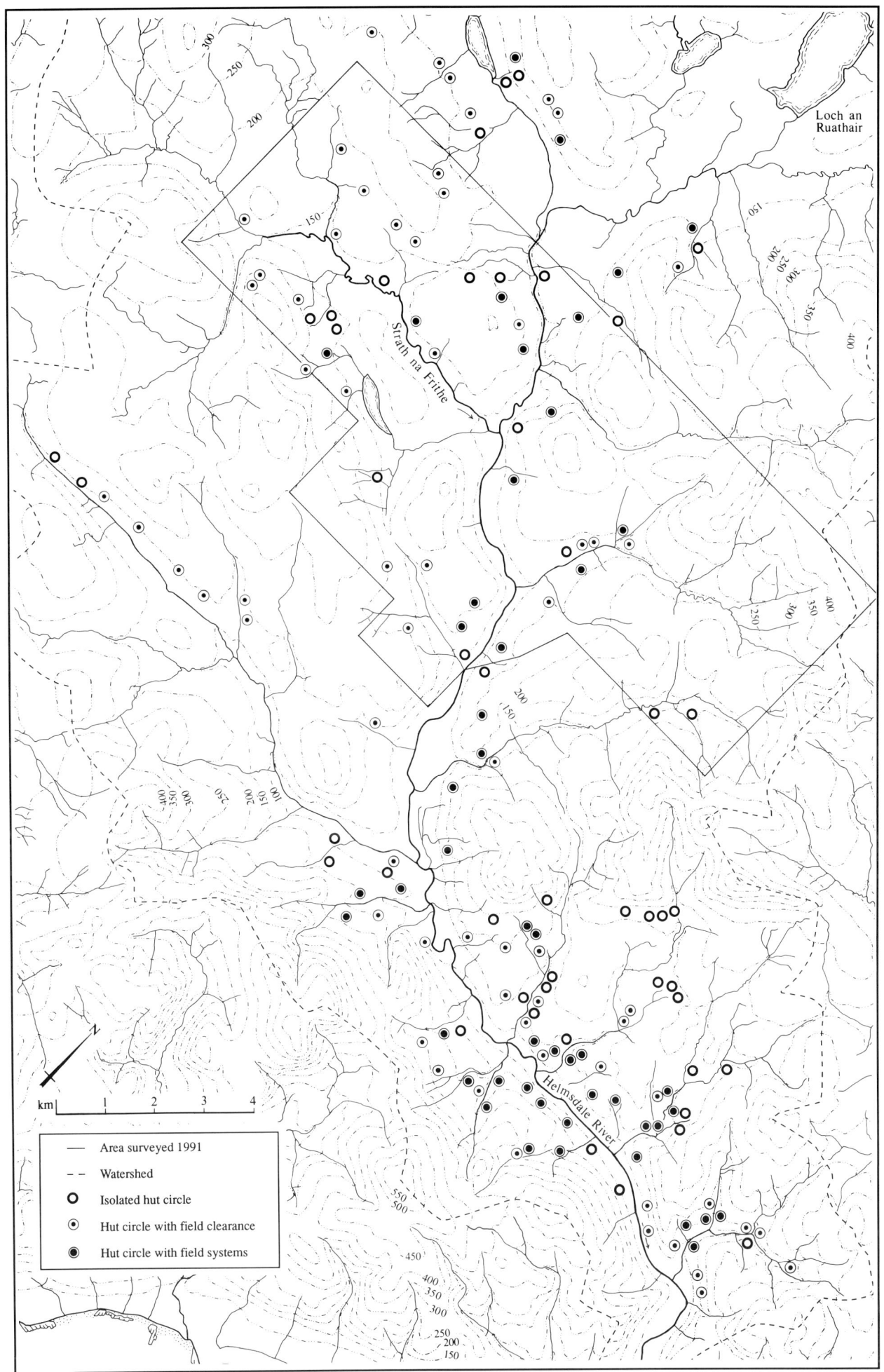

Figure 22.1 Strath of Kildonan study area. Distribution of hut-circle settlement.

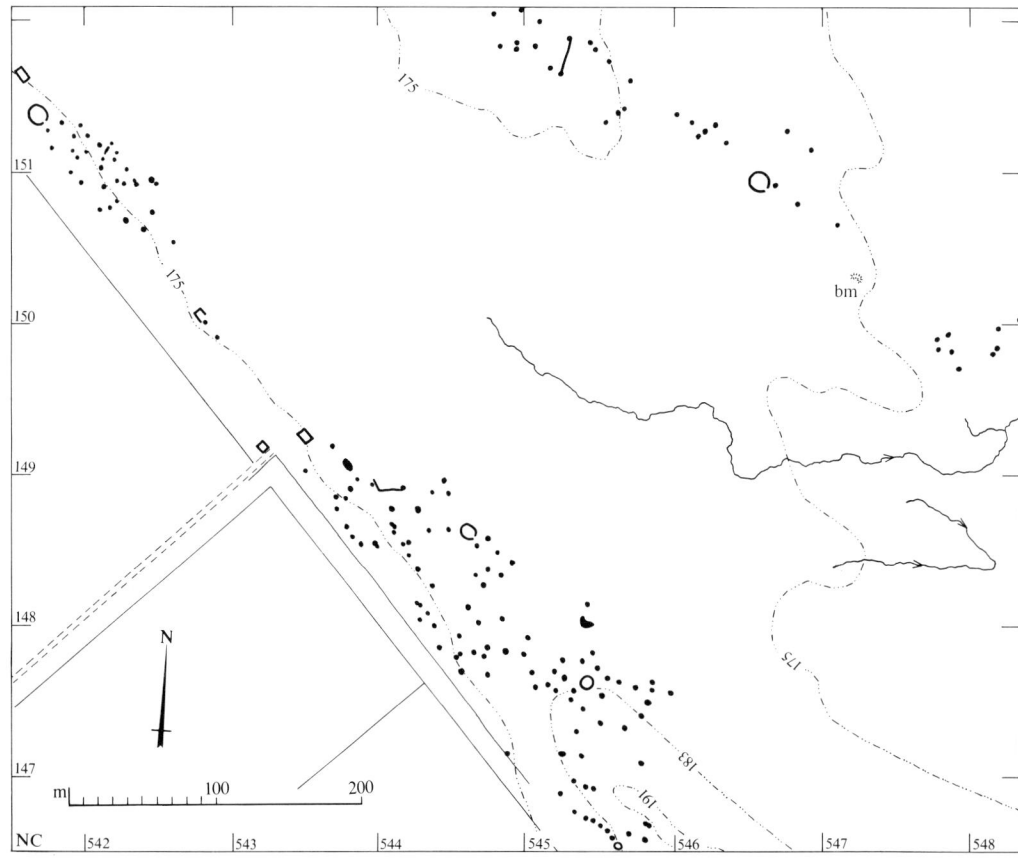

*Figure 22.2 Shinness Lodge, (NC 51NW 27 & 30; NC 51SW 14) near Lairg. Hut-circles with field clearance (Group 2).
Although outwith the study area, this site is typical of the group, and comprises four hut-circles disposed along two
ridges amongst a scatter of small cairns. A burnt mound lies to the north-east of the illustration.*

hut-circle at Cùl a'Bhaile, Jura remained as smaller units, at the upper edge of Bronze Age settlement (Stevenson 1985, 158). Activity at An Sithean continued over a longer, if intermittent, period than at Cùl a'Bhaile, and the two sites may be an equivalent of the Kildonan Group 3 and Group 2 sites respectively.

The distribution of sites on Figure 22.1 suggests a diversity in the intensity of the hut-circle settlement and land-use within the Strath of Kildonan. The lower reaches of the Strath show a particular density of hut-circle/field-system sites (Group 3) with a thinner, but even scatter up-river, suggesting that the Group 3 sites may disappear, or at least thin, in the upper reaches of the Strath na Frithe and up-river of Kinbrace. There is also a pattern of hut-circle/field-clearance sites (Group 2) infilling gaps between the concentrations of field-system sites, precluding a simplistic relationship between settlement and altitude. A near exclusivity of hut-circle/field-clearance sites along the main tributaries to the Helmsdale river is also notable indicating a less intensive exploitation of the side valleys. The most favourable locations in the Strath for hut-circle settlement are shown by the clusters of Group 3 sites, while the distributions of the Group 1 and 2 sites suggests a grading of settlement intensity elsewhere in the Strath. There are

some overlaps between the locations of Group 3 sites and the distributions of Neolithic monuments and to some extent brochs (RCAHMS 1993), suggesting that some of these areas may have been prime locations over long periods of time.

HUT-CIRCLES: DATING AND EXCAVATION

Dates for hut-circles within the Strath (Kilphedir GU-299 2370 ± 40 bp, range between SRR3 2100 ± 50 bp and GU10 1908 ± 60 bp, Fairhurst & Taylor 1971; Kilearnan Hill GU-1919 2935 ± 65 bp, GU-1917 2645 ± 100 bp, second date is not from a primary context, Haggarty forthcoming; Craggie Water, GU-2482 3200 ± 50 bp, date from a section through the hut-circle bank, Lowe pers comm) run from the late second millennium bc through to the beginning of the first millennium ad although dates from elsewhere in Scotland suggest a span from the very beginning of the second millennium bc. Modern excavations (Cnoc Stanger, Caithness, Mercer nd; Cùl a'Bhaile, Jura, Stevenson 1985; Tormore, Arran, Barber 1982) have shown that some hut-circles have been through repeated stages of construction, with, in some cases, abandonment

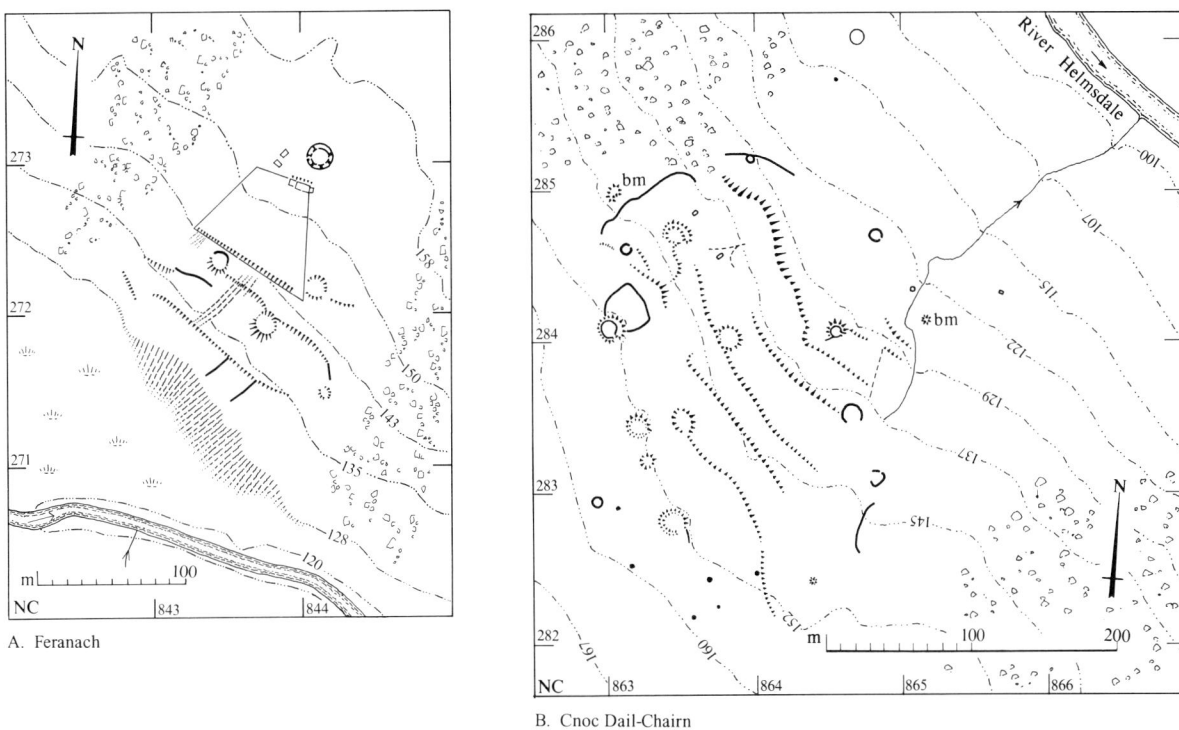

A. Feranach

B. Cnoc Dail-Chairn

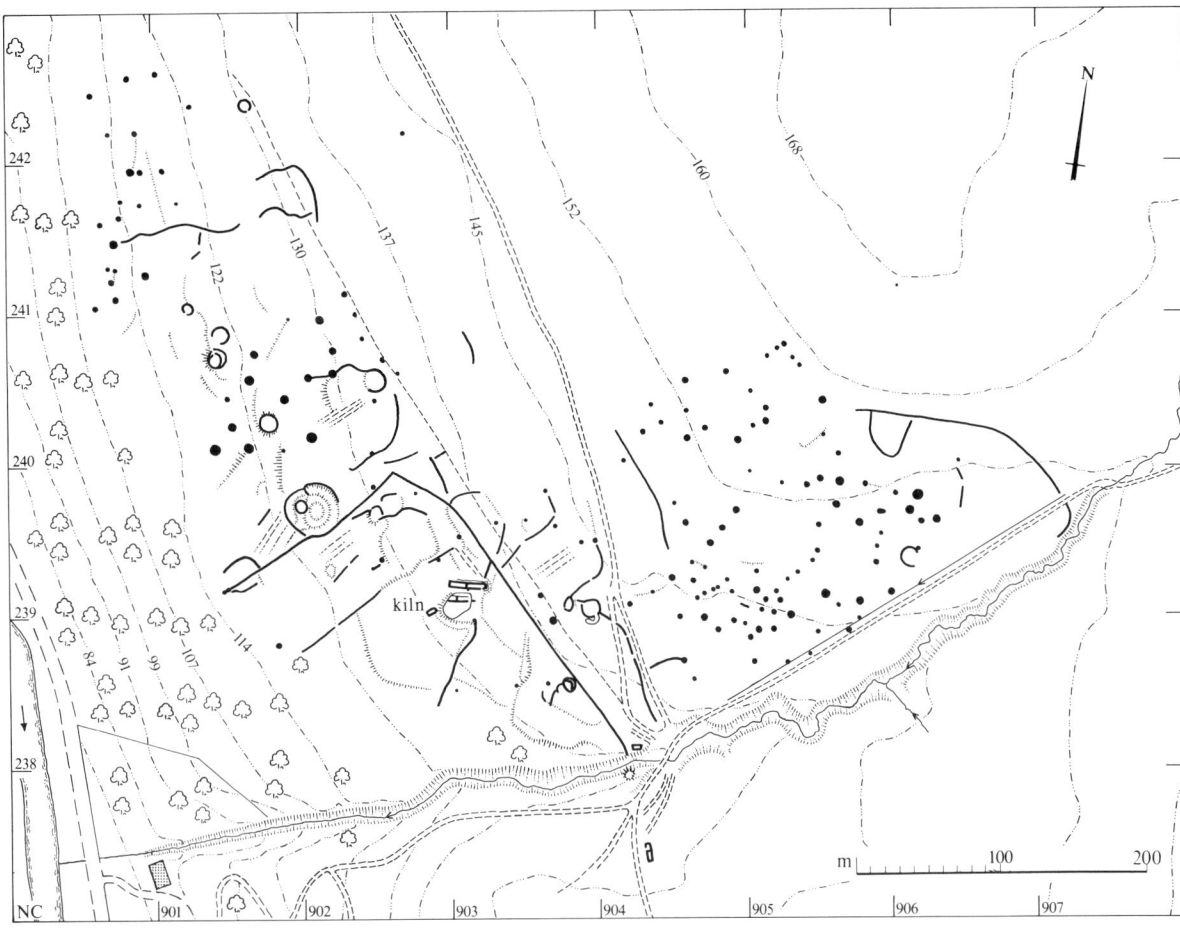

C. Ach na Fhionnfhuaraidh

Figure 22.3 Strath of Kildonan 1991 survey area. Hut-circles with field-systems (Group 3).

Figure 22.4 Allt Bad Ra'fin, Strath of Kildonan. Hut-circles, post-medieval building and cultivation remains. The southern-most hut-circle has a souterrain and an extended entrance passage.

between occupations. At Kilphedir the dates suggest two distinct phases of occupation, in the mid first millennium bc and at the end of the first millennium bc.

Within the context of the prime and peripheral areas identified in Kildonan, the evidence for intermittent occupation from some excavated hut-circle sites may be taken to suggest an ongoing cycle of expansion and contraction in later prehistoric settlement as the pressures which made a settlement viable ebbed and flowed. This pattern repeats itself in the catchment of the Shin to the south-west and may be applicable to much of Sutherland. Establishing, however, whether similar patterns represent common events is a problem still to be addressed, since local factors influencing the viability of a settlement may have out-weighed all but the most catastrophic economic,

environmental or social changes. That there is a shifting margin around the area that may have been regarded as viable for settlement is illustrated on many sites in the Strath, for example Auchnasheenish (NC 82NW 11, 13, 20, 21 & 22), where a farm, cultivation ridges, shielings, hut-circles, clearance cairns and burnt mounds occur in the same location, testifying to intermittent occupation and land-use from at least the second millennium BC to the eighteenth century AD, with the role of the site changing as the boundary of the margin shifted.

The identification of broad patterns of intensity of activity across a landscape may also be useful in assessing the applicability of environmental sequences at local and regional scales. Given the frequent spatial dislocation of coring sites and archaeological sites, linkages between sequences are often unsatisfactory. A case in point may be the Upper Suisgill excavation (Barclay 1987) where a core was taken from blanket bog some 2 km from the site (Andrews *et al* 1987). The predominant tree type in the Suisgill core is pine which is at odds with the published sites in Caithness and Sutherland in general and Kilearnan Hill in particular. The Kilearnan core suggests a birch and alder woodland within the valley and organic samples from the Kilearnan excavation also indicate the presence of hazel (Birnie *in* Haggarty forthcoming). The contrast between the cores suggests that the Suisgill core may be more representative of the uplands and interior of Sutherland, which features blanket bog and scattered pine, and is therefore unlikely to represent the activities on the excavated site which lies on the edge of the haughland of the Helmsdale River (Birnie *in* Haggarty forthcoming). The Suisgill core may be more relevant to a consideration of those sites in the upper and middle reaches of the Suisgill Burn which lie in less favourable locations.

CONCLUSIONS

An examination of the character and distributions of the land-use and settlement sites broadly dating to the second and first millennium bc has identified areas and sites which represent relatively intensive and prolonged activity. Other sites represent lower intensities of activity and may have been more sensitive to changes in the pressures on their viability. These sites may have been characterised by an intermittent pattern of occupation. The Group 3 sites, situated on and around the better land in the Strath, would have been more resilient in their response to increased stresses, although, of course, they have all failed at some point. Although identifying individual sites as peripheral elements within a wider settlement system is beyond the scope of the survey data, given the chronological span of the hut-circle settlements and the shifting boundary of the fringe, the areas where Group 1/2 sites predominate define what may always have been a periphery.

ACKNOWLEDGEMENTS

I am grateful to the following for permission to reference unpublished material and for discussing their sites: A Haggarty (Kilearnan Hill, excavations), C Lowe (Craggie Water), J Birnie (Kilearnan Hill, pollen), R J Mercer (Cnoc Stanger). The work at Kilearnan Hill and Craggie Water was sponsored by Historic Scotland. Fieldwork in the Strath of Kildonan in 1991 was undertaken by the Afforestable Land Survey of RCAHMS (D Cowley, P Dixon, S Halliday and G Ritchie) with survey and drawing work undertaken by G Brown, H Graham, J Green, A Leith, K Mcleod, I Parker and A Wardell. My thanks to colleagues at RCAHMS for comments on the text which has been edited by J B Stevenson and G S Maxwell. Most of the data used here has been compiled from the survey records of the Archaeology Division of the Ordnance Survey (in Inverness: J Barneveld, K Blood and J Macrae); a high quality record whose potential has yet to be fully realised.

REFERENCES

Andrews, M V, Blackham A M & Gilbertson D 1987 'A summary of the radiocarbon-dated vegetational history of the Suisgill area', *in* Barclay 1987.

Barber, J 1982 'Arran', *Current Archaeology*, 83, 358–62.

Barber, J W & Brown, M M 1985 'An Sithean, Islay', *Proceedings of the Society of Antiquaries of Scotland,* 114 (1984), 161–188.

Barclay, G J 1987 'Excavations at Upper Suisgill', *Proceedings of the Society of Antiquaries of Scotland*, 115 (1987), 159–198.

Birnie, J forthcoming 'Pollen analysis of sediments in association with excavations at Kilearnan Hill', *in* Haggarty forthcoming.

Fairhurst, H & Taylor, D B 1971 'A Hut-circle settlement at Kilphedir, Sutherland', *Proceedings of the Society of Antiquaries of Scotland*, 103 (1970–1), 65–99.

Haggarty, A forthcoming 'Excavations at Kilearnan Hill'.

McCullagh, R 1991 'Lairg', *in* Batey, C E & Ball, J (eds), *Discovery and Excavation in Scotland, 1991.* Council for Scottish Archaeology, 46–48.

Mercer, R J nd *Cnoc Stanger, Reay, Caithness.* Typescript excavation report.

RCAHMS 1993 *Strath of Kildonan: An Archaeological Survey.* Edinburgh: RCAHMS.

Stevenson, J B 1985 'The excavation of a hut circle at Cùl a'Bhaile', Jura, *Proceedings of the Society of Antiquaries of Scotland*, 114 (1984), 127–160.

23. Animal husbandry in a wetland: the case of Assendelft

Louise van Wijngaarden-Bakker

Abstract

The Iron Age habitation of the region of the Assendelver Polders in the western Netherlands consists of two separate colonisation phases. In a landscape that can be characterized as a prehistoric wetland the agricultural economy depended largely on animal husbandry. The possibilities and constraints for animal husbandry are examined. The influence of the foraging strategy of domestic herbivores, wet ground conditions, frost and snow are discussed, as are management logistics.

INTRODUCTION

In 1979 a large-scale research project was initiated by the Institute for Pre- and Protohistory of the University of Amsterdam. The main aim of the project was to analyze the settlement pattern in the region of the Assendelver Polders (Figure 23.1). Research was carried out on three levels, the intra-site, inter-site and regional level, and focused on the Iron Age and Roman periods. From the beginning of the project the study of the relationship between man, nature and the economic structure of the area was specified as a focal point of the research (Brandt *et al* 1984; van der Leeuw 1987). The first results of the palynological, palaeobotanical and zooarchaeological research are available (Groenman-van Waateringe 1987; 1988; Pals 1987; 1988; Seeman 1987; van Wijngaarden-Bakker 1988) as well as the general cultural data (Brandt *et al* 1987).

In this paper I will first give a short description of the palaeoeconomic situation in the research area, then focus on the possibilities and constraints for animal husbandry in a wetland area and lastly try to integrate the results of both studies.

GENERAL PALAEOECONOMIC SITUATION

During the middle and late Bronze Age (*circa* 1300 – 650

BC) the area of the Assendelver polders was uninhabited. In that period habitation was concentrated in the coastal zone and in West Friesland and may be briefly characterized by:

* large farms, 12 – 32 m long
* mixed farming
* animal husbandry based on cattle and sheep
* low percentage of pig, no horses
* arable farming: emmer, barley, linseed (Buurman 1988)

In the late Bronze Age the ground water level rises due to the Dunkirk Ia transgression and occupation of the west-Frisian area ceases. To the south, in the Assendelver area, the oligotrophic peat spread westwards forming 'peat cushions' in the surrounding reed peat.

In the early Iron Age a system of drainage gulleys developed at the edge of the peat. These drained the peat adequately for *circa* 75 years (650–525 BC), so that the area became habitable for the first time. Permanent settlement of the area was shown by the excavation of House Q, built on an oligotrophic peat cushion (Therkorn *et al* 1984). In brief, the habitation may be characterized by:

* smaller farms, maximum 20 m long

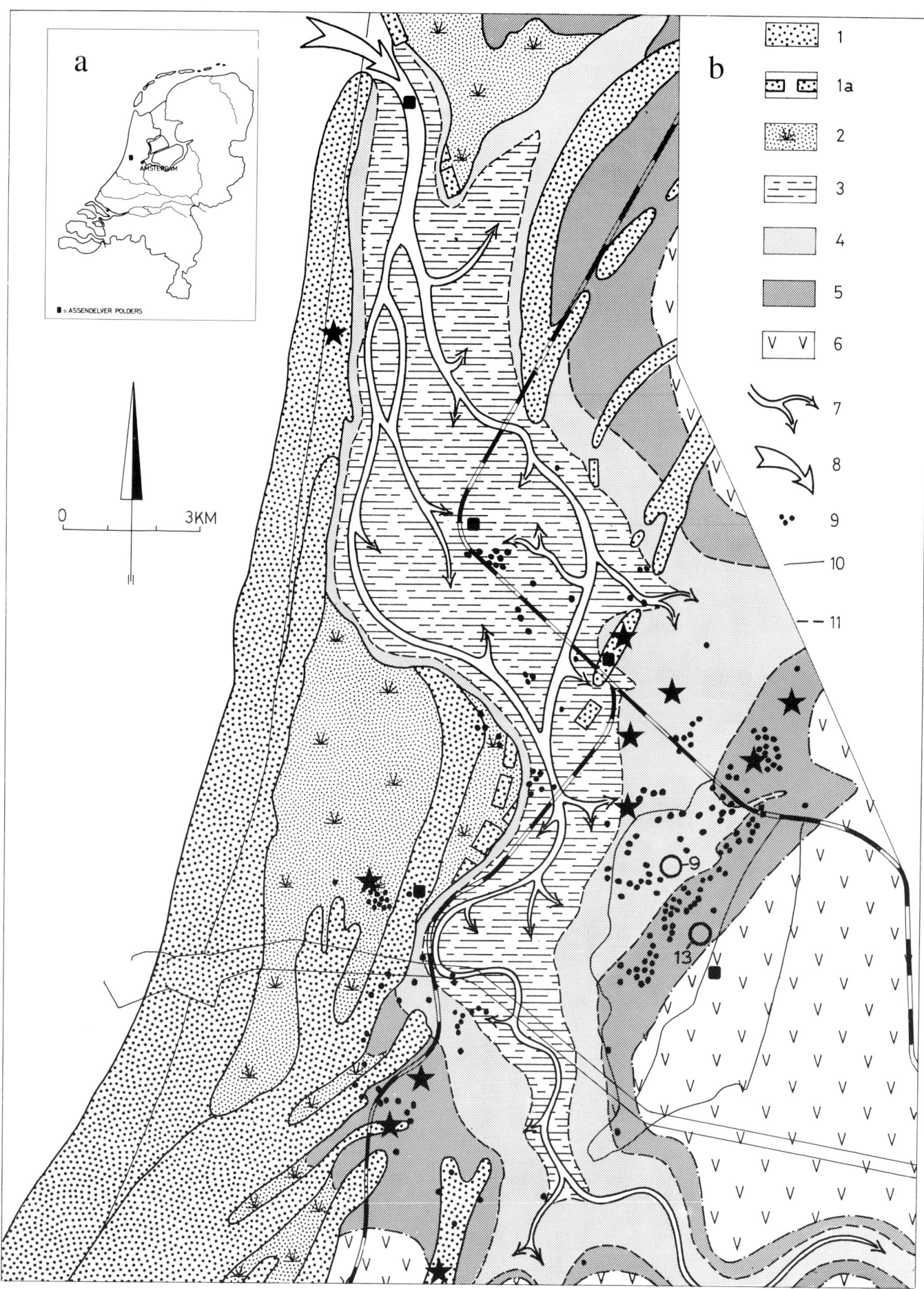

Figure 23.1. a) Location of the Assendelver Polders. b) The area of the Assendelver Polders at the beginning of the first century AD (after Brandt et al 1984).

1 *coastal barriers covered by dunes; 1a. idem, presumed. 2 emergent sandy estuarine deposits, beaches and other sandy soils between coastal barriers.*
3 *emergent intertidal region of Dunkirk I estuary (clay and sandy tidal flats). 4 high marshes of Dunkirk I estuary. 5 idem, peaty subsoil (reed swamp).*
6 *raised oligotrophic bog. 7 still functioning tidal channels of the Dunkirk I system. 8 tidal inlet. 9 sites. 10 modern coastline; outline of the Assendelver*
Polders. 11 presumed boundaries.

	EARLY IRON AGE		LATE IRON AGE	
	N	%	N	%
BOS	36	52	1052	74
OVIS/CAPRA	31	45	316	22
SUS	2	3	25	1.5
EQUUS	–	–	23	1.5
CANIS	–	–	16	1

Table 23.1 Frequency (based on fragment counts) of domestic mammals in excavations in the Assendelver Polders. Data from Seeman 1987 and van Wijngaarden-Bakker 1988.

* pastoralism: cattle 52%, sheep/goat 45%, pig 3% (Table 23.1)
* cultivation: gold-of-pleasure and import of barley
* stabling of cattle and goats
* no horse

This pattern is almost exactly matched by the results of a recent excavation in the peat area south of Rotterdam. As opposed to Site Q where a bone sample of only 69 fragments was available, at Midden Delfland a large bone sample could be analyzed. It showed roughly the same proportion of domestic animals, but here the horse was present and eaten (van Dijk 1992). In the stable parts of the farm houses layers of cattle dung were encountered. Dung of ovicaprid origin was identified as coming from goats on the basis of the presence of shoots of bog myrtle, a shrub apparently not eaten by sheep (Therkorn *et al* 1984).

By the end of the period the gullies that drained the peat area in the Assendelver Polders became larger, leading to frequent inundations. The area was no longer inhabitable. Around 200 BC large areas of saltmarsh in the Assendelver Polders had been covered with silt, so that the area became habitable again.

A second colonization phase took place in the Late Iron Age and this time the habitation lasted some 400 years (200 BC to AD 200). Characteristics of the habitation are:

* farms on natural levees, *circa* 20 m long
* gradually smaller 'farms', disappearance of stable part
* platforms on small artificial mounds ('terps')
* mixed farming with emphasis on animal husbandry: cattle 63–70%, sheep/goat: 20–22%, pig 2–7% (Table 23.1)
* horses present: eaten + ritual evidence
* small scale cultivation of barley
* millet, linseed and celtic bean grown in 'gardens'

The area becomes gradually wetter, and is once again abandoned in the second century AD.

POSSIBILITIES AND CONSTRAINTS FOR ANIMAL HUSBANDRY

The two periods of colonization of the Assendelver Polders in respectively the Early and the Late Iron Age show some similarities and some differences in agricultural practices. To analyze the situation I want to focus on the possibilities and constraints for animal husbandry in this wetland environment.

Wetlands are areas where water always plays an essential role. They have a high natural value from the point of economy, science, cultural history and recreation. In the Netherlands there are now 103 officially recognised wetlands (Swart *et al* 1989). A number of them are severely threatened by pollution, disturbance, desiccation and drainage. This has led to a protective program by several nature conservation agencies. As this program got under way, it was felt that protection of existing wetlands was not enough, but that new wetlands should also be created.

Apart from technical measures such as the control of water management, grazing by herbivores was considered an important tool for development of such areas (Vera 1988). An example of such a newly developed wetland are the Oostvaardersplassen in the most recently reclaimed part of the former Zuyderzee. Here a large area, consisting of dry reedland with local patches of brushwood and grassy terrain, is grazed by Heck cattle, Konik horses and red deer. Another wetland grazing experiment is currently under way in the Ilperveld, a peat area north of Amsterdam where Soay sheep have been introduced. Observation of the grazing behaviour of cattle, horses, deer and sheep can help us understand the interaction between grazers and the (prehistoric) wetland ecosystem. All of the above mentioned projects have shown the great importance of the foraging strategy of grazers.

Foraging strategy of domestic herbivores

Grazing has a definite effect on vegetation: some plants will not produce flowers and others such as grasses will produce numerous new shoots. In the latter case the period of supply of grasses with a high food content is lengthened and therefore the primary production and consequently the grazing capacity both increase (de Bie *et al* 1987).

Herbivores may be divided into:

* *grazers*: herbivores that feed on grasses with a high food content; high ability to digest cell walls. Example: cattle and horses.
* *intermediate feeders*: herbivores feeding on grasses, herbs and leaves. Example: sheep.
* *browsers*: herbivores feeding on leaves, twigs with a low percentage of cell walls. Example: goat (Figure 23.2).

Food intake of Heck cattle in the Oostvaarders Plassen has been studied by Drost (1986). Winter feeding of these

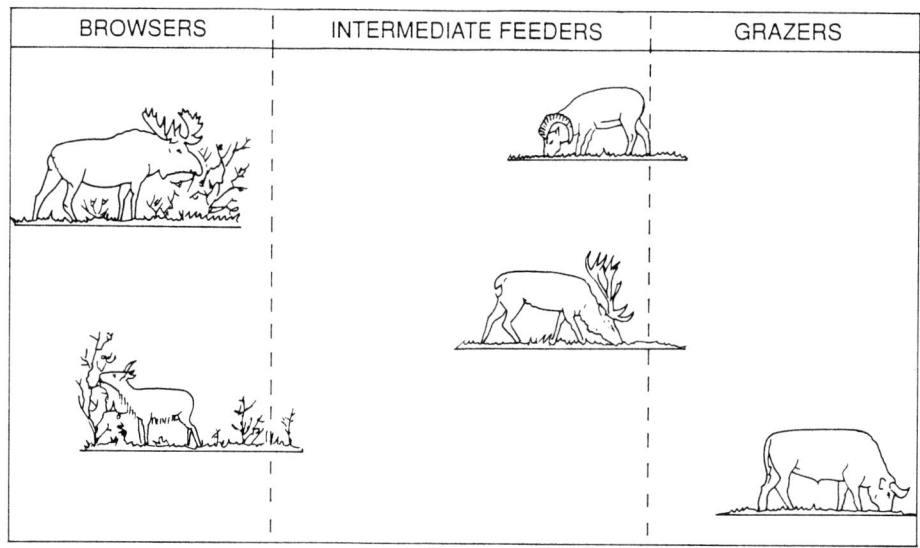

Figure 23.2 Food selection strategy of large ruminants (adapted from van de Veen & Lardinois 1991).

cattle has been kept to an absolute minimum. Year round grazing, however, was found to be possible, but depended on the presence of grassy terrain and brushwood (Table 23.2).

There are as yet no data available on the food intake of the Soay sheep in the wetland project at the Ilperveld, but so far the herd seems to be in good health with a high reproduction rate. Study of food intake of Mouflons introduced into a nature park in Czechoslovakia shows that they eat a varied diet as intermediate feeders (Table 23.3). The problem with sheep is that they are choosy: at the Ilperveld the sheep have eaten all the available specimens of the lesser butterfly orchid (*Platanthera bifolia*), a species that figures on the national list of protected plant species.

Cattle and sheep can easily graze on the same terrain. As intermediate grazers, sheep can survive on a poorer diet and can also graze closer to the ground as they have a split upper lip. Cattle eat by slinging their tongue around grasses and therefore need longer stems.

On the basis of pollen analysis the landscape around Assendelft may be characterized by the presence of grasses, sedges, bog myrtle, heather and marsh fern (Groenman-van Waateringe 1987; 1988). This vegetation did not present an important constraint on the food selection strategy of cattle and sheep.

Goats are browsers and thus need bushes and low trees for food. Botanical samples from the area around Assendelft show a high presence of bog myrtle (*Myrica gale*), a shrub that is especially appreciated by goats. Dung samples from House Q consisted almost entirely of bog myrtle, showing that either branches were purposely collected and fed to caprovids or that these animals were alternately kept outside and inside the stable. As sheep are reported not to eat bog myrtle, the dung was interpreted as coming from goats (Therkorn *et al* 1984). The same observation of dung

	SPRING	SUMMER	AUTUMN	WINTER
GRASSES	+ +		+ +	
REEDS		+ +	+	
HERBS		+	+	
TWIGS				+ +

Table 23.2 Food intake of Heck Cattle in experimental grazing project of the Oostvaardersplassen (after Drost 1986).

	SPRING	SUMMER	AUTUMN	WINTER
GRASSES/HERBS	44	56	13	8
LEAVES/TWIGS	18	7	11	29
BUSHES	23	32	20	27
BARK	3	–	–	3
MOSSES	10	1	1	21
FRUITS	2	4	55	12

Table 23.3 Food intake in percentages by Mouflon sheep in Czechoslovakia (after Pfeffer 1967).

containing bog myrtle was made at the Middle Iron Age site of Midden Delfland, in the peat area south of Rotterdam (van Dijk 1992).

The food of pigs consists of 90% vegetal and 10% animal matter (insects and worms). Extensive pig breeding is usually associated with a masting period in oak and beech woods. As the reconstructed environment of the

Assendelft area is essentially treeless, the food selection strategy of pigs would constitute a constraint on their presence in the area.

Wet ground conditions

For cattle, sheep and pigs the wetness of the soil would present no great problems. The grazing projects with Heck cattle and Soay sheep in wetlands have not produced any pathological conditions of the feet or hoofs. Pigs are anatomically adapted to wet and marshy terrain where their two lateral toes will spread out. Goats, in contrast, are not well adapted to humid terrains: in these conditions they often develop severe ailments of the hooves (French 1970). They need special care in the wet season.

Apart from pathological conditions of the feet, a wet environment may also influence the spreading of some illnesses, such as infestation with the liver fluke and some specific bacteria. So far there is no definite evidence that these zoonoses were present in the prehistoric Assendelver wetland.

Frost and snow

Frost in itself does not present a problem to cattle, sheep, goats or pigs, although goats may be rather more susceptible due to the absence of a woollen undercoat. Deep snow cover may present a problem. If the snow reaches higher than the length of the front leg, an animal may literally drown in the snow. For each species there is thus a critical depth of snow cover (Table 23.4).

A strongly frozen soil presents a problem to true grazers. Cattle will eventually resort to twigs, but as these have a very low food content, a prolonged frost could lead to starvation, unless the farmer provides supplementary feeding.

Data on the past and present climate of the Netherlands suggest that prolonged periods of heavy frost and/or deep snow cover are extremely rare, particularly in the coastal area. With the possible exception of goats, frost and snow do not seem to have been constraining factors for year round grazing.

Management logistics

The stabling of livestock formed part of the Iron Age economy of the Assendelver Polders. Conceivably, stabling only took place during the winter months. Hypothetically there are two possible regimes: night-and-day stabling or stabling only during the night. The first regime would involve the collection of sufficient winter fodder, while in the second case this would be unnecessary.

During the first colonization period at House Q goats were fed with branches of bog myrtle, but it is not known whether the branches were brought to the goats or whether the goats were taken to the shrubs. The collection of sufficient winter fodder for cattle may have presented a

	CRITICAL DEPTH OF SNOW
CATTLE	90 – 100 cm
SHEEP	25 – 30 cm
GOAT	30 – 40 cm
PIG	40 – 50 cm
HORSE	90 – 100 cm

Table 23.4 Critical depth of snow cover for domestic animals (after Pfeffer 1964).

	year	S	C	C/S ratio
Camelina sativa – gold-of-pleasure	1975	20	1150	57.5
Brassica rapa – rapeseed	1977	20	565	28.3
Avena sativa – oats	1975	130	2850	21.9
Vicia faba – celtic bean	1975	180	3000	16.6
Linum usitatissimum – linseed	1975	60	890	14.5
Hordeum vulgare – hulled barley	1977	125	1650	13.2
Panicum miliaceum – millet	1975	60	330	5.5
Triticum aestivum – bread wheat	1977	130	305	2.4
Hordeum distichum – two-row barley	1977	130	230	1.8
Triticum dicoccum – emmer	1976	60	37	0.6
Triticum spelta – spelt	1976	40	–	0.0

Table 23.5 Maximum yields of crop seeds from sown seeds in experimental cultivation (after van Zeist et al 1976; Bottema et al 1980). KEY: S = amount of seeds (g); C = crop (g).

serious logistical problem in a wetland situation. There is however ample evidence for local cultivation of gold-of-pleasure (*Camelina sativa*), which might well have been used as winter fodder. Experimental cultivation of different crops in a brackish environment gave by far the highest yields for gold-of-pleasure (Table 23.5).

In the second colonization period a local disappearance of stables and an appearance of platforms have been observed (Therkorn & Abbink 1987). It is difficult to assess whether the disappearance of stables is connected to a pastoral economy where there is less use for dung or to the possibility that collection of sufficient winter fodder became a constraining factor. The zooarchaeological data give evidence of a concentration on cattle and sheep keeping, with very few pigs and an occasional horse (Seeman 1987; van Wijngaarden-Bakker 1988).

The technological equipment for the use of secondary products such as milk was available: ceramic cheese forms have been found at several sites (Therkorn & Abbink 1987, 139). Also, a number of houses show a second hearth in the stable part of the house. This suggests some activity carried out there in summer. It is tempting to think of cheese making which involves some heating.

If the animals followed their more or less 'natural' reproductive cycle (Table 23.6), the period of lactation would fall in summer when the stables are otherwise empty. When resorting to year-round grazing the milking and cheese making could have been concentrated on the so-called platforms. Supplementary evidence comes from the ageing data for cattle which show a fairly high proportion of animals over four years old, again pointing to milk production. The ageing data for sheep do not show a clear cut pattern either for primary or for secondary products. There is however considerable artefactual evidence for spinning and weaving of wool.

CONCLUSION

Both the archaeological and the ecological data show a community well adapted to wetland circumstances. During the second colonization phase, especially, there is evidence for an intensification of animal husbandry with an increasing emphasis on cattle breeding and to a lesser extent on sheep. The fact that there is very little evidence for cultivation has led to the supposition that the animal husbandry in this environment may have led to the production of a surplus that could then be exchanged for grain from outside the wetland area. The surplus may have been in the form of secondary products, notably milk and wool. The marginality for cultivation of the area was thus counterbalanced by a high productivity of animal husbandry. This productivity can be linked to the grazing capacity of the wetland and might have been increasing due to the foraging strategy of cattle and sheep.

A point of special interest is the fact that despite important constraints on the keeping of goats in the wetland environment, these animals were present. This presence therefore seems to be more an effect of a cultural choice than of environmental factors.

REFERENCES

Bie, de S, Joenje, W & van Wieren, S E 1987 *Begrazing in de natuur.* Wageningen: PUDOC.

Bottema, S, Hoorn, T C van, Woldring, H & Gremmen, W H E 1980 'An agricultural experiment in the unprotected salt marsh, part II', *Palaeohistoria*, 22, 127–140.

Brandt, R W , Groenman – van Waateringe, W & Leeuw, S E van der (eds) 1987 *Assendelver polder papers I.* Amsterdam: Universiteit van Amsterdam.

Brandt, R W, Leeuw, S E van der & Wijngaarden-Bakker, L H van 1984 'Transformations in a Dutch estuary: research in a wet landscape', *World Archaeology*, 16(1), 1–17.

Buurman, J 1988 'Economy and environment in Bronze Age West Friesland, Noord Holland (from wetland to wetland)', *in* Murphy, P & French, C (eds), *The exploitation of wetlands*, Oxford: BAR British Series, 267–292. (= Brit Archaeol Rep Brit Ser, 186).

Dijk, J van 1992 *Melkboeren in Midden Delfland.* Doctoraalscriptie

	MATING	BIRTH	LACTATION
CATTLE	aug/sept	may/june	may–oct
SHEEP	oct/dec	march/april	march–aug
GOAT	nov/dec	april	april–sept
PIG	jan/feb	april/may	—
HORSE	april/july	march/june	—

Table 23.6 Reproductive cycle of domestic mammals.

Instituut voor Prehistorie Leiden.

Drost, H J 1986 'Begrazingsonderzoek in de Oostvaardersplassen. Runderen in het riet', *Landbouwkundig tijdschrift*, 98, 25–28.

French, M H 1970 *Observations on the goat.* (=FAO agricultural studies 80). Rome: FAO.

Groenman – van Waateringe, W 1987 'Palynology of the settlements on the levee, sites B-F', *in* Brandt, R W *et al* (eds), *Assendelver Polder Papers I*, Amsterdam: Universiteit van Amsterdam, 49–82.

Groenman – van Waateringe, W 1988 'Lokale bosbestanden en houtgebruik in West-Nederland in IJzertijd, Romeinse Tijd en Middeleeuwen', *in* Bloemers, J H F (ed), *Archeologie en oecologie tussen Rijn en Vlie*, Assen: van Gorcum, 133–153.

Leeuw, S E van der 1987 'Outline of the project and first results', *in* Brandt, R W *et al* (eds), *Assendelver Polder Papers I*, Amsterdam: Universiteit van Amsterdam, 1–22.

Pals, J P 1987 'Environment and economy as revealed by macroscopic plant remains', *in* Brandt, R W *et al* (eds), *Assendelver Polder Papers I*, Amsterdam: Universiteit van Amsterdam, 83–90.

Pals, J P 1988 'Milieu, landschap en landbouweconomie in Noord-Holland 1300v.Chr – 1300 na Chr.', *in* Bloemers, J H F (ed), *Archeologie en oecologie tussen Rijn en Vlie*, Assen: van Gorcum, 121–130.

Pfeffer, P 1964 'Le rôle des facteurs climatiques dans la dynamique des populations d'ongulés sauvages des steppes et déserts paléarctiques', *La terre et la vie. Revue d'écologie appliquée*, 18, 167–177.

Pfeffer, P 1967 'Le moufflon de Corse (*Ovis ammon musimon* Schreber, 1782); position systématique, écologie et éthologie comparées', *Mammalia. Supplément* , tome 31.

Seeman, M 1987 'Faunal remains of excavations during 1978 and 1979', *in* Brandt, R W *et al* (eds), *Assendelver Polder Papers I*, Amsterdam: Universiteit van Amsterdam, 91–98.

Swart, J, Braakhekke, W G & Kramer, T 1989 *Besturen van waterrijke gebieden. Het omgaan met wetlands door lagere besturen.* Utrecht: Natuur & Milieu.

Therkorn, L L & Abbink, A A 1987 'Seven levee sites: B, C, D, G, H, F and P', *in* Brandt, R W *et al* (eds), *Assendelver Polder Papers I*, Amsterdam: Universiteit van Amsterdam, 115–168.

Therkorn, L, Brandt, R W , Pals, J P & Taylor, M 1984 'An early Iron Age Farmstead: Site Q of the Assendelver Polders Project', *Proceedings of the Prehistoric Society*, 50, 351–373.

Veen, H van de & Lardinois, R 1991 *De Veluwe natuurlijk! Een herkansing en eerherstel voor onze natuur.* Haarrlem: Schuyt.

Vera, F 1988 *De Oostvaardersplassen. Van spontane natuuruitbarsting tot gerichte natuurontwikkeling.* Amsterdam: IVN/Grasduinen.

Wijngaarden-Bakker, L H van 1988 'Zoöarcheologisch onderzoek in de west-Nederlandse delta 1983–1987', *in* Bloemers, J H F (ed), *Archeologie en oecologie van Holland tussen Rijn en Vlie*, Assen: van Gorcum, 154–185.

Zeist, W van, Hoorn, T C van, Bottema, S & Woldring, H 1976 'An agricultural experiment in the unprotected salt marsh', *Palaeohistoria*, 18, 111–153.

24. Douara Cave, Syria: the botanical evidence from a Palaeolithic site in an arid zone

Frances S McLaren

Abstract

Excavations of a Middle Palaeolithic hearth found at the site of Douara Cave, Syria produced thousands of tiny *Prunus* endocarp (stone) fragments. The small size of the endocarp fragments suggested that the plum stones might have been smashed to obtain the oily kernels. The paper also explores some methods of food exploitation that the foragers may have used.

INTRODUCTION

Excavations of the Levantine Mousterian deposits at Douara Cave in the Palmyra Basin, Syria, produced not only faunal remains (Payne 1983) but also a large amount of macrobotanical remains. These plant residues offered the first opportunity to examine such an early plant assemblage since the macrobotanical material found associated with Peking Man at Zhoukoudian over fifty years ago (Oakley 1969). The Douara Cave macrobotanical residues have been investigated with three prime aims: initially to identify unequivocally the plants and establish reliable morphological criteria, secondly to reconstruct the environments surrounding the cave during the Mousterian occupation and finally to produce models for the possible exploitation of these plants by the foragers. This paper discusses some of the methods of food exploitation that the foragers may have employed. The nomenclature employed throughout this paper follows that used by Zohary & Hopf (1993).

THE EXCAVATIONS OF DOUARA CAVE

Excavations at Douara Cave 1 are just part of a number of geological and archaeological investigations into the Palaeolithic history of the Palmyra basin undertaken by the University of Tokyo since 1970. Douara Cave 1 is a Paleogene limestone and marl cave located on the lower part of the southern slope of Jabal ad Douara about 20 km north-east of Palmyra (Central Syria). The cave is about 170 m above a plain of central marshland situated in a transitional zone between mountains to the north and a steppic piedmont landscape to the south (Endo 1978).

A series of excavations at Douara Cave have produced evidence for the irregular occupation of the cave by Early Man during both the Middle Palaeolithic and Epipalaeolithic periods but to date no evidence of any Upper Palaeolithic activities has been found (Akazawa 1987; Endo 1978; Hirai 1987).

The large amount of ancient plant remains recovered during excavations of the Levantine Mousterian deposits included *Celtis australis* L. (but with no *Celtis tournefortii* Lam. present) together with a huge number of seeds of several members of the Boraginaceae identified (a) from seed morphology by Gordon Hillman and (b) from tissue histology of Boraginaceae, seed fragments and wood particles by Akiko Matsutani (1979; 1987).

In 1984 the archaeological investigations, directed by Professor Takeru Akazawa, concentrated on a series of Mousterian hearths. The hearth site measured about 6 m by 4 m and was encircled by limestone fragments. These hearths have been subjected to several physical dating methods (Nishimura 1979; Kobayashi *et al* 1987; Ninagawa *et al* 1987; Sakai 1987), which at present have resulted in a range of dates between 75,000 and 52,000 BP. An entire section of the hearth was removed during the excavations by Yoshi Nishiaka for closer laboratory excavation and examination by Gordon Hillman. The hearth deposits were formed of a series of very compact concretions of ash interspersed with charred plant remains (Hirai 1987). In one layer of the deposit Hillman discovered some thousands of tiny fruit endocarp (stone) fragments.

ARCHAEOLOGICAL EVIDENCE FOR THE USE OF TEMPERATE FRUIT AND NUT TREES

Archaeological evidence for the cultivation of fruit and nut trees is based on their appearance outside their natural range and the recognition of technological processing evidence such as fruit presses and large scale storage facilities. The earliest archaeological evidence in southwest Asia for the cultivation of temperate orchard fruits such as apples, pears, cherries and plums occurs in the Early Bronze Age long after the local adoption of settled agriculture (Zohary & Spiegel-Roy 1975; Spiegel-Roy 1986; Zohary & Hopf 1993; Miller 1991).

Absence of evidence for earlier cultivation of these tree fruits does not, of course, mean that wild arboreal fruits or nuts were not exploited by foragers. At the open air Epipalaeolithic site of Ohalo II by the Sea of Galilee (Kislev *et al* 1992) a large number of plant species were recovered. Nine edible fruits and nuts were identified including two nutshell fragments of an almond (*Amygdalus* sp.). At Hayonim, a Natufian cave and terrace site in Israel, a much smaller assemblage of macrobotanical residues were found but again almond shell fragments were identified (Miller 1991). At present no comparable wild plum stones have been reported from pre-agrarian sites in south west Asia.

In Europe plum stones have been recorded from the Pre-Neolithic (presumably Mesolithic) period onwards in the Upper Rhine and Danube regions (Werneck & Bertsch 1959). Zohary & Hopf (1993) suggested that these plums from European pre-agricultural sites were spontaneous forms of a hexaploid *P. domestica* identical to the small subglobose fruits still commonly found wild in Europe and Turkey today.

There are very few published systematic studies of ancient plums (Behre 1978; Körber-Grohne 1984). Long term selection and cultivation of European plums by horticulturists has today resulted in a large number of commonly named domesticated plum forms such as damsons, bullaces and greengages which are all included in *P. domestica* L. The wide variation in plum morphology has caused difficulties in arranging the wild and domesticated plums into a systematic classification since research by Linnaeus during the 18th century (McLaren 1995).

In a recent paper Zohary (1992) reviewed the position regarding the origin of *P. domestica*. Zohary felt that current genetic research pointed to a direct evolution of *P. domestica* out of the genetically variable *P. divaricata* stock. This lead him to propose that if direct evolution was the case then all the *P. divaricata* (=*P. cerasifera*) and *P. domestica* plums of Europe and west Asia belonged to a single polymorphic species, which should be known as *Prunus domestica* L.

MORPHOLOGICAL EXAMINATION OF THE DOUARA CAVE CHARRED FRUIT STONES

The charred fruit endocarps that Hillman recovered from the hearth, were tiny fragments measuring approximately 2 mm by 1 mm. Drupaceous fruits are made up of a thin outer pericarp forming the skin, containing a succulent fleshy mesocarp inside which is a hard stony endocarp known as a stone or pit surrounding a seed often misnamed a nut. The endocarp consists primarily of hard dense tissues largely of stone cells and like all solid plant tissues the chemical content of a stone is predominately carbohydrate.

Hillman's initial investigation began with light microscopic and Scanning Electron Microscopic (SEM) studies of a range of modern drupaceous fruits likely to have grown in the area at that time together with other hard walled fruits such as *Zilla spinosa* (Turra) Prantl which when charred and smashed might produce fragments of the Douara type. Microscopic examination of the Douara Cave fragments revealed that they retained seemingly diagnostic features which were closely similar to those of *Prunus domestica* (the domestic plum and its immediate wild ancestor(s)) but that none of them had the diagnostic features of *P. divaricata*. Clearly, in view of the early date of the remains, it is to the wild forms of *P. domestica* that he referred the specimens concerned. However, most of the fragments lacked all diagnostic features including the characteristic cell patterns of the inner surface of the endocarp. Thus, although he regarded many such specimens as having the general features of *Prunus* and/or *P. cerasus* (cherry), the bulk of them were unidentifiable.

The charred residues recovered from Douara cave have survived as a result of only partial degradation. Charring is a surface phenomenon and it appears that at least some components of the original chemical make-up of the stone (including long chain hydrocarbons) remain unaltered (Evans & Biek 1976; Winter 1983). It was, therefore, decided to subject the stones to a chemical analysis, the spectrometric technique of Infra Red analysis (IR), that had previously proved successful in the separation of early wheats and ryes (McLaren *et al* 1991; Hillman *et al* 1993).

CHEMICAL ANALYSIS OF ARCHAEOLOGICAL MATERIAL

Archaeological materials, including ancient macrobotanical remains, have a long period of unknown history. Despite being recovered from a cave, in a fairly arid environment, the fruit stones have probably been subjected to occasional alterations in their environment. For example, possible exposure to a long period of flooding or the action of soil fauna and flora could cause an alteration in the associated matrix. Post-excavation events such as the number of people handling the material and the storage media may all have had an effect. Unknown events could cause contamination of the botanical residue. Consequently, it was felt that these types of problems could limit the practical use of the more sensitive analytical techniques. To be of any use to the archaeobotanist, however, the analytical technique must be sensitive enough to unambiguously identify a single seed.

An individual IR spectrum represents a complex of organic and/or inorganic compounds. The spectra obtained from the charred seeds were used solely in a best match 'finger print' fashion and not to identify individual compounds.

When interpreting spectra obtained from amorphous residues such as those recovered from pottery or stone assemblages there is the possibility of misleading results because the residues may be derived from a mixture of one or more substances. The problem of interpreting the spectra derived from botanical remains is reduced because there are normally morphological or histological clues for guidance. For example, even a scrap of identifiable material will suggest that you are dealing with a fruit stone and not a grass seed and therefore limit the range of possible matches between spectra.

MATERIALS AND METHOD OF ANALYSIS

A series of standards were prepared for analysis by IR, based on Hillman's collection of modern fruit stones (McLaren *et al* 1991). The standards included not only plum and cherry stones but also a range of other stones including *Cornus mas* L., *Armeniaca vulgaris* Lam., *Zilla spinosa* (Turra) Prantl and *Ziziyphus spina-christi* (L.). Where possible the stones were extracted separately from the nut.

A total of nine ancient stone fragments from Douara Cave were subjected to IR analysis. A sample of associated hearth ash, containing only the minimum amount of charcoal dust was also examined but it proved impossible to obtain a good background spectrum (McLaren 1995).

RESULTS AND DISCUSSION ON THE IR ANALYSIS OF THE DOUARA CAVE PLUMS

The spectra of both modern and ancient stones were poor in quality when compared to those obtained from grains and seeds (McLaren 1995). Distinctions could be made between the IR spectra of the modern stones of plums and other fruits including *P. spinosa* although, as might be expected, a certain similarity was found between the spectra of members of *Prunus* genus (Figure 24.1).

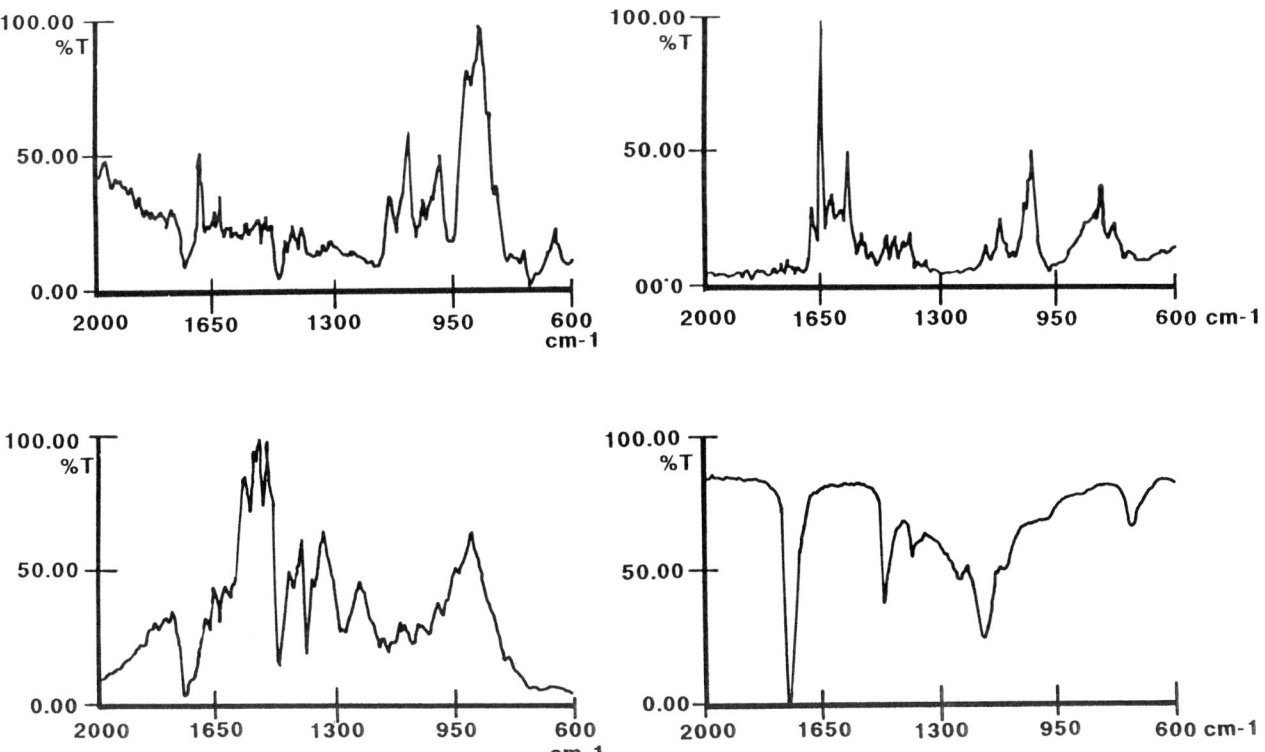

Figure 24.1 Infra Red absorption. Hexane extracts. Top left: IRFM443/90, modern stone of Prunus divaricata *Ledeb. (SP113(F)). Top right: IRFM902/239, modern stone of* Prunus domestica *L. (GCH2911(-)). Bottom left: IRFM893/237, modern stone of* Prunus domestica *L. subsp.* italica *(GCH3974). Bottom right: IRFM894/238, modern stone of* Prunus domestica *L. (GCH4135(-)).*

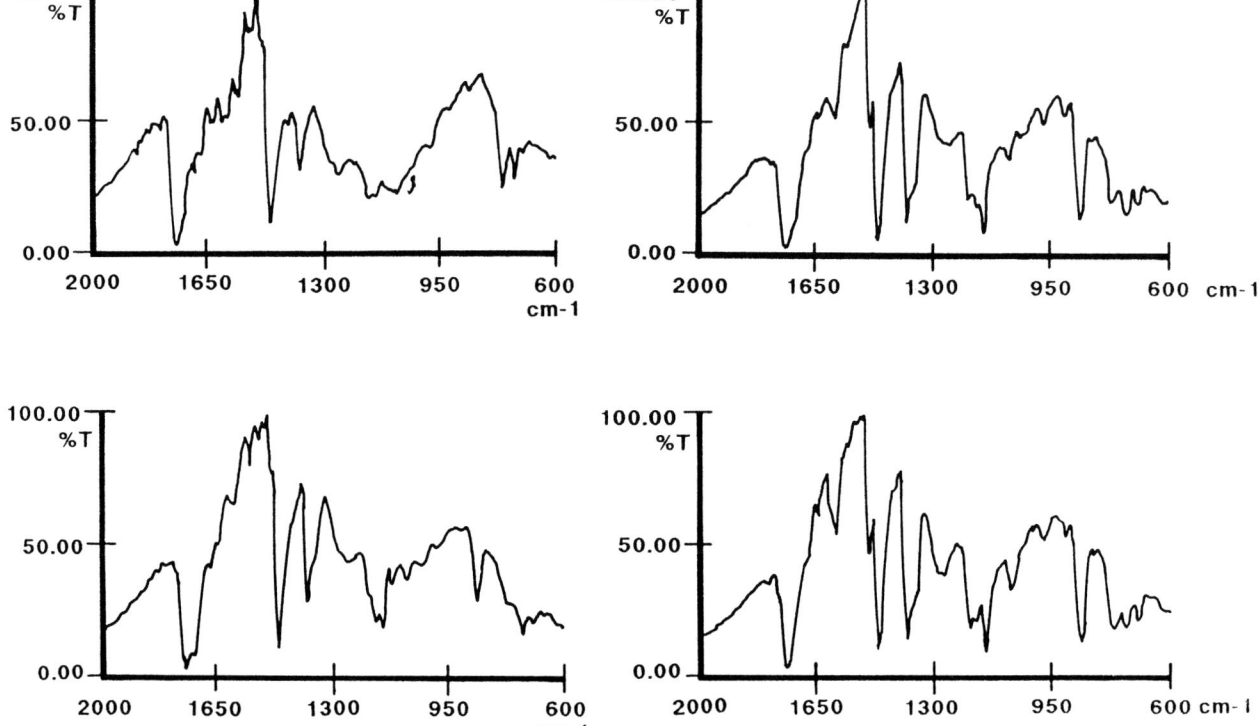

Figure 24.2 Infra Red absorption. Chloroform extracts. Top left: RFM463/90, modern stone of Prunus divaricata Ledeb. *(SP113(F)). Top right: IRFM445/91-6, ancient stone from Douara Cave (Stone 6). Bottom left: IRFM444/92-8, ancient stone from Douara cave (Stone 8). Bottom right: IRFM447/93-9, ancient stone from Douara Cave (Stone 9).*

The next step was to compare all the ancient stone spectra to ascertain if more than one species or subspecies were present. All the Mousterian stones examined show the same general pattern as the *Prunus domestica/ cerasifera/divaricata* complex (Figure 24.2). A mis-match between the modern stones and the ancient stones was observed in the 1049.8 cm^{-1} to 933.0 cm^{-1} range of the spectra (Figure 24.2). However, no such variation has yet been seen in spectra of different modern populations of the same species. The mis-match could be the result of either degenerative changes in the ancient material, or species/ subspecies differences.

Six of the Douara Cave stones showed closest resemblance to a domesticated strain *P. divaricata* subsp. *divaricata* (SP113 (f)). The closeness of the 'fingerprint' match is particularly evident in the chloroform spectra (Figure 24.2). A further two stones had distinct spectra which had a close similarity to the spectra of the *Prunus domestica/cerasifera/divaricata* complex so far examined. However, they showed no close match in all three solvent spectra to any set of modern plum or cherry stones so far examined.

Surprisingly, the hexane and chloroform extracts from Stone 1 produced a startling match with the spectra produced from a modern *P. domestica* subsp. *italica* (Figures 24.3 & 24.4). The separate extractions and IR analyses of the ancient stone and the greengage were carried

out over eighteen months apart which means that the possibility of laboratory cross contamination is unlikely. The propan-2-ol spectrum of this ancient stone is relatively poor in quality and does not, unfortunately, produce such a close match with the modern *P. domestica* subsp. *italica*. An interesting point about the greengage is that unlike most plums it often breeds true from seed (Masefield *et al* 1969). Further examination of the modern greengage populations are obviously required to confirm this identification and possibly the analysis of more of the Douara Cave specimens to see if any further possible greengages can be found.

THE DOUARA CAVE HUNTER-GATHERER EXPLOITATION STRATEGIES

The environmental evidence recoverable from a cave normally includes not only evidence of its own history but also details of the environments exploited by visitors to the cave.

Payne's (1983) analysis of the Douara Cave faunal remains suggested that this site had been occupied by the hunter-gatherers on a sporadic basis. The faunal remains could be divided into two main groups; bones primarily of large ungulates which were deposited as a result of human occupation and smaller bones which derived from the

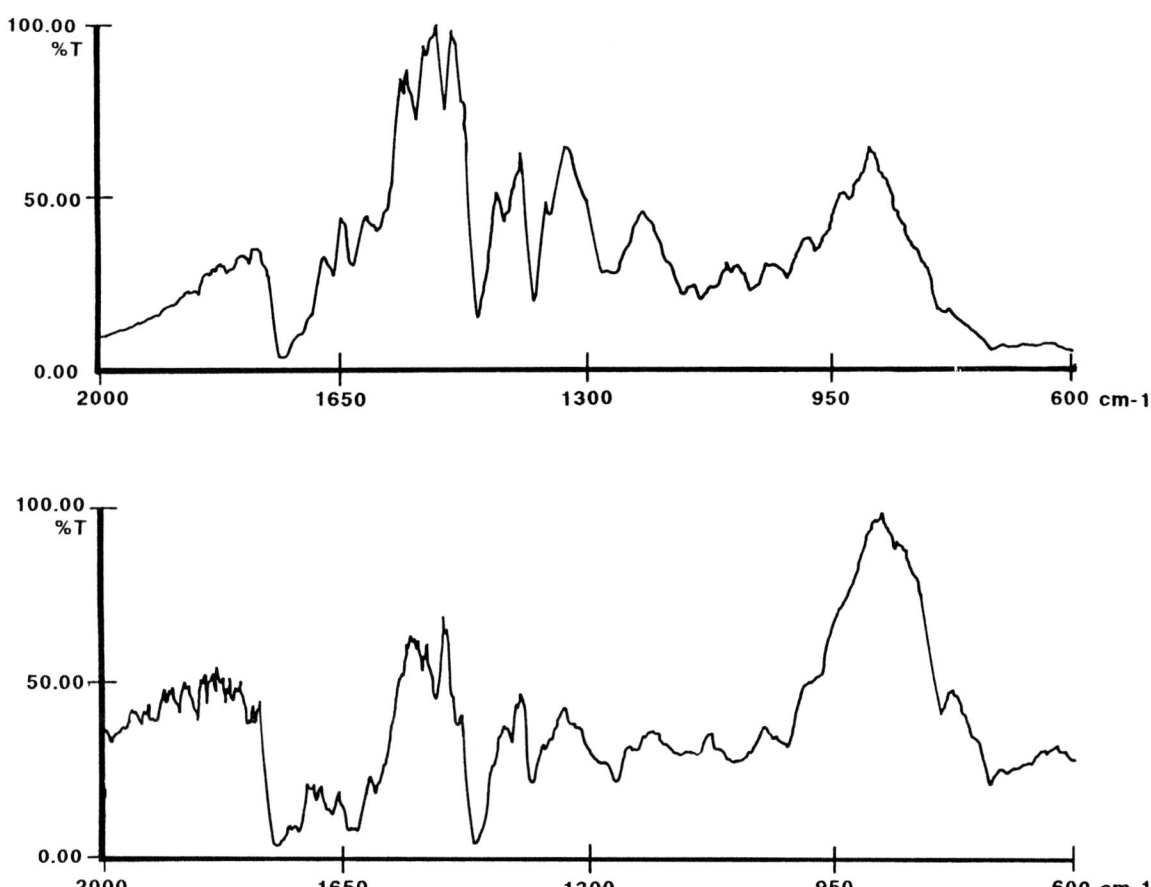

Figure 24.3 Infra Red absorption. Hexane extracts. Top: IRFM893/237, modern stone of Prunus domestica *L. subsp.* italica *(GCH3974). Bottom: IRFM527/113-1, ancient charred stone from Douara Cave (Stone 1).*

presence of non-human cave occupants such as bats and owls. These latter bones included rodent, lizard and small bird bones which were probably deposited in the cave as owl pellets and, therefore, give a relatively good indication of the ecology of about 3 km around the cave. Payne's studies (particularly of the rodent fauna) indicated an arid climate and the likelihood that the cave was in an open dry steppe environment similar to that found in the area today. Hillman's (Akazawa 1987; Hillman pers comm 1990) initial examination of the plant remains suggested that the climate might have been slightly wetter than is experienced in the area today. Hillman also observed the occurrence of halophyte plants in the area of the cave today which suggests the possible presence of nearby springs.

Hunter-gatherer exploitation strategy is based on food procurement which controls the pattern of movement throughout a territory. The patterns of food procurement tend towards tactics governed by risk minimization resulting in a complex range of methods of exploitation of a large relatively stable area (Cribb 1991). Habitation is usually dominated by seasonal base camps, each centred on a water resource. These base camps have access to a wide range of major plant and animal resources. Occasionally, additional use of special purpose short term campsites

may also have been required. Gorecki's (1991) recent ethnographic research in New Guinea of the contemporary usage made by hunter-gatherers of rock shelters and caves suggested that concepts of home base camps and temporary campsites were irrelevant to hunter-gatherers themselves. These studies indicated that hunter-gatherers had a variety of uses for rock shelters and caves but in general these sites were only occupied in the short term compared to open air sites. Akazawa (1987) suggested that one possible reason for the human occupation of Douara cave was to exploit the nearby flint sources.

If the foragers were in the area to take advantage of the flint sources, rather than food procurement, it is highly probable that they brought portable food with them. The faunal remains left by hunter-gatherers include evidence of the use of fairly portable food such as eggs including both small eggs and a large number of ostrich (Struthion-idae) eggs. It has been suggested that the ostrich egg shells could have been used as some type of container, possibly for water (Payne pers comm 1992). That the plum stones arrived at Douara Cave as a result of forager agency is not in doubt because the fragments were charred and also recovered from a hearth. Plums would be a fairly portable food, particularly if they had been dried.

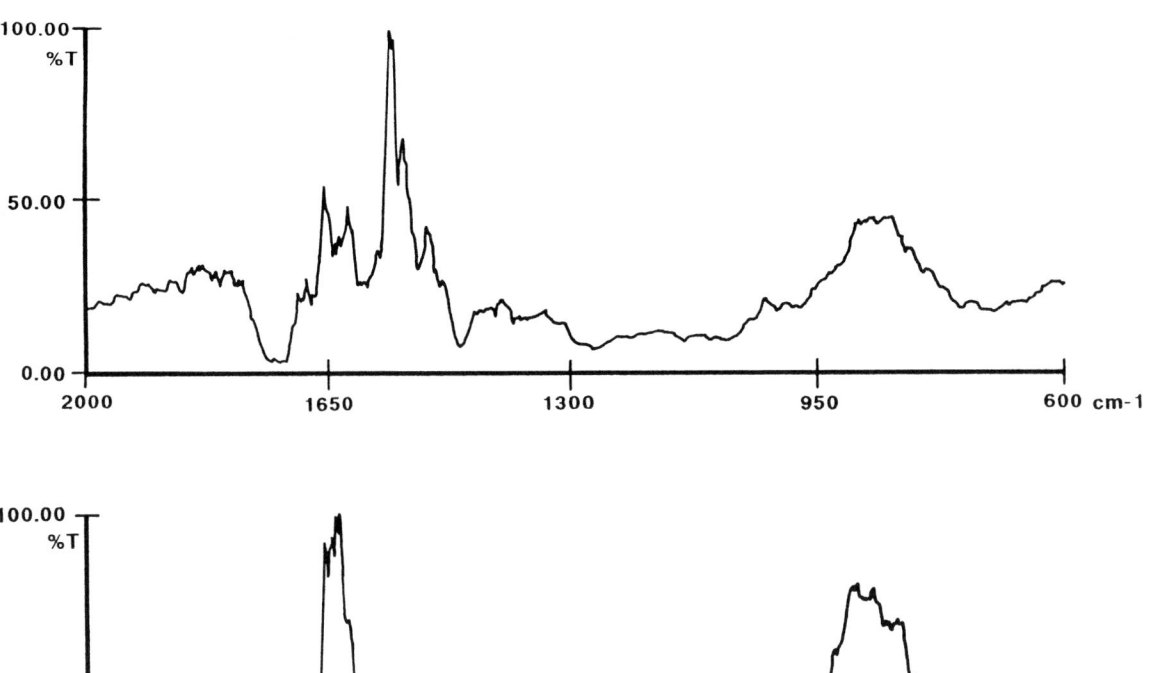

Figure 24.4 Infra Red absorption. Chloroform extracts. Top: IRFM899/237, modern stone of Prunus domestica *L. subsp.* italica *(GCH3974). Bottom: IRFM547/113-1, ancient charred stone from Douara Cave (Stone 1).*

THE ROLE OF TOXINS

A large number of plants including plums have pharmacologically active and/or toxic compounds. Plant compounds may effect human health, particularly if they are not part of a balanced diet. Were the plums from Douara Cave toxic? The likelihood of the presence of a plant toxin amygdalin in the Douara Cave plums seeds cannot be ignored. Like many other members of the Rosaceae family plums and cherries contain a cyanogenic glycoside compound known as amygdalin, which is an inactive sugar-cyanide complex. Amygdalin is normally found in the seed of the plum. The enzyme which breaks down the amygdalin is found in a separate area within the fruit and/or nut. Contact between the two is only established when cutting, peeling or crushing the fruit/nut. Once contact occurs the gas hydrogen cyanide (HCN) is produced (Montgomery 1980; Forbes & Watson 1992). Consumption of fresh plum kernels containing amygdalin can produce a variety of health problems. This varies from soft tissue diseases such as goitres, to paralysis, or instant death. The outcome depends on the amount of amygdalin present and the state of health of the human consumer, for example, whether or not they are suffering from malnutrition (Montgomery 1980).

It is reasonable to propose that prior to the establishment of tree cultivation, the presence of non-toxic fruits should be considered the exception rather than the rule. Only an occasional chance wilding of trees such as plums or almonds (*Amygdalis communis* L.) would produce fruits or nuts that were fit for human consumption without some form of processing. Furthermore, low hominid population numbers reduces still further the chance of a foraging group discovering such rare trees unless obvious morphological differences are visible. Distinct variation is only very occasionally seen and a notable example is the almond, whose blossom colour separates the deep pink flowered sweet almond and the white flowered bitter almond. Unfortunately, there is an intermediate colour range and in these cases the best method of finding edible sweet fruit is to observe which fruit the wildlife prefer (Daniel Zohary pers comm 1992). If a tree was established as non-toxic then the foragers could be certain that any fruit crops from the tree would always be edible (Dicenta & Garcia 1993).

At what level of concentration amygdalin would have occurred in these pre-domestic plums is open to conjecture. The amount of amygdalin present in fruits of the Prunoideae varies from the high concentrations found in the cherry laurel (*Prunus laurocerasus L.*) to the very low levels occurring in domesticated cherries. The broken plum stones were found in a hearth deposit, which may suggest that the seed was exploited as a food. Amygdalin is most likely to have been present in these seeds which would suggest that the Douara Cave inhabitants were carrying out some form of processing and detoxification of the seeds for food.

THE USE OF PLUMS AS FOOD

This leads to the question of what form of food processing or low level storage the foragers could be expected to practise on the fruits. How far back the use of plant food processing extends, particularly involving the controlled use of fire, has long been the subject of speculation (for a selected example see: Leopold & Ardley 1972; 1973; Dornstreisch 1973; Stahl 1984; 1989).

The plums could easily have been detoxified by pinching them to rupture the skin and release the gas. Processing them with water and/or heat is also simple. Any toxins would be either leached away or the hydrogen cyanide driven off on heating provided an open cooking system was used. The resulting product would then be safe for human consumption. If storage of the plums was also contemplated then dehydration, fermentation or the heating of a sugar source such as honey with the fruit solution (Needham & Evans 1987) to produce a preserve may have been practised. These methods of processing would serve as both a means of preservation and detoxification without having to resort to any complicated technology (Stanton 1971; 1985; 1987; Dennis 1987; Coultate 1989).

The Douara Cave plums were charred and came from a hearth; these factors allow comparison with ethnographic evidence for processing. The best recorded example is probably the exploitation of plums by the North American Indians (Balls 1962; Prescott Barrows 1971; Timbrook 1982). Plum kernels were prized as a valuable alternative to starchy acorns because of their high protein and oil content. The consequences of not processing the bitter fruit was well understood by the various North American Indian groups.

At least seven species of plums are known to have been exploited. Like some *P. divaricata* forms the wild North American fruits are characterised by having a large stone surrounded by very little flesh. The Californian Indians exploited a native plum, *Prunus ilicifolia* Walp. (islay). In general the islay kernel was used for food. Occasionally, the flesh was removed and the entire stone or pit was exploited. Various methods of roasting and leaching in preparation of the islay for food have been noted. In one method the fruit was gathered in August and initially sun dried. Subsequently the thin shells of the pits were easily broken and the extracted kernels ground in a mortar to the consistency of a meal. The islay meal was then leached in a sand basket. The cleaned meal was then boiled as a type of soup known as 'atole' (Prescott Barrows 1971). The Chumash Indians are recorded as employing an alternative

method of islay preparation when the kernels were to be stored. The pits were first boiled 'until done' and retained overnight. The stones were subsequently cracked and the extracted kernels were stored in baskets for later use (Timbrook 1982, 166).

In Western Asia today the so called 'wild' fruits of the cherry plums (*P. cerasifera*) are commonly referred to as orchard fruits and sold in local markets. The trees produce high summer fruits that have a taste that ranges from very sour to sweet and even tasteless soft juicy flesh which benefits from cooking. In Turkey these fruits are first pinched (to detoxify?) and subsequently dried (Hillman pers comm 1990). In the Caucasus they are traditionally cooked, dried and made into meal by mixing with either wheat or barley flour (Masefield *et al* 1969; Roach 1986).

Balls (1962) also reported that a beverage was produced from the flesh of islay. However, no further details of the method or type of drink was noted and at present no further details have been traced. The possibility of drink production including the use of fermentation cannot be ruled out for the Douara Cave plums. Recently van Zeist & Waterbolk-van Rooijen (1992) suggested that a small concentration of *Prunus mahaleb* L. stones recovered from the Early Bronze Age site of Tell Hammam et-Turkman in Northern Syria could have been used to produce a fermented drink. For the moment all we can say is that it is not impossible that equivalent roasting, leaching or fermentation could have been performed to detoxify the oil rich plum kernels at Douara Cave.

CONCLUSION

How far archaeologists will be able to trace early usages of plants and identify food production methods used by foragers is open to conjecture. It is unlikely that even the most sophisticated chemical technique could distinguish the differences between natural biodegradation activities and the deliberate harnessing of fermentation to produce safe, palatable and stable foodstuffs. Food processing experiments such as smashing the stones to extract the kernel, making fruit preserves, and natural fermentation may produce additional information.

Examination of these fruit stones has provided a key-hole view of exploitation of a valuable food resource in the Middle Palaeolithic. By its very nature a cave is a marginal site; however, excavations of a cave can provide data not only from the close environment of a cave but also from a wider setting. The faunal evidence suggests that the immediate environment was an open dry steppe but the plums show that the foragers exploited at least one fruit from a nearby forest or park environment.

ACKNOWLEDGEMENTS

Chemical analyses of ancient wheat and rye grains are

being supported through a Research Grant (No: GR/ H07597) from the Science-Based Archaeology committee of Britain's Science and Engineering Research Council (now through NERC) which is gratefully acknowledged.

I wish to express my grateful thanks to all those who have contributed to this study including Prof. Takeru Akazawa, The University Museum, Tokyo and Gordon Hillman for providing the Douara Cave endocarp remains for analysis. I am extremely grateful to Gordon Hillman for providing the modern fruit standards from his reference collection. I would like to acknowledge Professor Daniel Zohary, John Evans, Gordon Hillman, Dr. Ann Butler, Dr. Sarah Mason, and the numerous other people who have commented, discussed or read aspects of this paper. Finally special mention should be made of the staff of University of East London for allowing the research to be carried out at times that were often inconvenient to them.

REFERENCES

Akazawa, T 1987 'The ecology of the Middle Palaeolithic occupation at Douara Cave, Syria', *in* Akazawa, T & Sakaguchi, Y (eds), *Paleolithic site of Douara cave and paleogeography of Palmyra Basin in Syria; Part IV, 1984 excavations*, Tokyo: The University Museum, The University of Tokyo, 29, 155–166.

Balls, E K 1962 *Early uses of California plants*. California: California Natural History Guides No. 10. University of California Press, Berkeley.

Behre, K-E 1978 'Formenkreise von *Prunus domestica* L. von der Wikingerzeit bis in die frühe Neuzeit nach Fruchtsteinen aus Haithabu und Alt-Schleswig', *Bericht Deutschen Botanik Gessellschaft Bd.*, 91, 161–179.

Coultate, T P 1989 *Food: The chemistry of its components*. (2nd edition). London: Royal Society of Chemistry paperbacks.

Cribb, B 1991 *Nomads in archaeology*. Cambridge: Cambridge University Press, 20–22.

Dennis, C 1987 'Microbiology of fruits and vegetables', *in* Norris, J R & Pettipher, G L (eds), *Essays in agriculture and food microbiology*. Chichester: John Wiley & Sons Ltd, 227–260.

Dicenta, F & Garcia, J E 1993 'Inheritance of the kernel flavour in almond', *Heredity*, 70, 308–312.

Dornstreisch, M D 1973 'Food habits of early man: Balance between hunting and gathering', *Science*, 179, 306.

Endo, K 1978 'Stratigraphy and paleoenvironments of the deposits in and around the Douara cave site', *in* Hanihara, K & Sakaguchi, Y (eds), *Paleolithic site of Douara cave and paleogeography of Palmyra Basin in Syria: Part I, stratigraphy and paleogeography in the late quaternary*, Tokyo: The University Museum, The University of Tokyo, 14, 53–81.

Evans, J & Biek, L 1976 'Overcooked food residues on pot sherds', *in* Slater E A & Tate J A (eds), *Proceedings of the 16th International Symposium on Archaeometry*, Edinburgh: National Museum of Antiquities of Scotland.

Forbes, J C & Watson, R D 1992 *Plants in agriculture*. Cambridge: Cambridge Academic Press.

Gorecki, P P 1991 'Horticulturalists as hunter-gatherers: Rock shelter usage in Papua New Guinea', *in* Gamble, C S & Boismier, W A (eds), *Hunter-gatherer and pastoralist case studies, International Monographs in Prehistory*, USA: Ethnoarchaeological Series, 1, 237–262.

Hillman, G, Wales, S, McLaren, F, Evans, J & Butler, A 1993

'Identifying problematic remains of ancient plant foods: A comparison of the role of chemical, histological and morphological criteria', *World Archaeology* , 25, 1, 94–121.

Hirai, Y 1987 'Stratigraphy and sedimentological analysis of the Douara Cave deposits', *in* Akazawa, T & Sakaguchi, Y (eds), *Paleolithic site of Douara cave and paleogeography of Palmyra Basin in Syria: Part IV, 1984 excavations*, Tokyo: The University Museum, The University of Tokyo, 29, 49–6.

Kislev, M E, Nadel, D & Carmi, I 1992 'Epipalaeolithic (19,000 BP) cereal and fruit diet at Ohalo 11, Sea of Galilee, Israel', *Review of Palaeobotany and Palynology*, 73, 161–166.

Kobayashi, K, Yoshida, K, Nagai, H, Imamura, M, Yoshikawa, H, Yamashita, H, Okizaki, S, Yagi, S, Kobayashi, T & Honda, M 1987 '14C dating by Accelerator Mass Spectroscopy of carbonized plant remains from a Middle Palaeolithic hearth at Douara Cave, Syria', *in* Akazawa, T & Sakaguchi, Y (eds), *Paleolithic site of Douara cave and paleogeography of Palmyra Basin in Syria: Part IV, 1984 excavations*, Tokyo: The University Museum, The University of Tokyo, 29, 147–153.

Körber-Grohne, U 1984 'Über die Notwendigkeit einer Registrierung und Dokumentation wilder und primitiver Fruchtbäume, zu deren Erhaltung und zur Gewinnung von Vergleichsmaterial für paläoethnobotanische Funde', *in* van Zeist W & Casparie W A (eds) *Plants and ancient man: Studies in palaeoethnobotany*. Rotterdam: A A Balkema, 237–241.

Leopold, A C & Ardley, R 1972 'Toxic substances and the food habits of early man', *Science*, 176, 512–514.

Leopold, A C & Ardley, R 1973 'Response to M D Dornstreisch, Food habits of early man: Balance between hunting and gathering', *Science,* 179, 306–307.

McLaren, F S 1995 'Plums from the Palaeolithic site of Douara Cave: A chemical analysis', *in* Kroll, H & Pasternak, R (eds), *Res archaeobotanicae, Proceedings of the 9th symposium, International Workgroup of Palaeobotany, Kiel 1992*. Kiel: Institut für Ur- und Frühgeschichte der Christian-Albrecht-Universität, 195–218.

McLaren, F S, Evans, J & Hillman, G C 1991 'Identification of charred seeds from SW Asia', *in* Pernicka E & Wagner G A (eds), *Proceedings of the 26th International Symposium on Archaeometry, Archaeometry '90: Heidleberg*. Basel: Birkhäuser Verlag, 797–806.

Masefield, G B, Wallis, M, Harrison, S G & Nicholson, B E 1969 *The Oxford book of food plants*. Oxford: Oxford University Press.

Matsutani, A 1979 'Microscopic study of the amorphous silica and charcoal from the Douara cave', *in* Hanihara, K & Akazawa, T (eds), *Paleolithic site of Douara cave and paleogeography of Palmyra Basin in Syria: Part II, Prehistoric occurrences and chronology in Palmyra Basin,* Tokyo: The University Museum, The University of Tokyo, 16, 225–233.

Matsutani, A 1987 'Plant remains from the 1984 excavations at Douara Cave', *in* Akazawa, T & Sakaguchi, Y (eds), *Paleolithic site of Douara cave and paleogeography of Palmyra Basin in Syria: Part IV, 1984 excavations*, Tokyo: The University Museum, The University of Tokyo, 29, 117–122.

Miller, N F 1991 'The Near East', in van Zeist, W, Wasylikowa, K & Behre, K-E (eds). *Progress in Old World palaeoethnobotany: A retrospective view on the occasion of 20 years of the International Work Group for Palaeoethnobotany*, Rotterdam: A A Balkema, 133–160.

Montgomery, R D 1980 'Cyanogens', *in* Liener, I E (ed), *Toxic constituents of plant food stuffs*. London: Academic Press, 143–157.

Needham, S & Evans, J 1987 'Honey and dripping: Neolithic food plant residues from Runnymede Bridge', *Oxford Journal of Archaeology,* 6, 21–28.

Ninagawa, K, Yamamoto, I, Takahashi, N, Inoue, N, Yamashita,

Y, Wada, T & Sakai, H 1987 'Thermoluminescence of barite nodules from the Middle Paleolithic hearth at Douara Cave, Syria', *in* Akazawa, T & Sakaguchi, Y (eds), *Paleolithic site of Douara cave and paleogeography of Palmyra Basin in Syria: Part IV, 1984 excavations*, Tokyo: The University Museum, The University of Tokyo, 29, 133–145.

Nishimura, S 1979 'Fission track age of the baked pebble excavated at Douara cave', *in* Hanihara, K & Akazawa, T (eds), *Paleolithic site of Douara cave and paleogeography of Palmyra Basin in Syria: Part II, Prehistoric occurrences and chronology in Palmyra Basin*, Tokyo: The University Museum, The University of Tokyo, 16, 235–238.

Oakley, K 1969 *Framework for dating fossil Man*. (3rd edition) London: Weidenfield & Nicholson.

Payne, S 1983 'The animal bones from the 1974 excavations at Douara cave', *in* Hanihara, K & Akazawa, T (eds), *Paleolithic site of Douara cave and paleogeography of Palmyra Basin in Syria: Part III, animal bones and further analysis of archaeological materials*', Tokyo: The University Museum, The University of Tokyo, 21, 1–108.

Prescott Barrows, D 1971 'Desert Plants of the Coahuilla', *in* Heizer R F & Whipple M A (eds), *The California Indians: A source book*. Berkeley, Los Angeles: University of California Press, 306–314.

Roach, F A 1986 *Cultivated fruits of Britain: Their origin and history*. Oxford: Basil Blackwell Ltd.

Sakai, H 1987 'Paleomagnetic study of the Middle Paleolithic Hearth at Douara Cave', *in* Akazawa, T & Sakaguchi Y (eds), *Paleolithic site of Douara cave and paleogeography of Palmyra Basin in Syria: Part IV, 1984 excavations*, Tokyo: The University Museum, The University of Tokyo, 29, 123–131.

Spiegel-Roy, P 1986 'Domestication of fruit trees', *in* Barigozzi, C (ed), *The origin and domestication of cultivated plants* (Symposium organized by Centro Linceo Interdisciplinare di Scienze Matematiche e Loro Applicazioni, Rome, November 1985), Rome: Accademia Nazionale dei Lincei, 201–212.

Stahl, A B 1984 'Hominid dietary selection before fire', *Current Anthropology*, 25 (2), 151–168.

Stahl, A B 1989 'Plant food processing applications for dietary quality', *in* Harris, D R & Hillman, G C (eds), *Foraging and farming: The evolution of plant exploitation*. London: Unwin and Hyman, 171–194.

Stanton, W R 1971 'Some domesticated lower plants in South East Asian food technology', *in* Ucko, P J & Dimbleby, G W (eds), *The domestication and exploitation of plants and animals*. London: G Duckworth Ltd, 463–469.

Stanton, W R 1985 'Food fermentation in the tropics', *in* Wood, B J B (ed), *Microbiology of fermented foods* (volume 2). London: Elsevier Applied Science Publishers Ltd, 193–192.

Stanton, W R 1987 'Microbial processes in the production of food', *in* Norris, J R & Pettipher, G L (eds), *Essays in agriculture and food microbiology*. Chichester: John Wiley & Sons Ltd, 345–369.

Timbrook, J 1982 'Use of wild cherry pits as food by the Californian Indians', *Journal of Ethnobiology*, 2 (2), 162–176.

Werneck, H L & Bertsch, K 1959 'Zur Ur- und Fruhgeschichte der Pflaumen im oberen Rein- und Donauraume', *Angewandte Botanik*, 33, 19–33.

Winter, J 1983 'The characterization of pigments based on carbon', *Studies in Conservation*, 28, 49–66.

van Zeist, W & Waterbolk-van Rooijen, W 1992 'Two interesting floral finds from the third millennium B.C. Tell Hammam et-Turkman, Northern Syria', *Vegetation History and Archaeobotany*, 1, 157–161.

Zohary, D 1992 'Is the European plum *Prunus domestica* L. a *P. cerasifera* EHRH x *P. spinosa* L. allo-polyploid?', *Euphytica*, 60, 75–77.

Zohary, D & Hopf, M 1993 *Domestication of plants in the old world* (2nd edition). Oxford: Oxford University Press.

Zohary, D & Spiegel-Roy, P 1975 'Beginnings of fruit growing in the Old World' *Science*, 187, 319–327.

List of contributors

Timothy G Acott School of Earth Sciences, University of Greenwich, Medway Towns Campus, Chatham Maritime, Chatham, Kent, ME4 4AW, England.

Ibrahim M Al-Resseeni School of Geography, University of Leeds, Leeds, LS2 9JT, England.

Ian Armit Historic Scotland, Longmore House, Salisbury Place, Edinburgh, EH9 1SH, Scotland.

Mike Baillie School of Geosciences, Queen's University, Belfast, BT7 1NN, Northern Ireland.

Jonathan Bell Ulster Folk & Transport Museum, Cultra, Holywood, Co Down, BT18 OEU, Northern Ireland.

Julie Bond Dept of Archaeological Sciences, University of Bradford, Bradford, West Yorkshire, BD7 1DP, England.

Tony Brown Department of Geography, University of Exeter, Amory Building, Exeter, EX4 4RJ, England.

Lawrence A S Butler Department of Archaeology, University of York, Kings Manor, York, YO1 2EP, England.

Stephen Carter 27 Montague Street, Edinburgh, EH8 9QT, Scotland.

Geraint Coles Dept of Archaeology, University of Edinburgh, The Old High School, Infirmary Street, Edinburgh, EH1 1LT, Scotland.

David Cowley RCAHMS, John Sinclair House, 16 Bernard Terrace, Edinburgh, EH8 9NX, Scotland.

Sarah Crane School of Archaeological Studies, University of Leicester, Leicester, LE1 7RH, England.

Camilla Dickson† Division of Environmental and Evolutionary Biology, Graham Kerr Building, University of Glasgow, Glasgow, G12 8QQ, Scotland.

Kevin Edwards Department of Archaeology & Prehistory, University of Sheffield, Northgate House, West Street, Sheffield, S1 4ET, England.

John Grattan The Institute of Earth Sciences, The University of Wales, Aberystwyth, SY23 3DB, Wales.

Sheila Hamilton-Dyer 5 Suffolk Avenue, Shirley, Southampton, SO15 5EF, England.

Pauline Kneale School of Geography, University of Leeds, Leeds, LS2 9JT, England.

Finbar McCormick School of Geosciences, Queen's University, Belfast, BT7 1NN, Northern Ireland.

Adrian T McDonald School of Geography, University of Leeds, Leeds, LS2 9JT, England.

Frances S McLaren Department of Environmental Sciences, Faculty of Sciences & Health, University of East London, Romford Road, London, E15 4LZ, England.

John McManus School of Geography & Geology, University of St Andrews, St Andrews, Fife, KY16 9ST, Scotland.

Coralie Mills AOC (Scotland) Ltd, The Schoolhouse, 4 Lochend Road, Leith, Edinburgh, EH6 8BR, Scotland.

Terry O'Connor Department of Archaeological Sciences, University of Bradford, Bradford, West Yorkshire, BD7 1DP, England.

Deirdre O'Sullivan School of Archaeological Studies, University of Leicester, Leicester, LE1 7RH, England.

Clare de Rouffignac Central Archaeological Service, English Heritage, Fort Cumberland, Eastney, Portsmouth, PO4 9LD, England.

Ian Simpson Dept of Environmental Science, University of Stirling, Stirling, FK9 4LA, Scotland.

Sue Stallibrass Department of Archaeology, University of Durham, Science Laboratories, South Road, Durham, DH1 3LE, England.

John Tierney Archaeological Services Unit, Department of Archaeology, University College Cork, Cork, Ireland.

Richard Tipping Department of Environmental Science, University of Stirling, Stirling, FK9 4LA, Scotland.

Kevin Walsh Centre Camille Julian, Université de Provence – CNRS, 29 Avenue Robert Schumann, 13621 Aix-En-Provence, France.

Mervyn Watson Ulster Folk & Transport Museum, Cultra, Holywood, Co Down, BT18 OEU, Northern Ireland.

Graeme Whittington School of Geography & Geology, University of St Andrews, St Andrews, Fife, KY16 9ST, Scotland.

Louise van Wijngaarden-Bakker IPP, Nieuwe Prinsengracht 130, 1018 VZ, Amsterdam, Netherlands.

Rob Young School of Archaeological Studies, University of Leicester, Leicester, LE1 7RH, England.